Wave Energy Devices

Wave Energy Devices

Design, Development, and Experimental Studies

Srinivasan Chandrasekaran
Faisal Khan
Rouzbeh Abbassi

CRC Press is an imprint of the
Taylor & Francis Group, an **informa** business

MATLAB® is a trademark of The MathWorks, Inc. and is used with permission. The MathWorks does not warrant the accuracy of the text or exercises in this book. This book's use or discussion of MATLAB® software or related products does not constitute endorsement or sponsorship by The MathWorks of a particular pedagogical approach or particular use of the MATLAB® software.

First edition published 2022
by CRC Press
6000 Broken Sound Parkway NW, Suite 300, Boca Raton, FL 33487-2742

and by CRC Press
4 Park Square, Milton Park, Abingdon, Oxon, OX14 4RN

CRC Press is an imprint of Taylor & Francis Group, LLC

ISBN: 978-1-032-25075-5 (hbk)
ISBN: 978-1-032-25077-9 (pbk)
ISBN: 978-1-003-28142-9 (ebk)

DOI: 10.1201/9781003281429

Typeset in Times
by Newgen Publishing UK

Contents

Preface ix
Authors xi

Chapter 1 Ocean Waves and Wind Forces: Basics 1

1.1 Ocean Environment 1
1.2 Wave Theories 2
1.3 Stoke's Fifth-Order Theory 6
1.4 Wave Spectra 16
1.4.1 PM Spectrum for Wave Loads 16
1.4.2 Modified PM Spectrum (Two Parameters, H_s, ω_0) 16
1.4.3 International Ship Structures Congress (ISSC) Spectrum (Two Parameters, H_s, $\bar{\omega}$) 16
1.4.4 Joint North Sea Wave Project (JONSWAP) Spectrum (Five Parameters, H_s, ω_{0o}, γ, τ_a, τ_b) 17
1.5 Wind and Current 22
1.5.1 Wind Spectra 28
1.5.2 Current 31
Exercises 32

Chapter 2 Perforated Cylinders and Applications 33

2.1 Introduction 33
2.2 Force Reduction in the Inner Cylinder 33
2.3 Effect of Annular Spacing and Perforation Ratio on Force Reduction 37
2.4 Effect of Perforation Parameters on Force Reduction 41
2.5 Twin Perforated Cylinders 47
Exercises 59

Chapter 3 Floating Wave Energy Converter 61

3.1 Introduction 61
3.2 Wave-to-Wire Transfer 65
3.3 Numerical Modeling 66
3.4 Frequency-Domain Modeling 69
3.5 Time-Domain Modeling 74
3.5.1 Linear Time-Domain Model (LTD) 75
3.6 Computational Tools 77
3.7 Multibody Floating Wave Energy Converter 77
3.8 Hydrodynamic Coefficients of Floating Wave Energy Converters 91

3.9 Wave Power....94
3.9.1 Regular Waves....94
3.9.2 Irregular Waves....95
3.10 Performance Curves with Virtual Power Take-Off System....100
3.11 Analysis with Hydraulic Power Take-Off System....104
3.11.1 Governing Parameters....104
3.12 Practical Guide to Design of Hydraulic Power Take-Off System....109
3.13 Numerical Studies on Hydraulic Power Take-Off Systems....110
3.14 Floating Wave Energy Converter Response without Power Take-Off System....116
3.15 Floating Wave Energy Converter with a Novel Hydraulic Power Take-Off System....124
3.16 Failure Assessment of Floating Wave Energy Converter....134
Exercises....141

Chapter 4 Double-Rack Mechanical Wave Energy Converter....143

4.1 Introduction....143
4.2 Mechanical Wave Energy Converter....144
4.2.1 Equation of Motion....145
4.2.2 Power Take-Off System Design....147
4.2.3 Experimental Studies....152
4.3 Double Rack Mechanical Wave Energy Converter....155
4.3.1 Experimental Studies on Double Rack Mechanical Wave-Energy Converters....156
4.4 Failure Assessment....158
4.4.1 Conducting Design Failure Mode and Effect Analysis....160
4.5 Failure Mode and Effect Analysis of Mechanical Wave-Energy Converter....162
4.6 Deep-Ocean Wave Energy Converter: Conceptual Design....163
4.6.1 Geometric Design of the Device....163
4.7 Working Principle of Deep Ocean Wave Energy Converters....166
4.8 Experimental Investigations on Deep Ocean Wave Energy Converters....168
4.8.1 Cylindrical Float....172
4.8.2 Cylindrical Float Integrated with Fin....176
4.9 Multi-Utility Development Devices....185
4.9.1 Applications of TSUSUCA-DOLPHIN....196
4.9.2 Workable Alternatives....197
4.9.3 Cost Benefits....199
Exercises....201

Chapter 5 Offshore Wind Turbines ... 203

5.1 Introduction ... 203
5.1.1 Support Systems for Wind Turbines ... 204
5.2 Wind Power ... 206
5.3 Numerical Tools ... 207
5.4 Offshore Wind Turbine Classifications ... 208
5.5 Offshore Floating Wind Turbine: Components ... 209
5.6 Offshore Floating Wind Turbines ... 209
5.6.1 Single Point Anchor Reservoir Type ... 210
5.6.2 Tension-Leg Platform Type ... 212
5.6.3 Pontoon (Barge) Type ... 213
5.6.4 Semi-Submersible Type ... 214
5.6.5 Triceratops ... 215
5.6.6 Triceratops-Supported Wind Turbines ... 219
5.7 Experimental and Numerical Analyses ... 220
5.8 Mathematical Background ... 222
5.8.1 Blade Momentum Theory ... 222
5.8.2 Blade Element Theory ... 225
5.9 Aero-Elastic Model ... 227
5.9.1 Kane Method ... 228
5.9.2 Wind Load ... 228
5.9.3 Wind Shear Effect ... 229
5.10 Normal Turbulence Model ... 229
5.11 Wind Spectrum ... 230
5.12 Numerical Analysis of Triceratops-Supported Wind Turbine ... 230
5.13 Responses under Operable Loads and Parked Conditions ... 236
5.14 Dynamic Tension Variation of Tethers ... 238
Exercises ... 241

References ... 243

Index ... 255

Preface

The book *Wave Energy Devices: Design, Development, and Experimental Studies* is an intensive effort, backed up with the very rich experiences in teaching, research, and industrial consultancies of the authors with international exposure. The contents are distributed in five chapters, each focusing on developing expertise and strengthening the research abilities in the wave energy harvest. Detailed discussions on the design and development of floating wave energy converters, mechanical wave energy converters, and offshore wind turbines are testimonies of research-backed development on the new arena of wave energy. There has been a positive kick-start of high-value research contributions by scientists worldwide, based on the detailed experimental and numerical studies presented in various chapters. Industrial-backed experience possessed by the authors in designing and developing offshore wind turbines in alignment with the international guidelines are added factors to this book. Although a classroom mode of presentation is not adopted, the contents discussed in the step-by-step procedures help us to understand the design and development of wave energy devices from scratch. The inclusion of the conceptual development of multi-utility development devices like TSUSUCA is a stimulator for young researchers to diversify their ideas for commercial success. The use of perforated cylinders, though finding a wider application in coastal structures, wave energy harvesting will result in force reduction on members. The basic concepts of ocean waves and wind will help graduate engineers and researchers strengthen their numerical modeling techniques while helping practicing engineers revisit the mathematical models. Failure assessment, discussed in a few chapters, helps achieve a complete product development of wave energy devices using simple but effective reliability tools. A variety of mechanical components, their sizing, design principles, and final product development help readers achieve a workflow in the design of wave energy devices.

The authors sincerely thank various researchers who contributed to the chapters by sharing their findings and knowledge. The lead editor, in particular, thanks the Chairman of the Center for Continuing Education and the administrative authorities of the Indian Institute of Technology Madras for their generous support extended in conducting the experiments, identifying industrial partners for technology transfer, and assisting in filing patents of the developed wave energy devices. The editors also extend their sincere gratitude to the administrative authorities of the School of Engineering, Macquarie University, Australia, and the Process Safety Center, Texas A&M University, United States. The association of the authors with the schools of engineering of international repute strengthened the editors' confidence in writing this book and they sincerely thank the authors for providing rich research and an academic environment in which to complete this book. The contributions made by Dr. N. Madhavi, Freelance Technical Designer, San Diego, United States, in Chapter 2 are sincerely acknowledged.

The contributions made by Dr. VVS. Sri Charan, Consultant, Oscilla Power, United States, in Chapter 3 are also acknowledged. The research input and technical contributions shared by Dr. Harender Sinhmar, Department of Mechanical Engineering, Shiv Nadar University (Institution of Eminence), Delhi NCR, India and Mr. Deepak, CR, Scientist, National Institute of Ocean Technology, India are further acknowledged.

Authors

Srinivasan Chandrasekaran is Professor in the Department of Ocean Engineering, Indian Institute of Technology Madras. He commands a rich experience in teaching, research, and industrial consultancy of about 29 years. He has supervised many sponsored research projects and offshore consultancy assignments, both in India and abroad. His active research areas include dynamic analysis and design of offshore structures, structural health monitoring of ocean structures, risk and reliability, fire resistant design of structures, use of functionally graded materials in marine risers, and health, safety and environmental management in process industries. He was a visiting fellow under the Ministry of Italian University Research invitation to the University of Naples Federico II for two years. During his stay in Italy, he researched the advanced nonlinear analysis of buildings under earthquake loads and other impact loads with experimental validation on full-scaled models. He has authored about 170 research papers in peer-reviewed international journals and refereed conferences organized by professional societies worldwide. He has authored eighteen textbooks, which various publishers of international repute publish. He is an active member of several professional bodies and societies, both in India and abroad. He has also conducted about twenty distance-education programs on various engineering subjects for the National Program on Technology-Enhanced Learning, Government of India. He is a vibrant speaker and delivered many keynote addresses in international conferences, workshops, and seminars organized in India and abroad.

Faisal Khan is Professor and Director of Mary O'Connor Process Safety Center at Texas A&M University, USA, and a Professor and the Canada Research Chair (Tier I) of Offshore Safety and Risk Engineering. He is the founder of the Centre for Risk Integrity and Safety and Engineering. He is actively involved with global energy industries on the issue of safety, risk, and asset integrity. He is the recipient of many National and Regional awards, including the SPE award for his contribution to health, safety, and Risk Engineering. He has authored over 500 research articles in peer-reviewed journals and conferences on safety, risk, and reliability engineering. He has authored eights books on the subject area.

Rouzbeh Abbassi is Associate Professor and Discipline Leader of the Civil Engineering Program at the School of Engineering of Macquarie University, Sydney, New South Wales, Australia. He is a member of Sustainable Energy Research Centre at Macquarie University. He has expertise in environmental, safety, and risk engineering applications to different engineering operations such as onshore and offshore renewable energy systems. He actively collaborates with various national and international research and industry partners on multimillion dollar projects in this area, such as the Blue Economy Cooperative Research Centre. He has published over 150 research papers in peer-reviewed journals and conferences in the field of his expertise.

1 Ocean Waves and Wind Forces

Basics

1.1 OCEAN ENVIRONMENT

Wave-structure interaction, which dominates the geometric design of the floating offshore structures, is a well-known and clearly understood phenomenon. They were successfully practiced in the design principles of ships and other large floating structures like floating docks. Wave-structure interaction causes a complicated hydrodynamic behavior in open sea conditions. The wave loads developed on floating offshore structures are much higher than those developed by wind action. Wave loads that develop due to water particle motion also depend upon the size of the structural member. Due to the smaller influence of the slender members with diameter to wavelength ratio (D/L), less than 0.2 on the wave field, the wave loads can be calculated using Morison's equation. In the case of floating structures with large-diameter members, their presence affects the wave field, and hence diffraction theory should be used to estimate these loads. The response analysis of offshore floating structures under wave loads depends on the type of waves considered (Nagavinothini and Chandrasekaran, 2020; Chandrasekaran and Nagavinothini, 2020). The large floating offshore structures shall also be under the influence of wind and current, in addition to the waves. The mast of the offshore wind turbine will be highly wind-sensitive and more susceptible to low-frequency oscillations. Offshore wind turbines resting on large floating structures exhibit a complex behavior under the coupled action of wind and waves. The platform's response increases with the increase in the exposed area of the structure and the wind velocity. In addition to the wind effects, the current causes a varying pressure distribution around the structural members resulting in steady drag force.

Environmental loads continuously interact with offshore structures in waves, wind, current, or seaquakes. Continuous exposure to these loads can also induce fatigue damage to the structure over time as these loads are also cyclic. Among all these loads, a primary concern is wave and wind loads; the former is a high-frequency phenomenon compared to the latter. The basic concept of floating offshore structures should therefore be FORM-dominant; resistance to load should be incurred from the geometric shape and not by the strength of the material. For a broader convenience of commissioning and de-commissioning, large floating structures are designed as positively buoyant. Excessive buoyancy keeps them afloat, enabling easy installation, but poses serious challenges to their structural response.

DOI: 10.1201/9781003281429-1

1.2 WAVE THEORIES

Waves play a critical role in the design and development of floating wave energy converters (FWEC). The harnessing principle depends on the wave characteristics and the range of sea states. Wave analysis can be performed either by the design wave approach or the statistical approach. A regular wave is defined using wave amplitude and period (*H*, *T*). The water particle kinematics is calculated as a function of sea-surface elevation using the potential theory; waves are assumed to be long-crested. Various wave theories are developed to include a wide range of wave parameters (Lungren 1963). The most common theory is Airy's linear wave theory. On the other hand, the statistical approach recommends using the appropriate wave spectra, which are site-specific. It is useful to assess the dynamic response of the FWEC accurately; it helps arrive at a response spectrum that defines the maximum expected response within a particular interval of time.

Generally, ocean waves are random but can be represented as a regular wave described by a deterministic approach. As the waveform in each cycle in a regular wave is the same, wave theories describe the characteristics of a typical cycle; it remains invariant for other cycles. Two significant parameters are the period (*T*) and Height (*H*) of the waves and water depth (*d*). A sample time history of a regular wave is shown in Fig. 1.1. Airy's wave theory assumes linearity between the kinematic quantities and the wave height, making the wave theory simple. Airy's theory assumes a sinusoidal waveform of wave height (*H*), which is small in comparison to the wavelength (λ) and water depth (*d*). Sea surface elevation, wave number, horizontal and vertical water particle velocities, and accelerations are given as below:

$$\eta(x,t) = \frac{H}{2}\cos(kx - \omega t) \tag{1.1}$$

$$k = \frac{2\pi}{\lambda} \tag{1.2}$$

$$\dot{u}(x,t) = \frac{\omega H}{2}\frac{\cosh(ky)}{\sinh(kd)}\cos(kx - \omega t) \tag{1.3}$$

$$\dot{v}(x,t) = \frac{\omega H}{2}\frac{\sinh(ky)}{\sinh(kd)}\sin(kx - \omega t) \tag{1.4}$$

$$\ddot{u}(x,t) = \frac{\omega^2 H}{2}\frac{\cosh(ky)}{\sinh(kd)}\sin(kx - \omega t) \tag{1.5}$$

$$\ddot{v}(x,t) = -\frac{\omega^2 H}{2}\frac{\sinh(ky)}{\sinh(kd)}\cos(kx - \omega t) \tag{1.6}$$

Airy's linear wave theory is valid up until the mean sea level only. However, due to the variable submergence effect, the submerged length of the members will be continuously changing and will attract additional forces at any given instant of time. To compute the water particle kinematics up to the actual level of submergence,

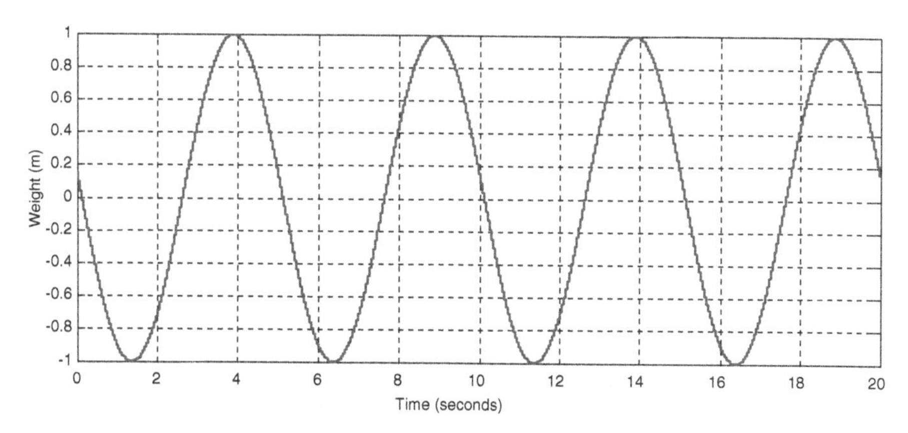

FIGURE 1.1 Typical regular wave (2m, 5s)

TABLE 1.1
Classification of Waves according to Relative Depth

Classification	*d/L*	*Kd*	tan*h* (*kd*)
Deep water	0.5 to infinity	∏ to infinity	-1
Transitional	(1/20) to 0.5	(∏/10) to ∏	tan*h* (*kd*)
Shallow water	0 to (1/20)	0 to (∏/10)	-*kd*

several suggests have been made (Chakrabarthi and Tam, 1975; Chakrabarti et al., 1976; Hogben and Standing, 1974; Chandrasekaran, 2015; 2016b; Chandrasekaran and Jain, 2016; Chandrasekaran et al., 2004). One of the limitations of Airy's linear theory is that the velocity potential does not satisfy the Laplace equation but satisfies the dynamic free-surface boundary conditions. Therefore, in many physical situations, Airy's linear theory, even with the stretching modifications, is not adequately describing the water particle kinematics completely. Waves are progressively definable by their amplitude and period at a given water depth, but this simple treatment is inadequate for larger wave heights. Hence, a few higher-order wave theories are used to obtain appropriate water particle kinematics. These theories become nonlinear and allow the formulation of waves that are not purely sinusoidal. Table 1.1 shows the classifications of waves according to the relative water depth.

Wheeler suggested the following modifications in the horizontal water particle velocity and acceleration to include the actual level of submergence of the member:

$$\begin{aligned} \dot{u}(x,t) &= \frac{\omega H}{2} \frac{\cosh\left(ky\left[\dfrac{d}{d+\eta}\right]\right)}{\sinh(kd)} \cos(kx-\omega t) \\ \ddot{u}(x,t) &= \frac{\omega^2 H}{2} \frac{\cosh\left(ky\left[\dfrac{d}{d+\eta}\right]\right)}{\sinh(kd)} \sin(kx-\omega t) \end{aligned} \tag{1.7}$$

Chakrabarti suggested the modifications as given below:

$$\dot{u}(x,t)=\frac{\omega H}{2}\frac{\cosh(ky)}{\sinh(k(d+\eta))}\cos(kx-\omega t)$$
$$\ddot{u}(x,t)=\frac{\omega^2 H}{2}\frac{\cosh(ky)}{\sinh(k(d+\eta))}\sin(kx-\omega t) \tag{1.8}$$

Variations in the horizontal water particle velocity for the intermediate water depth condition, based on the modifications, are shown in Fig. 1.2. As seen from the figure, horizontal water particle velocity variation increases with the increase in the wave amplitude. The same variation is observed in deep and shallow water conditions as well. At a wave height less than 1 m, the variations due to Wheeler's and Chakrabarti's modifications are found to be negligible (Fig. 1.2).

Based on the structural waveform, waves can be classified as regular waves and irregular or random waves. Initially, the transfer of energy causes capillary waves, which grow to form irregular waves with different amplitudes and periods depending on wind speed and direction. Suppose the wind blowing over the ocean surface has a constant wind velocity. In that case, the generated irregular wavelets will grow to a fully developed sea with constant wave amplitude and period. Such waves are referred to as regular waves, and these waves exhibit a sinusoidal motion. Since the actual wave formation is complex, the following mathematical formulations model the ocean surface with regular waves.

- Airy's two-dimensional small-amplitude linear wave theory
- Stoke's theory
- Solitary wave theory
- Cnoidal theory
- Stream function theory

Airy's theory, which assumes linearity between the wave height and kinematic quantities, is commonly used in the theories listed above. The regular waves are usually defined by their wave height (H) and wave period (T), as shown in Fig. 1.3. Airy's small-amplitude linear wave theory is valid for deep water conditions where $(d/gT^2) > 0.8$ and Stoke's theory can be used when $(H/gT^2) > 0.04$.

The most straightforward is Airy's linear wave theory or small-amplitude wave theory. According to this theory, the waveform has a sinusoidal profile and provides the kinematic and dynamic amplitudes as a linear function of wave amplitude. Thus, the normalized amplitude value is unique and invariant to the wave amplitude and thus helps represent the response of the offshore structures as a normalized value. As a function of wave height, the normalized responses are called transfer function or the Response Amplitude Operator (RAO), which is one of the common ways to express the extreme response of the structures.

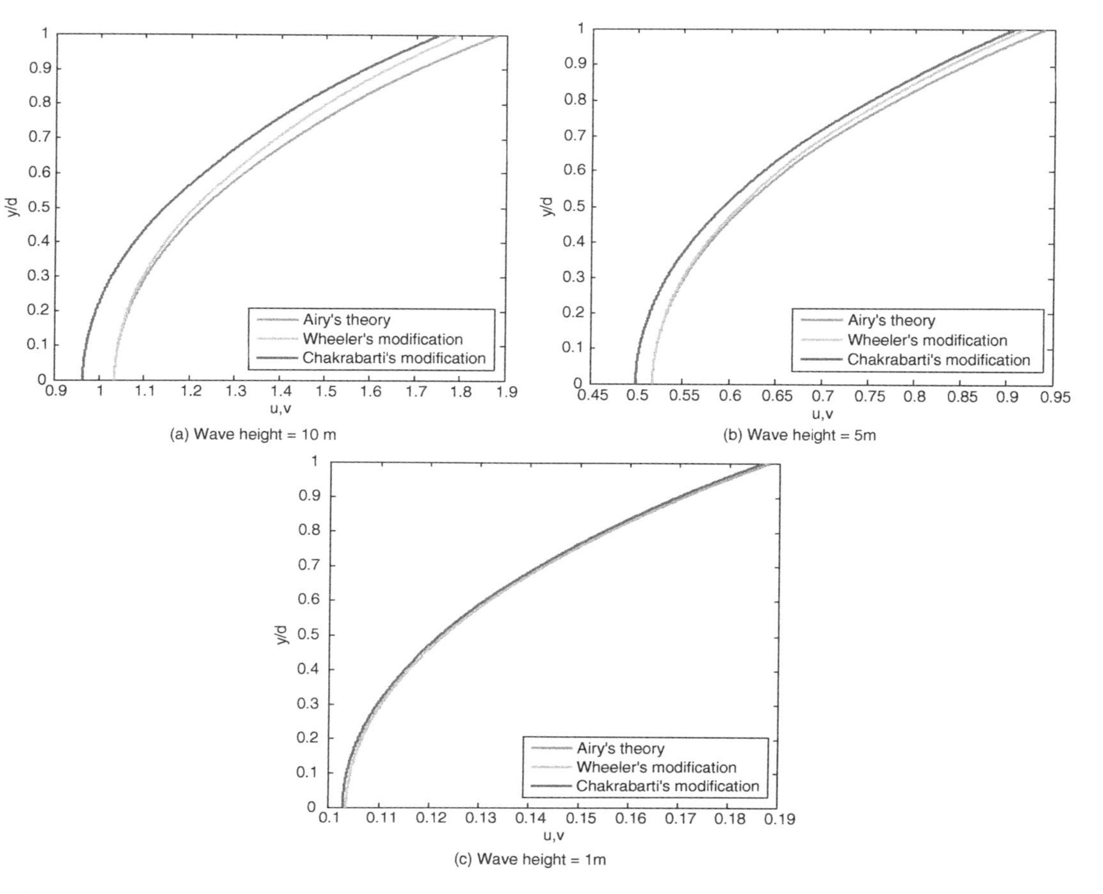

FIGURE 1.2 Horizontal water particle velocity variation (intermediate depth)

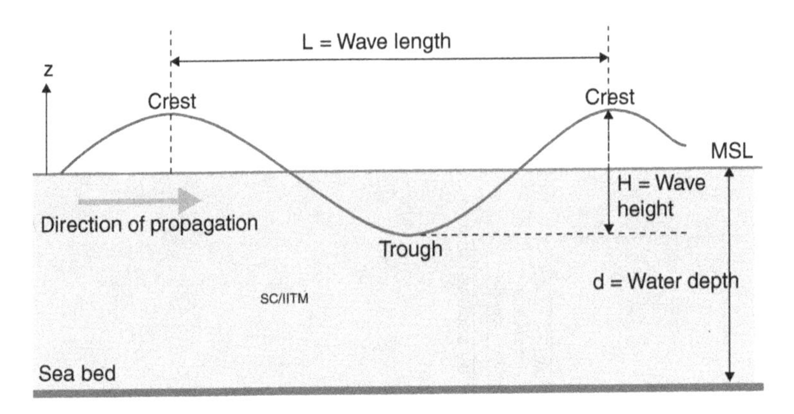

FIGURE 1.3 Wave parameters

1.3 STOKE'S FIFTH-ORDER THEORY

Stokes (1847) suggested an analytical solution for steady water waves, first derived to second order for finite depth and third order for infinite depth. Later, the third order was extended for finite depth to the fifth order for infinite depth using the inverse plane method (Stokes, 1880; De, 1955). However, variations in the Stokes solution exist due to three main reasons (Zhao et al., 2016). The primary difference between the solutions arises from the definition of the perturbation parameter. Several researchers have defined it based on wave steepness (Bretschneider, 1960; Fenton, 1985). The following relationship gives wave steepness:

$$wave\,steepness = \frac{kH}{2} \tag{1.9}$$

Skjelbreia and Hendrickson (1960) defined it as components of the non-dimensional variables $\left[\frac{U_b \sinh(kh)}{c}, k\eta, \frac{kc_{(0)}u_m}{g}, \frac{u_b}{c_{(o)}}\right]$, where (u_b) is the water particle velocity at the bottom, η is the mean sea level elevation, (u_m) is water particle velocity at mean sea level, h is the water depth, c is the wave celerity, and $c_{(o)}$ is the wave celerity defined by linear theory. A common recommendation made by researchers is to use wave steepness as the perturbation parameter. The second main difference arises from the definition of depth. A few researchers measured depth (h) from the bottom to the still water level (SWL); other researchers used the mean water depth, measured from the sea bottom to the mean water level (MWL). One common recommendation to avoid this discrepancy was to use the still water depth, but the solution will become more rigorous. The third and major reason for the discrepancies is that the unique solution to Stokes theory cannot be obtained only using three parameters (H, T, d); current should also be specified to determine the wave celerity. Stokes (1847) proposed two methods: one with a zero uniform current to determine the wave

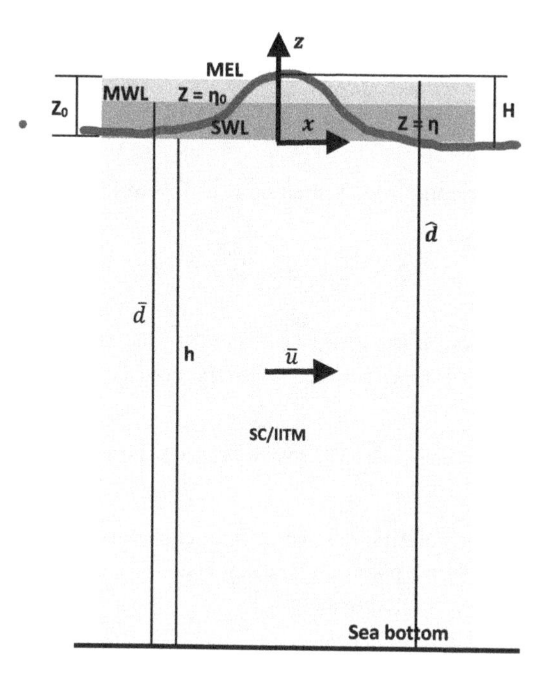

FIGURE 1.4 Wave nomenclature

celerity, and the other is to set a zero-volume flux transport under waves. A common recommendation was to include a uniform current, which can predict wave celerity correction (Fenton, 1985).

Let us consider a steady progressive wave with the wavelength (L), wave amplitude (H), and period (T), as shown in Fig. 1.4. Current ($\bar{u}$) is considered to be uniform throughout the complete water depth. Still water level (SWL) is the undisturbed water surface, while mean water level (MWL) is differed by (η_0) from the SWL, as shown in the figure. MWL is higher than SWL. Mean energy level (MEL) is measured (z_0) from the SWL. The sea surface elevation (z) is measured from the SWL upwards, while (h) is the still water depth, measured from the seabed to the SWL, (η) indicates the free surface elevation, and (z_0) is the total head measured from the SWL up to the MEL. The difference between the MWL and MEL is noted as the set-down effect of ocean waves.

For the fluid assumed as inviscid and incompressible and for an irrotational flow, the potential function should satisfy Laplace's equation:

$$\varphi_{xx} + \varphi_{zz} = 0 \quad \{for -h < z < \eta\} \tag{1.10}$$

At ($z = -h$), the boundary condition applicable is no-fluid passing through, as it is the seabed. Hence,

$$\varphi_z = 0 \tag{1.11}$$

At $z = \eta\ (x,t)$ at the free surface, the kinematic boundary condition applicable is as follows:

$$\varphi_z = \eta_t + \varphi_x \eta_x \tag{1.12}$$

The dynamic boundary condition applicable is as follows:

$$\varphi_t + \frac{1}{2}\left[\varphi_x^2 + \varphi_z^2\right] + g\eta = gz_0 \tag{1.13}$$

Now, the potential function (as given by Eq. 1.10) and the boundary condition (as given by Eq. 1.11) can be approximated using the Fourier series, as follows:

$$\varphi = \bar{u}x + \frac{c}{k}\sum_{n=1}^{\infty}\left\{\varphi_n \cos h\left[nk(h+z)\right]\sin(n\theta)\right\} \tag{1.14a}$$

Where c (= ω/k) is wave celerity, ω is wave frequency, k is wave number, θ (= kx-ωt) is the phase angle and φ_n are potential coefficients.

Let the derivatives be defined as below:

$$\frac{\partial}{\partial x} = \frac{\partial}{\partial \theta}\frac{\partial \theta}{\partial x} = k\frac{\partial}{\partial \theta} \tag{1.14b}$$

$$\frac{\partial}{\partial t} = \frac{\partial}{\partial \theta}\frac{\partial \theta}{\partial t} = -\omega\frac{\partial}{\partial \theta} \tag{1.14c}$$

By substituting Eq. (1.14a) in Eq. (1.12), we get:

$$k\left(1-\frac{\bar{u}}{c}\right)\frac{\partial \eta}{\partial \theta} = \sum_{n=1}^{\infty}\left\{nk\varphi_n \cos h\left[nk(h+\eta)\right]\cos(\mathrm{n}\theta)\frac{\partial \eta}{\partial \theta} - n\varphi_n \sin h\left[nk(h+\eta)\right]\sin(n\theta)\right\} \tag{1.15}$$

Integrating Eq. (1.15) with respect to (θ),

$$k\left(1-\frac{\bar{u}}{c}\right)\eta = \sum_{n=1}^{\infty}\left\{\varphi_n \sin h\left[nk(h+\eta)\right]\cos(n\theta) + constant\right\} \tag{1.16}$$

For the no-wave condition, the integration constant in Eq. (1.16) will be zero. Hence, Eq (1.16) reduces to the following form:

$$k\left(1-\frac{\bar{u}}{c}\right)\eta = \sum_{n=1}^{\infty}\left\{\varphi_n \sin h\left[nk(h+\eta)\right]\cos(n\theta)\right\} \tag{1.17}$$

By substituting Eq. (1.14a) in Eq. (1.13), we get:

$$-c^2\sum_{n=1}^{\infty}\left\{n\varphi_n \cos h\left[nk(h+\eta)\right]\cos(n\theta)\right\}$$
$$+\frac{1}{2}c^2\left\{\frac{\bar{u}}{c}+\sum_{n=1}^{\infty}\left\{n\varphi_n \cos h\left[nk(h+\eta)\right]\cos(n\theta)\right\}\right\}^2 \tag{1.18}$$
$$+\frac{1}{2}c^2\left\{\sum_{n=1}^{\infty}\left\{n\varphi_n \sinh\left[nk(h+\eta)\right]\sin(n\theta)\right\}\right\}^2+g\eta=gz_0$$

Further, the free surface can also be approximated as follows:

$$\eta=\eta_0+\sum_{n=1}^{\infty}\left\{\eta_n\cos(n\theta)\right\} \tag{1.19}$$

Where η_n is Fourier coefficients of the free surface. The authors acknowledge the explanations and solution procedures given by the researchers (Zhao et al., 2016) towards Stokes' equation. More detailed explanations and validation of the solution can be seen in Zhao et al. (2016); for the benefit of the readers, the complete solution is presented here. The solution to the above equations is represented in two parts: The first part is to build a set of algebraic equations for the Fourier coefficients of free-surface and the potential function; the second part is to solve these equations using the perturbation method.

$$\varphi_n=\left[1-\frac{\bar{u}}{c}\right]\phi_n \tag{1.20}$$

Substituting this in Eqs. (1.17 to 1.18), and excluding the presence of current, we get:

$$k\eta=\sum_{n=1}^{\infty}\left\{\phi_n \sinh\left[nk(h+\eta)\right]\cos(n\theta)\right\} \tag{1.21}$$

When current is included along with the wave celerity term, then the equation takes the following form:

$$-(c-\bar{u})^2\sum_{n=1}^{\infty}\left\{n\phi_n \cos h\left[nk(h+\eta)\right]\cos(n\theta)\right\}$$
$$+\frac{1}{2}(c-\bar{u})^2\left\{\sum_{n=1}^{\infty}\left\{n\phi_n \cos h\left[nk(h+\eta)\right]\cos(n\theta)\right\}\right\}^2 \tag{1.22}$$
$$+\frac{1}{2}(c-\bar{u})^2\left\{\sum_{n=1}^{\infty}\left\{n\phi_n \sin h\left[nk(h+\eta)\right]\sin(n\theta)\right\}\right\}^2+g\eta=gz_0-\frac{1}{2}\bar{u}^2$$

The above set of equations (1.21–1.22) give two implicit sets of expressions for the Fourier coefficients of the free-surface elevation, in terms of the modified potential function, ϕ_n. It contains infinite Fourier series and is truncated to five terms, hence Stokes' fifth-order solution. As seen in Eq. (1.21), the hyperbolic functions of the free-surface elevation are expanded, as the Taylor series and the Fourier series include only five terms. Subsequently, the free-surface expressions are substituted on Eq. (1.22), leading to the solution of free-surface elevation in the potential function. After simplification and truncating up to five terms, the following set of equations are obtained for the free-surface elevation, up to the SWL:

$$\text{Let } s_1 = \sinh(kh) \,\&\, c_1 = \cosh(kh)$$

$$k\eta_0 = \frac{1}{2}\phi_1^2 c_1 s_1 + 2\phi_2^2 c_1 \left(2s_1^3 + s_1\right) + \left(\frac{1}{2}\phi_1^2 \phi_2 c_1 + \frac{1}{8}\phi_1^4 c_1\right)\left(8s_1^3 + 3s_1\right) \tag{1.23}$$

$$\begin{aligned} k\eta_1 &= \phi_1 s_1 + \phi_1 \phi_2 \left(3s_1^3 + 2s_1\right) + \frac{3}{8}\phi_1^3 \left(3s_1^3 + 2s_1\right) + \phi_2 \phi_3 \left(20s_1^5 + 25s_1^3 + 6s_1\right) \\ &+ \phi_1 \phi_2^2 \left(25s_1^5 + 29s_1^3 + 6s_1\right) + \frac{1}{8}\phi_1^2 \phi_3 \left(100s_1^5 + 107s_1^3 + 18s_1\right) \\ &+ \left(\frac{1}{6}\phi_1^3 \phi_2 + \frac{5}{192}\phi_1^5\right)\left(125s_1^5 + 136s_1^3 + 24s_1\right) \end{aligned} \tag{1.24}$$

$$k\eta_2 = 2\phi_2 c_1 s_1 + \frac{1}{2}\phi_1^2 c_1 s_1 + \left(\phi_1 \phi_3 c_1 + \phi_1^2 \phi_2 c_1 + \frac{1}{6}\phi_1^4 c_1\right)\left(8s_1^3 + 3s_1\right) \tag{1.25}$$

$$\begin{aligned} k\eta_3 &= \phi_3 \left(4s_1^3 + 3s_1\right) + \left(\phi_1\phi_2 + \frac{1}{8}\phi_1^3\right)\left(3s_1^3 + 2s_1\right) + 2\phi_1\phi_4 \left(10s_1^5 + 11s_1^3 + 2s_1\right) \\ &+ \frac{1}{2}\phi_1 \phi_2^2 \left(25s_1^5 + 29s_1^3 + 6s_1\right) + \frac{1}{4}\phi_1^2 \phi_3 \left(100s_1^5 + 107s_1^3 + 18s_1\right) \\ &+ \left(\frac{1}{8}\phi_1^3 \phi_2 + \frac{5}{384}\phi_1^5\right)\left(125s_1^5 + 136s_1^3 + 24s_1\right) \end{aligned} \tag{1.26}$$

$$k\eta_4 = \left(4\phi_4 c_1 + 2\phi_2^2 c_1\right)\left(2s_1^3 + s_1\right) + \left(\phi_1 \phi_3 s_1 + \frac{1}{2}\phi_1^2 \phi_2 c_1 + \frac{1}{24}\phi_1^4 c_1\right)\left(8s_1^3 + 3s_1\right) \tag{1.27}$$

$$\begin{aligned} k\eta_5 &= \phi_5 \left(16s_1^5 + 20s_1^3 + 5s_1\right) + 20\phi_1 \phi_4 \left(10s_1^5 + 11s_1^3 + 2s_1\right) + \phi_2\phi_3 \left(20s_1^5 + 25s_1^3 + 6s_1\right) \\ &+ \frac{1}{2}\phi_1 \phi_2^2 \left(25s_1^5 + 29s_1^3 + 6s_1\right) + \frac{1}{8}\phi_1^2\phi_3 \left(100s_1^5 + 107s_1^3 + 18s_1\right) \\ &+ \left(\frac{1}{24}\phi_1^3\phi_2 + \frac{1}{384}\phi_1^5\right)\left(125s_1^5 + 136s_1^3 + 24s_1\right) \end{aligned} \tag{1.28}$$

$$g\eta_0 = gz_0 - \frac{1}{2}\bar{u}^2 - \frac{1}{4}\left(c - \bar{u}\right)^2 \left(\phi_1^2 + 4\phi_2^2\right) \tag{1.29}$$

$$g\eta_1 = (c-\bar{u})^2 \left\{ \phi_1 c_1 + \phi_1\phi_2 c_1\left(s_1^2-1\right) + \frac{1}{8}\phi_1^3 c_1 s_1^2 + \phi_2\phi_3 c_1\left(s_1^4+s_1^2\right) + \frac{1}{8}\phi_1^2\phi_3 c_1\left(28s_1^4+3s_1^2\right) + \frac{1}{6}\phi_1^3\phi_2 c_1\left(10s_1^4+3s_1^2\right) + \frac{1}{192}\phi_1^5 c_1\left(25s_1^4+12s_1^2\right) \right\} \tag{1.30}$$

$$g\eta_2 = (c-\bar{u})^2 \left\{ 2\phi_2\left(2s_1^2+1\right) + \frac{1}{4}\phi_1^2\left(2s_1^2-1\right) + \frac{1}{2}\phi_1\phi_3\left(16s_1^4+6s_1^2-3\right) + 2\phi_1^2\phi_2\left(2s_1^4+s_1^2\right) + \frac{1}{12}\phi_1^4\left(4s_1^4+3s_1^2\right) \right\} \tag{1.31}$$

$$g\eta_3 = (c-\bar{u})^2 \left\{ 3\phi_3 c_1\left(4s_1^2+1\right) + \phi_1\phi_2 c_1\left(5s_1^2-1\right) + \frac{3}{8}\phi_1^3 c_1 s_1^2 + 2\phi_1\phi_4 c_1\left(18s_1^4+5s_1^2-1\right) + \frac{1}{2}\phi_2^2\phi_1 c_1\left(21s_1^4+9s_1^2\right) + \frac{27}{4}\phi_1^2\phi_3 c_1\left(4s_1^4+s_1^2\right) + \frac{1}{8}\phi_1^3\phi_2 c_1\left(75s_1^4+27s_1^2\right) + \frac{1}{128}\phi_1^5 c_1\left(75s_1^4+36s_1^2\right) \right\} \tag{1.32}$$

$$g\eta_4 = (c-\bar{u})^2 \left\{ 4\phi_4\left(8s_1^4+2s_1^2+1\right) + \phi_2^2\left(8s_1^4+2s_1^2+1\right) + \frac{1}{2}\phi_1\phi_3\left(40s_1^4+24s_1^2-3\right) + 2\phi_1^2\phi_2\left(3s_1^4+2s_1^2\right) + \frac{1}{12}\phi_1^4\left(4s_1^4+3s_1^2\right) \right\} \tag{1.33}$$

$$g\eta_5 = (c-\bar{u})^2 \left\{ 5\phi_5 c_1\left(16s_1^4+12s_1^2+1\right) + 2\phi_1\phi_4 c_1\left(31s_1^4+13-1\right) + \phi_2\phi_3 c_1\left(52s_1^4+39s_1^2-3\right) + \frac{1}{2}\phi_2^2\phi_1 c_1\left(45s_1^4+25s_1^2\right) + \frac{1}{8}\phi_1^2\phi_3 c_1\left(220s_1^4+75s_1^2\right) + \frac{1}{24}\phi_1^3\phi_2 c_1\left(175s_1^4+75s_1^2\right) + \frac{1}{384}\phi_1^5 c_1\left(125s_1^4+60s_1^2\right) \right\} \tag{1.34}$$

The above algebraic equations representing the free-surface elevation in terms of the potential functions (Eqns. 1.23 to 1.34) can be solved using the perturbation method. Let us expand the free-surface elevation (η_n), potential function (ϕ_n) and $(c-\bar{u})^2_{(n)}$ as series (n = 1, 2,3, 4, 5), taking (ε) as the perturbation parameter that satisfies the following condition:

$$\text{For } H = 0, \varepsilon = 0 \tag{1.35}$$

$\eta_{nj}, \phi_{n,j}, c^2_{(j)}$ are called perturbation coefficients, validating the following condition:

For $n > j$, $(j = 1,2,3,4,5)$, these coefficients will be zero. Let us assume the following series for these coefficients:

$$\eta_n = \varepsilon\eta_{n1} + \varepsilon^2\,\eta_{n2} + \varepsilon^3\,\eta_{n3} + \varepsilon^4\,\eta_{n4} + \varepsilon^5\,\eta_{n5} \tag{1.36}$$

$$\phi_n = \varepsilon\phi_{n1} + \varepsilon^2\,\phi_{n2} + \varepsilon^3\,\phi_{n3} + \varepsilon^4\,\phi_{n4} + \varepsilon^5\,\phi_{n5} \tag{1.37}$$

$$\left(c - \bar{u}\right)^2 = c^2_{(0)} + \varepsilon c^2_{(1)} + \varepsilon^2 c^2_{(2)} + \varepsilon^3 c^2_{(3)} + \varepsilon^4 c^2_{(4)} \tag{1.38}$$

By substituting Eqns. (1.36 to 1.38) in Eqns. (1.23 to 1.34), and equating the same order of (ε), one can obtain the separate algebraic perturbation equations. As the number of perturbation coefficients is larger than that of the perturbation equations, one of them is normalized to unity ($k\eta_{11} = 1$), and other values, namely ($k\eta_{12}, k\eta_{13}, k\eta_{14}, k\eta_{15}$) are left undetermined. Using the terminology, $s_1 = \sinh(kh)$ & $c_1 = \cosh(kh)$, the following are the set of expressions for $c^2_{(i)}$:

$$c^2_{(0)} = \frac{g}{k}\frac{s_1}{c_1} \tag{1.39}$$

$$c^2_{(1)} = 0 \tag{1.40}$$

$$c^2_{(2)} = \frac{g}{k}\frac{s_1}{c_1}\left[1 + \frac{3}{2s_1^2} + \frac{9}{8s_1^4}\right] \tag{1.41}$$

$$c^2_{(3)} = \frac{g}{k}\frac{s_1}{c_1}\left[2 + \frac{3}{s_1^2} + \frac{9}{4s_1^4}\right]\mathrm{k}\eta_{12} \tag{1.42}$$

$$c^2_{(4)} = \frac{g}{k}\frac{s_1}{c_1}\left\{\left[1 + \frac{3}{2s_1^2} + \frac{9}{8s_1^4}\right]\left[\left(\mathrm{k}\eta_{12}\right)^2 + 2\mathrm{k}\eta_{13}\right] + \left[\frac{5}{4} + \frac{17}{4s_1^2} + \frac{135}{32s_1^4} + \frac{27}{32\,s_1^6} - \frac{27}{128\,s_1^8} + \frac{81}{512\,s_1^{10}}\right]\right\} \tag{1.43}$$

The perturbation coefficients for the free-surface elevation are given below (Zhao et al., 2016):

$$\eta_{11} = \frac{1}{k} \tag{1.44}$$

$$\eta_{22} = \frac{1}{2k}\frac{c_1}{s_1}\left\{1 + \frac{3}{2s_1^2}\right\} \tag{1.45}$$

$$\eta_{23} = \frac{c_1}{s_1}\left\{1 + \frac{3}{2s_1^2}\right\}\eta_{12} \tag{1.46}$$

$$\eta_{24} = \frac{1}{2k}\frac{c_1}{s_1}\left\{1 + \frac{3}{2s_1^2}\right\}\left[(\mathrm{k}\eta_{12})^2 + 2\mathrm{k}\eta_{13}\right] + \frac{1}{k}\frac{c_1}{s_1}\left[\frac{17}{24} + \frac{25}{48s_1^2} - \frac{21}{16s_1^4} - \frac{81}{64s_1^6} - \frac{27}{128s_1^8}\right] \tag{1.47}$$

$$\eta_{25} = \frac{c_1}{s_1}\left\{1 + \frac{3}{2s_1^2}\right\}\left[\eta_{14} + (\mathrm{k}\eta_{13})\eta_{12}\right] + \eta_{12}\left[\frac{17}{6} + \frac{25}{12s_1^2} - \frac{21}{4s_1^4} - \frac{81}{16s_1^6} - \frac{27}{32s_1^8}\right] \tag{1.48}$$

$$\eta_{33} = \frac{1}{k}\left[\frac{3}{8} + \frac{9}{8s_1^2} + \frac{9}{8s_1^4} + \frac{27}{64s_1^6}\right] \tag{1.49}$$

$$\eta_{34} = \eta_{12}\left[\frac{9}{8} + \frac{27}{8s_1^2} + \frac{27}{8s_1^4} + \frac{81}{64s_1^6}\right] \tag{1.50}$$

$$\eta_{35} = \frac{1}{k}\left[\frac{9}{8} + \frac{27}{8s_1^2} + \frac{27}{8s_1^4} + \frac{81}{64s_1^6}\right]\left[(k\eta_{12})^2 + k\eta_{13}\right] + \frac{1}{k}\left(\frac{1}{6s_1^2 + 5}\right) \left[\frac{3393}{128} + \frac{459}{64}s_1^2 \frac{1845}{64s_1^2} - \frac{2025}{256s_1^4} - \frac{20277}{512s_1^6} - \frac{7857}{512s_1^8} - \frac{19197}{2048s_1^{10}} - \frac{3645}{4096s_1^{12}}\right] \tag{1.51}$$

$$\eta_{44} = \frac{1}{k}\frac{c_1}{s_1}\left(\frac{1}{6s_1^2 + 5}\right)\left[\frac{53}{3} + 2s_1^2 + \frac{365}{24s_1^2} + \frac{51}{4s_1^4} + \frac{351}{64s_1^6} + \frac{135}{128s_1^8}\right] \tag{1.52}$$

$$\eta_{45} = \frac{1}{k}\frac{c_1}{s_1}\left(\frac{1}{6s_1^2 + 5}\right)\eta_{12}\left[\frac{106}{3} + 8s_1^2 + \frac{365}{6s_1^2} + \frac{51}{s_1^4} + \frac{351}{16s_1^6} + \frac{135}{32s_1^8}\right] \tag{1.53}$$

$$\eta_{55} = \frac{1}{k}\left(\frac{1}{6s_1^2 + 5}\right)\left(\frac{1}{8s_1^2 + 5}\right)\left[\frac{37715}{128} + \frac{19895}{192}s_1^2 + \frac{125}{8}s_1^4 + \frac{29985}{64s_1^2} + \frac{349595}{798s_1^4} + \frac{142085}{512s_1^6} + \frac{13455}{128s_1^8} + \frac{47925}{2048s_1^{10}} + \frac{10125}{4096s_1^{12}}\right] \tag{1.54}$$

The perturbation coefficients for the potential function are given below (Zhao et al., 2016):

$$\phi_{11} = \frac{1}{s_1} \tag{1.55}$$

$$\phi_{12} = \frac{1}{s_1}k\eta_{12} \tag{1.56}$$

$$\phi_{13} = \frac{1}{s_1}\left[k\eta_{13} - \left(\frac{9}{8} + \frac{15}{8s_1^2} + \frac{3}{4s_1^4}\right)\right] \tag{1.57}$$

$$\phi_{14} = \frac{1}{s_1}\left[k\eta_{14} - k\eta_{12}\left(\frac{27}{8} + \frac{45}{8s_1^2} + \frac{9}{4s_1^4}\right)\right] \tag{1.58}$$

$$\begin{aligned}\phi_{15} = \frac{1}{s_1}\Bigg\{k\eta_{15} - \Bigg\{&\left(\frac{27}{8} + \frac{45}{8s_1^2} + \frac{9}{4s_1^4}\right)\left[\left(k\eta_{12}\right)^2 + k\eta_{13}\right]\\ &- \left[\frac{5}{24} + \frac{1}{24s_1^2} - \frac{221}{192s_1^4} - \frac{669}{512s_1^6} - \frac{117}{512s_1^8} + \frac{27}{256s_1^{10}}\right]\Bigg\}\Bigg\}\end{aligned} \tag{1.59}$$

$$\phi_{22} = \frac{3}{8s_1^4} \tag{1.60}$$

$$\phi_{23} = \frac{3}{4s_1^4}k\eta_{12} \tag{1.61}$$

$$\phi_{24} = \frac{3}{8s_1^4}\left[\left(k\eta_{12}\right)^2 + k\eta_{13}\right] + \frac{1}{4s_1^2}\left[1 - \frac{29}{24s_1^2} - \frac{21}{4s_1^4} - \frac{27}{8s_1^6} - \frac{27}{64s_1^8}\right] \tag{1.62}$$

$$\phi_{25} = \frac{3}{4s_1^4}\left[\left(k\eta_{14}\right) + \left(k\eta_{12}\,k\eta_{13}\right)\right] + k\eta_{12}\left[\frac{1}{s_1^2} - \frac{29}{24s_1^4} - \frac{21}{4s_1^6} - \frac{27}{8s_1^8} - \frac{27}{64s_1^{10}}\right] \tag{1.63}$$

$$\phi_{33} = -\frac{1}{16s_1^5} + \frac{9}{64s_1^7} \tag{1.64}$$

$$\phi_{34} = k\eta_{12}\left(-\frac{3}{16s_1^5} + \frac{27}{64s_1^7}\right) \tag{1.65}$$

$$\begin{aligned}\phi_{35} = &\left(-\frac{3}{16s_1^5} + \frac{27}{64s_1^7}\right)\left[\left(k\eta_{12}\right)^2 + k\eta_{13}\right] + \frac{1}{8s_1^3\left(6s_1^2 + 5\right)}\\ &\left[s_1^2 + \frac{87}{4} + \frac{1059}{32s_1^2} - \frac{813}{64s_1^4} - \frac{2835}{64s_1^6} - \frac{5589}{256s_1^8} - \frac{1215}{512s_1^{10}}\right]\end{aligned} \tag{1.66}$$

$$\phi_{44} = \frac{1}{\left(6s_1^2 + 5\right)}\left[\frac{5}{96s_1^4} - \frac{3}{8s_1^6} - \frac{9}{256s_1^8} - \frac{135}{512s_1^{10}}\right] \tag{1.67}$$

$$\phi_{45} = \frac{1}{\left(6s_1^2 + 5\right)}k\eta_{12}\left[\frac{5}{24s_1^4} - \frac{3}{2s_1^6} - \frac{9}{64s_1^8} + \frac{135}{128s_1^{10}}\right] \tag{1.68}$$

$$\phi_{55} = \frac{1}{8s_1^3\left(6s_1^2+5\right)\left(8s_1^2+5\right)}\left[-\frac{3}{8}+\frac{121}{16s_1^2}-\frac{131}{16s_1^4}-\frac{279}{16s_1^6}+\frac{135}{256s_1^8}+\frac{2025}{512s_1^{10}}\right] \tag{1.69}$$

By substituting Eqns. (1.36) to (1.38) in Eqns. (1.23) and (1.29) and neglecting terms higher than the fifth order, one can get the following expressions for set up and total head, respectively.

$$\begin{aligned}\eta_0 = \frac{1}{k}\frac{c_1}{s_1}\Bigg\{&\left(\frac{1}{2}\varepsilon^2\right)+\varepsilon^3\left(k\eta_{12}\right)+\varepsilon^4\left[\frac{1}{2}\left(k\eta_{12}\right)^2+\left(k\eta_{13}\right)+\left(-\frac{1}{8}+\frac{3}{8s_1^4}+\frac{9}{32s_1^6}\right)\right]\\&+\varepsilon^5\left[k\eta_{13}+\left(k\eta_{12}\right)\left(k\eta_{13}\right)+k\eta_{12}\left(-\frac{1}{2}+\frac{3}{2s_1^4}+\frac{9}{8s_1^6}\right)\right]\Bigg\}\end{aligned} \tag{1.70}$$

$$\begin{aligned}Z_0 = &\frac{u^{-2}}{2g}+\frac{\varepsilon^2}{k}+\left[\frac{C_1}{2s_1}+\frac{1}{4s_1c_1}\right]+\frac{\varepsilon^3}{k}\left(k\eta_{12}\right)\left[\frac{c_1}{s_1}+\frac{1}{2s_1c_1}\right]\\&+\frac{\varepsilon^4}{k}\Bigg\{[\left(k\eta_{12}\right)^2+2k\eta_{13}]\left[\frac{c_1}{2s_1}+\frac{1}{4s_1c_1}\right]+\frac{c_1}{2s_1}\left[-\frac{1}{4}+\frac{3}{4s_1^4}+\frac{9}{16s_1^{16}}\right]\\&+\frac{1}{4c_1s_1}\left[-\frac{5}{4}-\frac{9}{4s2_1^2}-\frac{3}{8s_1^4}+\frac{9}{8s_1^6}\right]\Bigg\}+\frac{\varepsilon^5}{k}\Bigg\{\left(\frac{c_1}{s_1}+\frac{1}{2s_1c_1}\right)\left(k\eta_{14}+\left(k\eta_{12}k\eta_{13}\right)\right)\\&+\left(k\eta_{12}\right)\left[\frac{c_1}{s_1}\left[-\frac{1}{2}+\frac{3}{2s_1^4}+\frac{9}{2s_1^6}\right]+\frac{1}{2c_1s_1}\right]\left[-\frac{5}{2}-\frac{9}{2s_1^4}-\frac{3}{4s_1^6}+\frac{9}{8s_1^6}\right]\Bigg\}\end{aligned} \tag{1.71}$$

The solution for Stokes' fifth-order equation is given as:

$$\begin{aligned}\eta = \eta_0 &+\varepsilon\cos(\theta)\left(\eta_{11}+\varepsilon\eta_{12}+\varepsilon^2\eta_{13}+\varepsilon^3\eta_{14}+\varepsilon^4\eta_{15}\right)\\&+\varepsilon^2\cos(2\theta)\left(\eta_{22}+\varepsilon\eta_{23}+\varepsilon^2\eta_{24}+\varepsilon^3\eta_{25}\right)\\&+\varepsilon^3\cos(3\theta)\left(\eta_{33}+\varepsilon\eta_{34}+\varepsilon^2\eta_{35}\right)+\varepsilon^4\cos(4\theta)\left(\eta_{44}+\varepsilon\eta_{45}\right)\\&+\varepsilon^5\cos(5\theta)\,\eta_{55}\end{aligned} \tag{1.72}$$

$$\begin{aligned}\varphi = \bar{u}x+\frac{c-\bar{u}}{k}\{&\sin(\theta)\cosh\left(k(h+z)\right)\varepsilon\left(\phi_{11}+\varepsilon\phi_{12}+\varepsilon^2\phi_{13}+\varepsilon^3\phi_{14}+\varepsilon^4\phi_{15}\right)\\&+\sin(2\theta)\cosh\left(2k(h+z)\right)\varepsilon^2\left(\phi_{22}+\varepsilon\phi_{23}+\varepsilon^2\phi_{24}+\varepsilon^3\phi_{25}\right)\\&+\sin(3\theta)\cosh\left(3k(h+z)\right)\varepsilon^3\left(\phi_{33}+\varepsilon\phi_{34}+\varepsilon^2\phi_{35}\right)\\&+\sin(4\theta)\cosh\left(4k(h+z)\right)\varepsilon^4\left(\phi_{44}+\varepsilon\phi_{45}\right)+\sin(5\theta)\cosh\left(5k(h+z)\right)\varepsilon^5\,\phi_{55}\}\end{aligned} \tag{1.73}$$

$$\left(c-\bar{u}\right)^2 = c_{(0)}^2+\varepsilon c_{(1)}^2+\varepsilon^2c_{(2)}^2+\varepsilon^3c_{(3)}^2+\varepsilon^4c_{(4)}^2 \tag{1.74}$$

1.4 WAVE SPECTRA

Wave data are approximately collected for 20 minutes every three hours and assumed to represent the stationary sea state between the measurements. The sea state is defined by significant wave height, significant wave period, peak period, and wave direction. Data collection is done by visual investigation and instrumental observations by buoys, radars, lasers, and satellites. The sea state in the short term is typically 3 hours and is assumed as a zero-mean, ergodic Gaussian process. It can be defined completely by a wave spectrum. While the JONSWAP spectrum is recommended for the North Sea, the Pierson-Moskowitz (PM) spectrum is recommended for open sea conditions. In the long term, variation of sea state is slower than the short-term fluctuations. It is approximated by a series of stationary, non-zero-mean Gaussian processes specified by the significant wave height (H_s) and peak wave period (T_p). The wave spectrum describes the energy distribution of different frequencies of a sea state. The spectrum should be selected based on the frequency characteristics of the wave environment (Boaghe et al., 1998; Isaacson et al., 2000). The following are a few relevant spectra applicable in the design and development of wave energy devices.

1.4.1 PM Spectrum for Wave Loads

The PM spectrum is a one-parameter spectrum used for fully developed sea conditions generated by relatively moderate winds over large fetch.

$$S^{+}(\omega)=\frac{\alpha g^{2}}{\omega^{5}}\exp\left[-1.25\left(\frac{\omega}{\omega_{0}}\right)^{-4}\right] \tag{1.75}$$

Where, α is the Phillips constant $\cong 0.0081$.

1.4.2 Modified PM Spectrum (Two Parameters, H_s, ω_0)

The modified PM spectrum is a two-parameter spectrum that was developed to account for the wave height. This spectrum is suitable for fully developed sea conditions, and it is usually employed to describe the tropical storm waves generated by hurricanes. It has a greater frequency bandwidth.

$$S^{+}(\omega)=\frac{5}{16}H_{s}\frac{\omega_{0}^{4}}{\omega^{5}}\exp\left[-1.25\left(\frac{\omega}{\omega_{0}}\right)^{-4}\right] \tag{1.76}$$

1.4.3 International Ship Structures Congress (ISSC) Spectrum (Two Parameters, H_s, $\bar{\omega}$)

The ISSC spectrum is a slight modification of the Bretschneider spectrum, and it is recommended for fully developed sea conditions. This spectral equation is true only for a narrow-banded spectrum, and the wave elevation follows a Gaussian distribution.

$$S^{+}(\omega) = 0.1107 H_s \frac{\omega^{-4}}{\omega^5} \exp\left[-0.4427\left(\frac{\omega}{\bar{\omega}}\right)^{-4}\right] \tag{1.77}$$

$$\bar{\omega} = \frac{M_1}{M_0}$$

Where M_1 and M_0 are spectral moments.

1.4.4 Joint North Sea Wave Project (JONSWAP) Spectrum (Five Parameters, H_s, ω_0, γ, τ_a, τ_b)

The JONSWAP spectrum is a modified form of the PM spectrum, and it is recommended for use in the reliability analysis. This spectrum is applicable only for limited fetch, and it is used to describe the winter storm waves of the North Sea.

$$S^{+}(\omega) = \frac{\bar{\alpha} g^2}{\omega^5} \exp\left[-1.25\left(\frac{\omega}{\omega_0}\right)^{-4}\right] \gamma^{a(\omega)} \tag{1.78}$$

where γ is the peakedness parameter. The value of 3.3 yields a mean spectrum for a specified wind speed and a given fetch length. The variation depends upon the duration of the wind and the stage of growth and decay of the storm. This value follows a normal probability distribution.

$$a(\omega) = \exp\left[-\frac{(\omega - \omega_0)^2}{2\bar{\sigma}^2 \omega_0^2}\right] \tag{1.79}$$

where $\bar{\sigma}$ is spectral width parameter or shape parameter and is given by:

$$\bar{\sigma}_a = 0.07, \omega \le \omega_0 \tag{1.80}$$

$$\bar{\sigma}_b = 0.09, \omega > \omega_0 \tag{1.81}$$

The modified Phillips constant is given by:

$$\bar{\alpha} = 3.25 \times 10^{-3} H_s^2 \omega_0^4 \left[1 - 0.287 \ln(\gamma)\right] \tag{1.82}$$

$$\gamma = 5 \, for \frac{T_p}{\sqrt{H_s}} \le 3.6 \tag{1.83}$$

$$= \exp\left[5.75 - 1.15\frac{T_p}{\sqrt{H_s}}\right] \text{ for } \frac{T_p}{\sqrt{H_s}} > 3.6 \quad (1.84)$$

$$H_s = 4\sqrt{m_0} \quad (1.85)$$

where γ varies from 1 to 7. As shown in Fig. 1.5, the wave spectra plot compares significant mean wind speed 20 m/s, wave height 5 m, and period 10 seconds. It is seen from the figure that the modified PM spectrum and the ISSC spectrum have the same spectral distribution, while the JONSWAP spectrum shows a higher energy peak.

The MATLAB® program for the wave spectra plot is given below. The variables are mean wind speed, significant wave height and time period.

```
%%WAVE SPECTRA plot ---- spectral density versus frequency ratio
%%Jonswap spectrum
hs=5; %wave height in m
t=10; %time period in seconds
v=3; %peakedness factor chosen between 1 to 7
g=9.81; %gravitational constant
w=0:0.000001:3; %frequency is the varying component
n=length(w);
wo=(2*pi)/t;
```

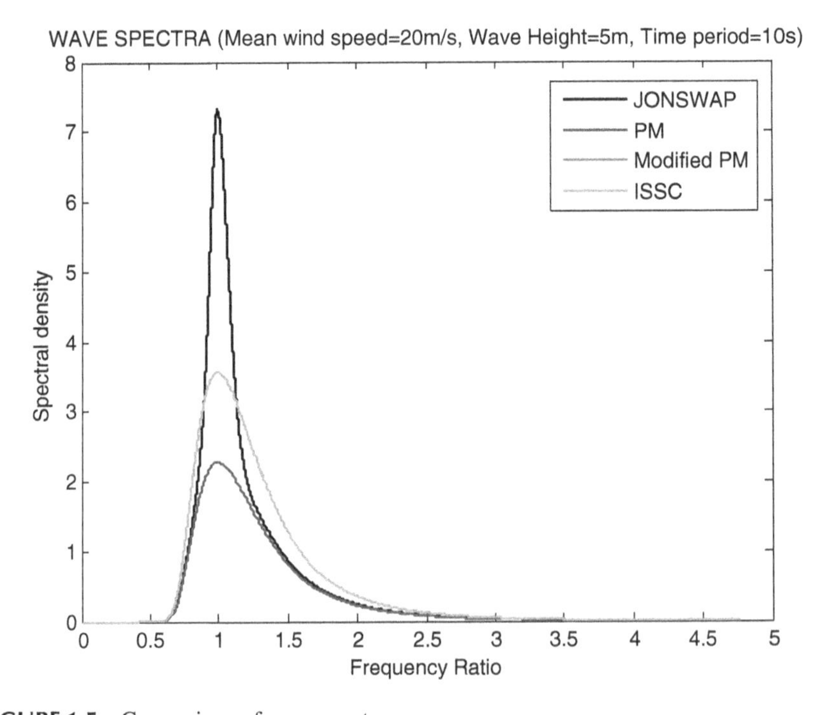

FIGURE 1.5 Comparison of wave spectra

```
alpha=3.25*(10^-3)*(hs^2)*(wo^4)*(1-(0.287*(log(v))));
for i=1:n
if w(i)<=wo
   sigma(i)=0.07;%spectral width parameter
else
   sigma(i)=0.09;
end
x(i)=-((w(i)-wo)^2)/(2*(sigma(i)^2)*((wo)^2));
y(i)=-1.25*((w(i)/wo)^(-4));
aw(i)=exp(x(i));
z(i)=exp(y(i))*(v^aw(i))*alpha*(g^2);
s(i)=z(i)/(w(i)^5);
p(i)=w(i)/wo;
i=i+1;
end
%%PM spectrum
hs=5; %wave height in m
t=10; %time period in seconds
g=9.81; %gravitational constant
v=20; %mean wind speed in m/s
wo=(2*pi)/t;
w=0:0.0001:3; %frequency is the varying component
n=length(w);
for i=1:n
x(i)=-1.25*((w(i)/wo)^(-4));
a(i)=exp(x(i));
b(i)=1/((w(i))^5);
s1(i)=0.0081*a(i)*b(i)*(g)^2;
p1(i)=w(i)/wo;
i=i+1;
end
%%Modified PM spectrum
hs=5; %wave height in m
t=10; %time period in seconds
wo=(2*pi)/t;
w=0:0.0001:3; %frequency is the varying component
n=length(w);
for i=1:n
y(i)=(-1.25)*((w(i)/wo)^(-4));
a(i)=exp(y(i));
b(i)=(wo^4)/((w(i))^5);
s2(i)=0.3125*((hs^2)*a(i)*b(i));
p2(i)=w(i)/wo;
i=i+1;
end
```

```
%%ISSC spectrum
hs=5; %wave height in m
t=10; %time period in seconds
wo=(2*pi)/t;
w=0:0.0001:3; %frequency is the varying component
n=length(w);
for i=1:n
y(i)=-1.2489*((w(i)/wo)^(-4));
a(i)=exp(y(i));
b(i)=(wo^4)/((w(i))^5);
s3(i)=0.3123*((hs^2)*a(i)*b(i));
p3(i)=w(i)/wo;
i=i+1;
end
plot(p,s,'k'); %jonswap spectrum
hold on;
plot(p1,s1,'r');%Modified PM spectrum
hold on;
plot(p2,s2,'b'); %Bretschneider spectrum
hold on;
plot(p3,s3,'g'); %ISSC spectrum
xlabel('Frequency Ratio');
ylabel('Spectral density');
title('WAVE SPECTRA (Mean wind speed=20m/s, Wave Height=5m, Time
period=10s)');
```

In reality, ocean waves combine a set of waves with different frequencies and directions; they appear as irregular or random waves (Chandrasekaran and Anubhab, 2004; 2005). Random waves are represented by wave energy density spectra, which describe the energy content of the ocean waves. It is spread over a wide frequency ranging from zero to infinite value, but waves are concentrated on a narrow band. Their statistical parameters characterize random waves. Sea states are represented by the significant wave height (Hs) and zero-crossing periods (Tz). Different sea states used to describe random waves are summarized in Table 1.2.

Several spectral models are available for use and are derived based on the observed properties of the ocean; hence, they are empirical. The frequency characteristics of

TABLE 1.2
Characteristics of Random Sea States

Sea State Description	Significant Wave Height H_s (m)	Zero Crossing Period T_z (s)	Wind Velocity (m/s)
Moderate	6.5	8.15	15
High	10	10	35
Very high	15	15	45

the real sea conditions influence the spectral formulation. The most commonly used spectral formulae include the PM spectrum, the JONSWAP spectrum, the ISSC spectrum, the Bredsneidger spectrum, and the Ochi-Hubble spectrum (Chandrasekaran, 2015; 2017; 2019b; 2020). Each spectrum model distributes the wave energy differently across the frequency band, so the structure's response will vary for the same wave height if different spectra are used. The commonly used spectral models in offshore structural design are the Pierson Moscowitz (PM) spectrum and the JONSWAP spectrum. The PM spectrum applies to various regions such as the Gulf of Mexico, offshore Brazil, Western Australia, offshore Newfoundland, and Western Africa, both in operational and survival conditions. The JONSWAP spectrum applies only to the North Sea for operational and survival conditions. The PM spectrum, representing the wave energy distribution under different frequencies, is suitable for open sea conditions. This spectrum is neither fetch-limited nor duration-limited. It is a two-parameter spectrum, developed under moderate wind conditions existing over large fetch and is given by the following relationship:

$$S^{+}(\omega) = \frac{1}{2\pi}\frac{H_s^2}{4\pi T_z^2}\left(\frac{2\pi}{\omega}\right)^2 \exp\left(-\frac{1}{\pi T_z^4}\left(\frac{2\pi}{\omega}\right)^4\right) \tag{1.86}$$

where H_s is the significant wave height, T_z is the zero-crossing period, and ω is the frequency. The spectral plot shows that the wave energy is concentrated on a narrow band. The typical wave energy PM spectra under different sea states are shown in Fig. 1.6. The sea state can be described using a suitable scale based on sea conditions,

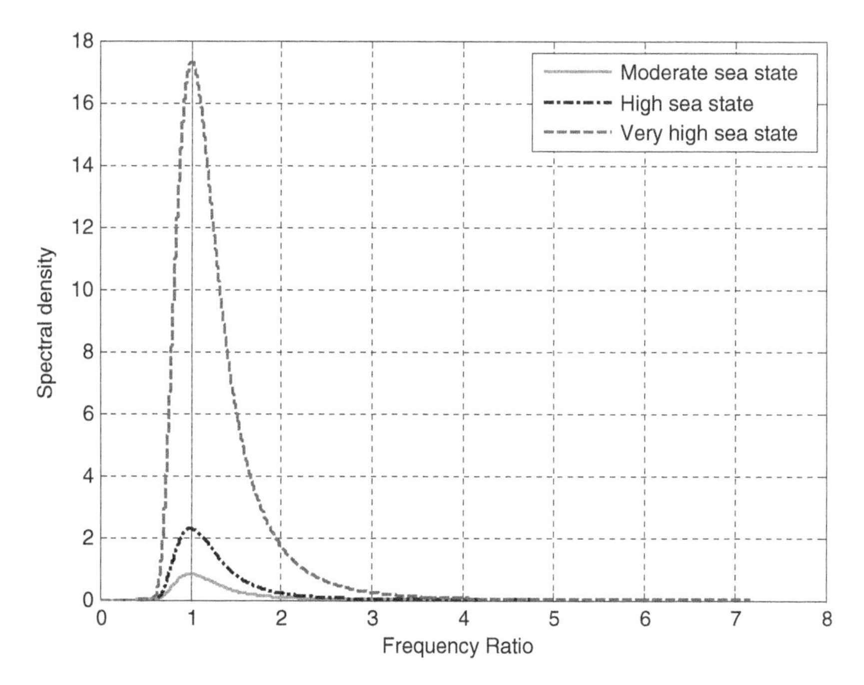

FIGURE 1.6 PM spectrum for different sea states

wave height, and wind speed. One such empirical measure for expressing the sea state is the Beaufort scale, developed by Irish Royal Navy officer Francis Beaufort in the nineteenth century. As per the World Meteorological Organization, the scale is standardized into thirteen classes, from zero to twelve. Since the description of the sea surface is more useful than the wind speed and wave heights alone, the World Meteorological Organization and MetService started defining the sea state using the Douglas sea state, developed in 1920 by Captain H. P. Douglas, which describes the roughness of the sea from smooth to very high (Chandrasekaran and Merin, 2016; Chandrasekaran and Madhavi, 2014a; 2014b; 2014c; 2014d; 2015a; 2015b; 2015c; 2015d; 2016; Chandarsekaran and Vishruth, 2013; Chandrasekaran et al., 2013; 2014a; 2014b; 2014c; Chandrasekaran and Abhishek, 2010).

1.5 WIND AND CURRENT

In addition to wave loads, offshore wave energy devices are also subjected to wind and current loads. The low-frequency motions of the device get more excited in the presence of wind. The fluctuating wind component, called the gust component, induces low-frequency oscillations. The magnitude of oscillation increases with the increase in the device's exposed area and the wind velocity. Another common occurrence introduced by the wind effect in the open sea is current. Current adds varying pressure distribution around the structural member, resulting in a steady drag force. The combined action of wind, waves, and current respond significantly to the floating offshore wave energy devices.

A significant wave generation source is the wind; hence, the floating wave energy device analysis in the wave-alone environment is not realistic. The dynamic wind effect will be significant on the superstructure of wind energy devices. Therefore, the impact of both mean wind and the gust component should be considered in the analysis. In the design of offshore structures, an average wind speed occurring over a one-hour duration is taken as steady wind speed. It is typically measured at 10 m above the mean sea level (MSL). Wind loads cause offset, resulting in the set-down of the deck as well. It induces different moments in pitch degree-of-freedom, which influences tension variation in tethers or the mooring lines. Thus, a strong coupling between surge, heave, and pitch is inherent in floating structures.

A wind spectrum can describe random wind blowing over a structure. Various spectral formulations such as Davenport spectra show significant differences at lower frequencies in the spectral density plots (Davenport, 1961; Harris, 1971; Kaimal, 1972; Simiu, 1971). The Davenport spectrum has more moderate energy content at lower frequencies than other wind spectra (Zaheer and Islam, 2008; 2012; 2017). Besides, the Davenport spectrum, being developed for land-based conditions, may not represent the wind velocity fluctuations at low frequencies in the offshore environment. Therefore, the most preferred wind spectrum for the analysis of offshore structures is the American Petroleum Institute (API) spectrum. It shows a higher energy content in the lower frequencies compared to other spectral formulations. The equation for the API spectrum is given by:

$$\frac{\omega S_u^+(\omega)}{\sigma_u(z)^2} = \frac{\theta}{(1+1.5\theta)^{5/3}} \tag{1.87}$$

where θ is frequency ratio or derivable variable $\left[\theta = \frac{\omega}{\omega_p}\right]$, ω_p is the peak frequency, z_s is the surface height (20 m), $\sigma_u(z)^2$ is the variance of $U(t)$ at a reference height, z is reference height (=10 m), and $S_u^+(\omega)$ is the spectral density (Chandrasekaran, 2014c; 2016; 2017; 2019a; Chandrasekaran et al., 2021). Variance $\sigma_u(z)^2$ at the reference height is given by:

$$0.01 \le \frac{\omega_p^2}{\bar{U}_z} \le 0.1 \tag{1.88}$$

$$\sigma_u(z) = \begin{cases} 0.15\bar{U}_z \left(\frac{z_s}{z}\right)^{0.125} & (if \rightarrow 2 \le z_s) \\ 0.15\bar{U}_z \left(\frac{z_s}{z}\right)^{0.275} & (if \rightarrow 2 > z_s) \end{cases} \tag{1.89}$$

The API wind spectral density plot for different wind velocities is shown in Fig. 1.7, which shows a significant energy concentration at lower frequencies. Unlike the wave

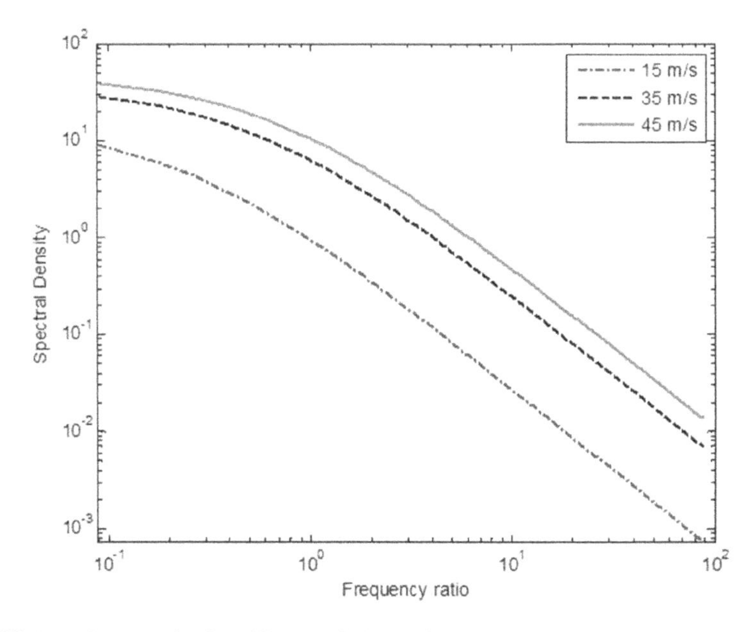

FIGURE 1.7 Spectra plot for different wind velocities

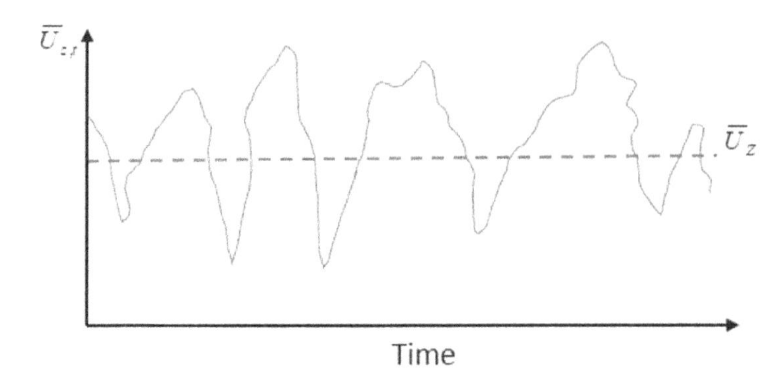

FIGURE 1.8 Dynamic variation of wind velocity

spectra, the wind spectra are found to be broad-banded. Current occurs commonly in the open sea due to the wind effect on water, tidal motion, temperature differences, density gradients, and salinity variations. The presence of current modifies the water particle kinematics and the apparent wave period. It also imposes drag forces on structures. The current velocity varies with depth; the highest value is observed near the MSL. The wind-generated current is applied in the same direction as that of the wave and wind loads.

The wind is the air circulation in the atmosphere, occurring due to unequal heating of the Earth's surface between the tropics and poles. Wind movement occurs due to the transfer of heat energy, which is profoundly affected by the Earth's rotation. The deflection of wind due to the rotation of the Earth is called the Coriolis effect. The wind is responsible for the generation of ocean waves and currents. The wind loads on offshore structures are quite complex due to the mean and time-varying wind velocity components, as shown in Fig. 1.8.

The time-varying component contributes to the dynamic effect of wind load on the structures. However, structures are designed by considering the wind loads as static by incorporating the dynamic component indirectly using the gust factor, which is the currently accepted method for computing wind effects on the structure. The time-varying component of wind velocity is termed wind gust, and it depends upon the mean wind velocity, height from the ground level, and the surface roughness of the exposed area. The wind direction depends upon the pressure gradient and the Coriolis force. The wind velocity is constant at greater heights and is referred to as gradient wind speed. At lower altitudes, eddy formation due to the surface roughness of the Earth causes the wind gusts to vary continuously with space and time. Wind speed of above 20.0 m/s and fetch length greater than 30.0 km at 10.0 m above the sea surface is assumed as stable wind condition, where the wind time history is stationary.

The wind is usually separated into mean and dynamic components for analytical considerations. The mean wind velocity and associated dynamic components are always calculated over a particular period due to the variation of the wind velocity concerning space and time. The energy spectrum of wind can be developed from the spectral analysis of wind velocity time history over such an extended period. The wind spectrum consists of short-period components due to gust, while the long-period component occurs due to the change in pressure. There is a spectral gap of about 360 seconds to 1.60 hours. It results from the lack of contribution from both wind gusts and pressure change. The pressure effect and the wind gusts change the wind velocity slowly and rapidly about the mean, respectively. An hour-average wind period is used for analysis, and the design wind speeds are usually specified by considering its return period. The commonly used wind spectra for offshore locations are the Kaimal spectrum and the API spectrum. The more straightforward equation used to represent the distribution of wind speed to height is the power-law distribution, which is given by,

$$\bar{U}_Z = \bar{U}_{10}\left(\frac{z}{10}\right)^{1/7} \tag{1.90}$$

where $\bar{U}_{10}$ is the basic design wind speed at 10.0 m above the MSL, $\bar{U}_Z$ is the wind speed at height z from the MSL (API, 2000). The spatial variation of mean wind velocity in an offshore location is shown in Fig. 1.9. It can be seen from the figure that the wind load acts on the exterior and transparent portions of the deck of the platform.

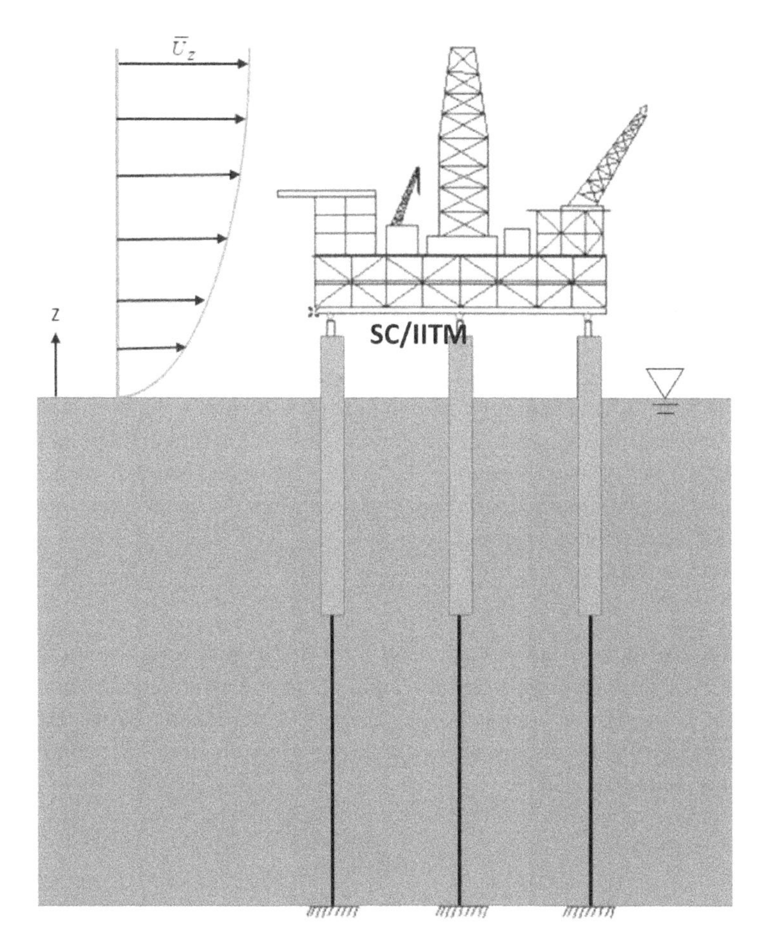

FIGURE 1.9 Spatial variation of mean wind velocity

From the design wind velocity, the wind force acting horizontally on a surface can be calculated using the following equation:

$$F_H = \frac{1}{2} C_D \rho_a U_z^2 A_p \tag{1.91}$$

Where F_H is the horizontal wind force, C_D is the drag coefficient, ρ_a is the wind density, and A_p is the projected area. The drag coefficient value depends upon the shape of the sections under consideration and the wind velocity. The wind force should be calculated for all the members separately and then added together in an offshore deck.

Wind forces on floating offshore structures are caused by complex fluid-dynamics phenomena, which are generally difficult to calculate with high accuracy. Wind pressure and wind force on a plate, placed orthogonal to the wind flow direction, are given as follows:

$$p_w = \frac{1}{2} \rho_a C_w v^2 \tag{1.92}$$

$$F_w = p_w A \tag{1.93}$$

where, ρ_a is the mass density of air (1.25 kg/m^3), and C_w is wind pressure coefficient. It is important to note that the mass density of air increases due to the water spray (splash) up to a height of 20 m above the MSL. If the plate is positioned at an angle (θ) with respect to the wind flow, then the appropriate projected area, normal to the flow direction, should be used in the above equation. The wind pressure coefficient (C_w) is experimentally determined in the wind tunnel under the controlled stationary wind-flow conditions. It depends on the Reynolds number, which has a typical value of 0.7 to 1.2 for cylindrical members.

The natural wind has two components: i) mean wind component (which is a static component); and ii) fluctuating, gust component (which is a dynamic component). The gust component is generated by the turbulence of the flow field in all three spatial directions. For offshore locations, the mean wind speed is much greater than the gust component. This means that in most of the design cases, the static analysis will suffice. The wind velocity is given by:

$$v(t) = \bar{v} + v(t) \tag{1.94}$$

where, $\bar{v}$ is the mean wind velocity, and $v(t)$ is the gust component. The spatial dependence of the mean component is only along with the height, while the gust component is considered homogeneous in both space and time. Wind forces in the directions parallel (drag force) and normal to the wind direction (lift force) are given by the following relationships:

$$\begin{aligned} F_D &= \frac{1}{2} \rho C_D \bar{v}_z A \\ F_L &= \frac{1}{2} \rho C_L \bar{v}_z A \end{aligned} \tag{1.95}$$

The gust component of wind velocity can be obtained by multiplying a gust factor with sustained wind speed. The average gust factor (F_g) is 1.35 to 1.45; variation of the gust factor and height is negligible. The sustained wind speed, which is used in the design, is the one-minute average wind speed. The product of sustained wind speed and the gust factor will give the *fastest mile velocity*. A 200-year sustained wind velocity of 125 miles per hour is commonly recommended for offshore structures.

The gust component of wind is modeled probabilistically, which can also be approximated by an aerodynamic admittance function. The aerodynamic admittance function is used for two reasons: i) to bypass the rigorous random analysis; and ii) to increase the possibility of an accurate measurement of this function through wind-tunnel experiments. Expanding the terms in Eq. (1.95), we get:

$$\begin{aligned} F_w(t) &= \frac{1}{2}\rho_a C_w v^2 A \\ &= \frac{1}{2}\rho_a C_w A[\overline{v} + v(t)]^2 \\ &= \frac{1}{2}\rho_a C_w A[\overline{v}^2 + (v(t))^2 + 2\overline{v}\, v(t)] \\ &\text{by neglecting higher powers of gust component,} \\ &\cong \overline{F}_w + \rho_a C_w A\overline{v}\, v(t) \end{aligned} \tag{1.96}$$

The wind force is expressed as a sum of the mean and the gust components in the above equation. The wind is considered an ergodic process, whose one-sided power spectral density function is expressed as a function of the wind spectrum as below:

$$S_F^+(\omega) = [\rho_a C_w A\overline{v}]^2 S_U^+(\omega) \tag{1.97}$$

Substituting and rearranging the terms, we get:

$$S_F^+(\omega) = \frac{4[\overline{F}_w]^2}{[\overline{v}]^2}\left[\chi\left\{\frac{\omega\sqrt{A}}{2\pi\overline{v}}\right\}\right]^2 S_U^+(\omega) \tag{1.98}$$

In the above equation, force and the response spectra are connected by the aerodynamic admittance function, which varies as below:

$$\begin{aligned} &\text{for } \frac{\omega\sqrt{A}}{2\pi\overline{v}} \Rightarrow 0, \quad \chi\left\{\frac{\omega\sqrt{A}}{2\pi\overline{v}}\right\} \Rightarrow 1 \\ &\text{for } \frac{\omega\sqrt{A}}{2\pi\overline{v}} \Rightarrow \infty, \quad \chi\left\{\frac{\omega\sqrt{A}}{2\pi\overline{v}}\right\} \Rightarrow 0 \end{aligned} \tag{1.99}$$

The aerodynamic admittance function is found empirically, as proposed below (Davenport, 1961):

$$\chi(x) = \left\{\frac{1}{[1 + (2x)^{4/3}]}\right\} \tag{1.100}$$

1.5.1 Wind Spectra

Wind spectra for the analysis of offshore structures are expressed in terms of the circular frequency for a reference height of 10 m.

$$S_u^+(\omega) = fG_u^+(f) \tag{1.101}$$

where S_u^+(É) is the wind spectral density, and f is the frequency.

The Davenport spectrum is focused on the high-frequency portion of the wind data, with the adjustment of a single site-specific parameter. This parameter describes the power in a frequency band as a complex function of average wind velocity.

$$\frac{\omega S_u^+(\omega)}{\delta \overline{U}_p^2} = \frac{4\theta^2}{\left(1+\theta^2\right)^{4/3}} \tag{1.102}$$

The Harris spectrum is used to calculate the steady wind forces from the time-averaged wind speed. It accounts for the spatial correlation of gust effects and the mean wind velocity variation. It is important to note that this spectrum is not recommended for frequencies of less than 0.1 Hz.

$$\frac{\omega S_u^+(\omega)}{\delta \overline{U}_p^2} = \frac{4\theta}{\left(2+\theta^2\right)^{5/6}} \tag{1.103}$$

A derivable variable θ is given by:

$$\theta = \frac{\omega L_u}{2\pi \overline{U}_{10}} = \frac{\delta L_u}{\overline{U}_{10}}, \quad 0 < \theta < \infty \tag{1.104}$$

where L_u is integral length scale (= 1200 m for the Davenport spectrum and 1800 m for the Harris spectrum), δ is the surface drag coefficient (= 0.001), and $\overline{U}_{10}$ is the mean wind speed at the height of 10 m. For large floating structures, the following spectra are recommended (Dyrbye and Hassen, 1997).

The Kaimal spectrum gives a better fit to the empirical observations of atmospheric turbulence.

$$\frac{\omega S_u^+(\omega)}{\sigma_u^2} = \frac{6.8\theta}{\left(1+10.2\theta\right)^{5/3}} \tag{1.105}$$

where σ_u^2 is the variance at a reference height of 10 m. The derivable variable is given by:

$$\theta = \frac{\omega}{\omega_p} \tag{1.106}$$

where ω_p is the peak frequency.

The API (2000) spectrum is given by:

$$\frac{\omega S_u^+(\omega)}{\sigma_u(z)^2} = \frac{\left(\omega/\omega_p\right)}{\left[1+1.5\left(\omega/\omega_p\right)\right]^{5/3}} \tag{1.107}$$

$$0.01 \le \frac{\omega_p z}{\overline{U}(z)} \le 0.1 \tag{1.108}$$

Usually, a value of 0.025 is obtained instead of the values computed from the above equation. Standard deviation and speed are given by:

$$\sigma_u(z) = \begin{cases} 0.15\overline{U}(z)\left(\dfrac{Z_s}{Z}\right)^{0.125} & : Z \le Z_s \\ 0.15\overline{U}(z)\left(\dfrac{Z_s}{Z}\right)^{0.275} & : Z > Z_s \end{cases} \tag{1.109}$$

where Z_s is the thickness of the surface layer, which is about 20 m. The spectral density plot showing different wind spectra for the mean wind speed of 20 m/s and period 12 s at a reference height of 10 m is given in Fig. 1.10.

The MATLAB® program for plotting the wind spectra is given below. The variables are mean wind speed and time period.

```
%%wind spectra plot---- spectral density versus theta
%%davenport spectrum
um=20; %mean wind speed at a height of 10 m
del=0.001; %surface drag coefficient
lu=1200; %integral length for davenport spectrum in m
w=0.001:0.001:10; %frequency is the varying component
theta=(w*lu)/(2*pi*um);
a=4*((theta).^2);
b=(1+(theta.^2)).^(4/3);
x=a./b;
y=(x*del*(um^2));
su=y./w;
%%Harris spectrum
um=20; %mean wind speed at a height of 10 m
del=0.001; %surface drag coefficient
lu=1800; %integral length for Harris spectrum in m
w=0.001:0.001:10; %frequency is the varying component
theta1=(w*lu)/(2*pi*um);
a=4*theta1;
```

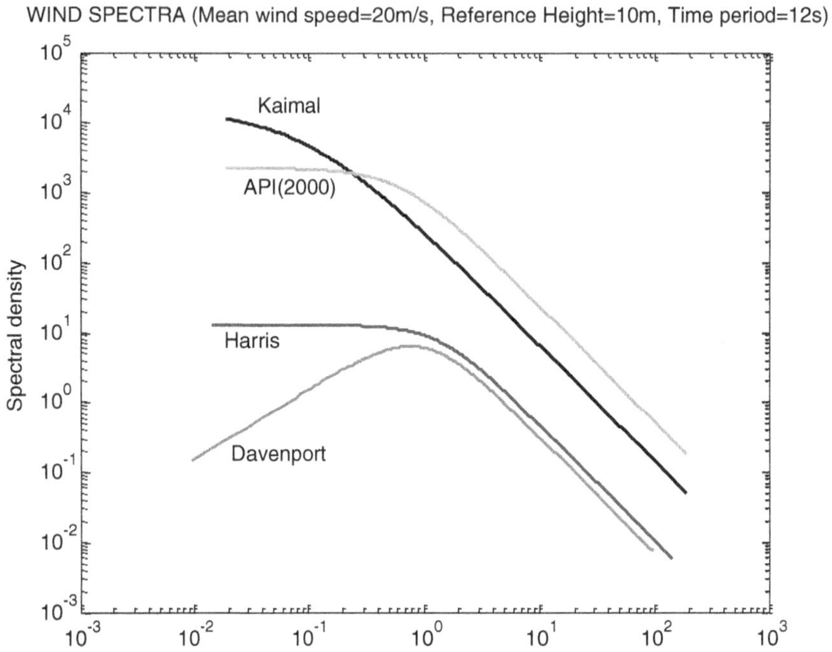

FIGURE 1.10 Wind spectra

```
b=(2+(theta1.^2)).^(5/6);
x=a./b;
y=x*del*(um^2);
su1=y./w;
%%Kaimal spectrum
t=12; %time period in seconds
w=0.01:0.001:100; %frequency is the varying component
wp=(2*pi)/t; %frequency in radians per second
z=10; %reference height is 10 m
zs=20; %the surface height usually taken as 20 m
uz=(wp*z)/0.025;
if z<=zs
  sigma=0.15*uz*((zs/z)^0.125);
else
  sigma=0.15*uz*((zs/z)^0.275);
end
theta2=w./wp;
a=6.8*theta2;
b=(1+(10.2*theta2)).^(5/3);
x=a./b;
y=x.*(sigma^2);
```

```
su2=y./w;
%%API(2000) spectrum
t=12; %time period in seconds
w=0.01:0.001:100; %frequency is the varying component
wp=(2*pi)/t; %frequency in radians per second
z=10; %reference height is 10 m
zs=20; %the surface height usually taken as 20 m
uz=(wp*z)/0.025;
if z<=zs
  sigma=0.15*uz*((zs/z)^0.125);
else
  sigma=0.15*uz*((zs/z)^0.275);
end
theta3=w./wp;
b=1+(1.5*(theta3)).^(5/3);
x=theta3./b;
y=x.*(sigma^2);
su3=y./w;
loglog(theta,su,'r','linewidth',2);%Davenport Spectrum
hold on;
loglog(theta1,su1,'b:','linewidth',2);%Harris spectrum
hold on;
loglog(theta2,su2,'k-.', 'linewidth',2);%Kaimal
hold on;
loglog(theta3,su3,'c--','linewidth',2);
xlabel('Derivable variable');
ylabel('Spectral density');
title('WIND SPECTRA (Mean wind speed=20m/s, Reference Height=10m,
Time period=12s)');
```

1.5.2 Current

Current occurs due to the transfer of water mass from one place to another, generally, due to the wind forces, temperature difference, tides, change in density, salinity, and river discharge. The landmasses, continental shelves, and the rotation of the Earth also modify the ocean currents. The action of current results in drag forces on offshore structures and thus modifies its response. Tidal current velocity is similar to the sinusoidal motion of ocean waves with long periods, and the maximum tidal current occurs near the mid-tide. Extreme wind conditions lead to the formation of surface wind drag currents along with the waves. However, the direction of current does not always act in the same direction as waves. For engineering conservative design purposes, the wind-generated currents are assumed to move in the same direction as waves. The wind-generated current velocity variation concerning depth is considered to be linear for design purposes, varying from 1.0% to 3.0% of sustained wind speed at the MSL to zero at the sea bottom. Typical tidal current and wind drag current profiles are shown in Fig. 1.11. It should be noted that the wave-current interactions

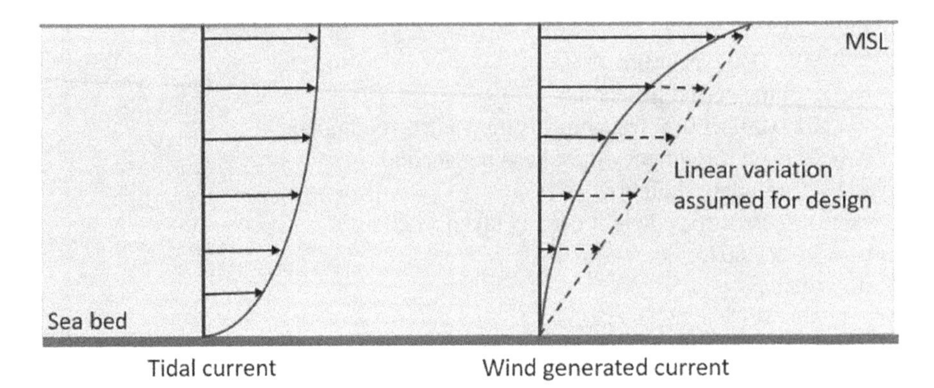

FIGURE 1.11 Current velocity variation

also govern the design of offshore platforms in ultra-deep waters because the current acting along the direction of the wave increases the wavelength. The current velocity should be added vectorially to the water particle velocity in the drag force component while computing the wave force using Morison's equation.

EXERCISES

1. What are the concerns in estimating the wave forces with large-diameter members?
2. Write a brief note on offshore wind turbines.
3. What do you understand by FORM-dominant design? How is it useful in offshore compliant structural design?
4. Explain how wave force on a cylindrical member is computed using Airy's wave theory.
5. Why is Airy's wave theory called a linear wave theory?
6. Plot the variation of water particle kinematics along the water depth. Use MATLAB® and explain the variables used in the computation.
7. How are waves classified according to relative depth?
8. What are the stretching modifications suggested by various researchers? Explain with a plot using MATLAB®.
9. Write a brief note on Stokes' fifth-order wave theory.
10. How is Stokes' fifth-order theory different from the third-order theory? What are the major changes made and how do these changes affect wave load on members?
11. Plot different wind spectra and compare their applications to offshore members.
12. What is the effect of current on waves?
13. Plot different wave spectra using MATLAB® and compare the results.
14. Explain the sea state description in terms of wave height and period.
15. What is an offset and how is it coupled with set-down effect?
16. Define fastest mile velocity.
17. Write short notes on the aerodynamic admittance function.

2 Perforated Cylinders and Applications

2.1 INTRODUCTION

Perforated structural forms are found to be effective in the force reduction under direct wave impact (Chandrasekaran and Abhishek, 2010); apart from reducing the forces, it is also one of the most practical methods to retrofit the coastal and offshore structures (Chandrasekaran and Madhavi, 2015a; Chandrasekaran et al., 2015a). Wang and Ren (1994) analyzed the wave kinematics around a protected impermeable pile with the porous outer cylinder. For the smaller annular spacing, they showed that long-period waves excite larger forces on the inner cylinder than short-period waves. The existence of an exterior porous cylinder reduces the hydrodynamic forces on the inner cylinder caused by the direct wave impact. William and Li (1998) showed a significant reduction in the wave field, which causes a reduction in the hydrodynamic forces experienced by the inner cylinder circumscribed by the semi-porous outer cylinder. Perforation ratio is one of the key factors that influence the reduction of hydrodynamic loads experienced by the inner cylinder (Williams and Li, 2000; Williams et al., 2000). Chandrasekaran and Vishruth (2013) quantified the response reduction of the column member of offshore TLP by encompassing the inner cylinder with the outer perforated member through experimental investigations. Response reduction is attributed to water particle kinematics and the cylinder's depth (Chandrasekaran and Madhavi, 2014a). Recent numerical studies highlighted parameters influencing the variations of water particle kinematics for which maximum force reduction in the damaged member shall be envisaged for the operational sea states (Chandrasekaran and Madhavi, 2014b).

2.2 FORCE REDUCTION IN THE INNER CYLINDER

Scaled models of PVC cylinders with the outer and inner diameter of 315 mm and 115 mm, respectively, are experimentally investigated to study the effect of perforation and the diameter of the perforation on the force reduction. Fig. 2.1 shows the model used for the study. The diameter of the perforations and the perforation ratio are varied appropriately. Table 2.1 shows the details. Steel frames consisting of ISMC channels and angles are used for clamping the inner cylinder and the outer cylinder on a single rigid frame. Strain gauges are placed along the inner cylinder to determine the forces acting on the cylinder during the passage of waves. Variations of strain and the inner cylinder during the passage of waves are recorded using the data acquisition system at the appropriate sampling rate. Strain gauges are placed closer, near the free surface, to obtain the variation of forces with higher accuracy.

DOI: 10.1201/9781003281429-2

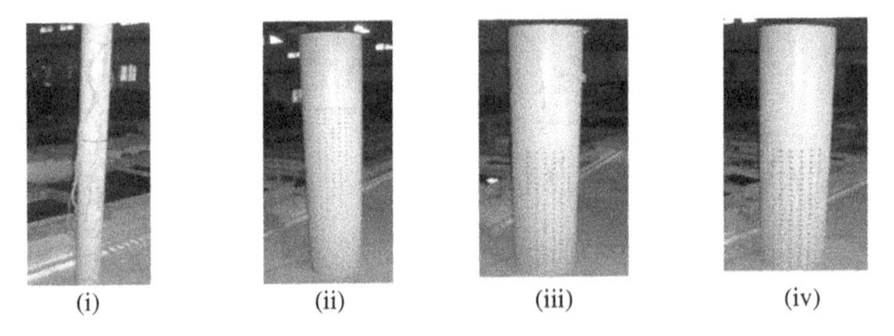

i) inner cylinder; ii) outer cylinder (A); iii) Outer cylinder (B); iv) Outer cylinder (C)

FIGURE 2.1 perforated cylinders considered for the study

TABLE 2.1
Details of Cylinders Used for the Experiment

	Designation of Porous Cylinders		
Details of Perforations	**A**	**B**	**C**
Diameter of the perforation	10 mm	15 mm	20 mm
Length of perforation	1450 mm	1050 mm	1050 mm
Number of perforations along length	41	26	24
Number of perforations along circumference	28	24	22
Perforation ratio	6.3%	10.6%	16%
	Inner Cylinder	**Outer Cylinders (A,B,C)**	
Diameter	110 mm	315 mm	
Length	1900 mm	1930 mm	
Thickness	4.4 mm	8.7 mm	

For the chosen perforation ratio of the outer cylinder, force variations are plotted for different combinations of wave height and wave periods. Fig. 2.2 shows the variations. It is seen that the variations are not proportional to the wave steepness index; there is a significant force reduction for long-period waves in comparison to that of short-period waves. The scaled model 1:100 is numerically investigated using the CFD tools, which have capabilities enabled in different modules to simulate viscous drag and turbulence effects caused by perforations (Chandrasekaran and Madhavi, 2016). A numerical model of the perforated member to a scale of 1:100 is shown in Fig. 2.3.

The length of the domain is taken in such a manner to reduce the reverse flow effect in the simulation, while the width of the block is taken equivalent to the width of the wave flume. The height of the domain is taken as the characteristic length of the inner cylinder that is effective during wave-structure interaction. A domain is chosen such that it can accommodate a minimum of two wavelengths upstream. The scaled model corresponds to the minimum frequency of 3.12 Hz, 0.312 Hz for the prototype

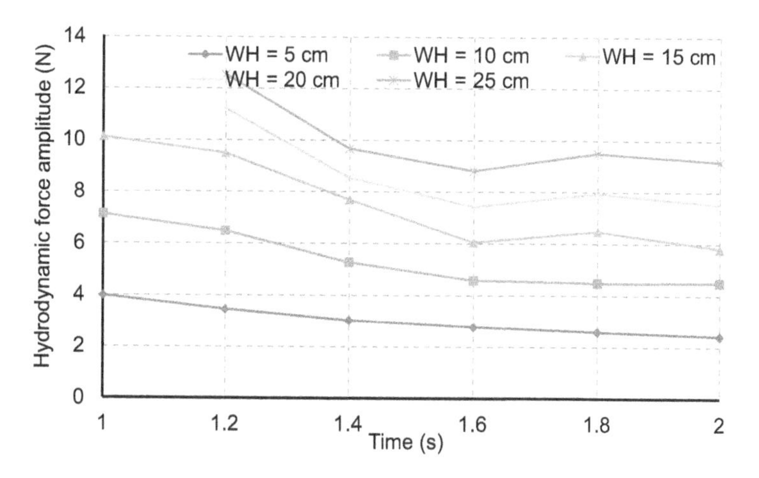

FIGURE 2.2 Force variations on inner cylinder with perforated outer cover

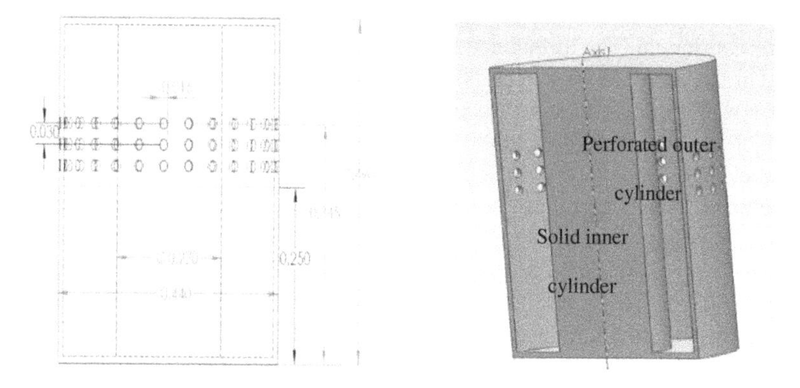

FIGURE 2.3 Numerical model of perforated outer cylinder with solid inner cylinder

by dispersion relationship; it corresponds to an incident wavelength up to 638 m for the prototype. Fig. 2.4 shows the domain of the perforated cylinder, generated with volumetric control. A surface and a volume mesh are generated using a surface remesher, defined by the base size, while the trimmer is used to generate the volume mesh. The efficiency of the generated mesh is relatively high as it can produce high-quality grids that consist of hexahedral cells. A core mesh is thus generated, and the cells are trimmed based on the surface mesh. Prism layer meshing is used along with the trimmer model to improve the accuracy of the flow solution. It generates orthogonal prismatic cells adjacent to the wall boundaries. In the present numerical simulation, mesh parameters are set globally and altered at specific boundaries; volumetric controls generate the desired mesh after assigning the appropriate regions.

Implicit Unsteady is chosen for each physical time-step involving the number of iterations to converge at the desired solution for that given instant of time. The volume of fluid is used to simulate the behavior of two fluids (air and water) within

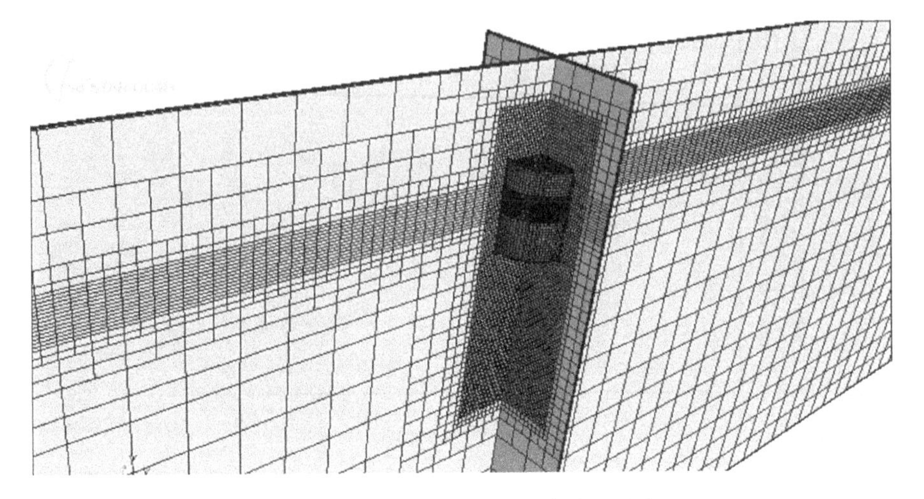

FIGURE 2.4 Domain of perforated cylinder for numerical modeling

the same continuum. The turbulence effects are simulated by using the K–omega turbulence in the Reynolds-averaged Navier–Stokes (RANS) turbulence. The transport equations that are solved are for the turbulent kinetic energy (k) and a quantity called (ω), which is defined as the specific dissipation rate, that is, the dissipation rate per unit turbulent kinetic energy ($\omega \sim \varepsilon / k$). The action of gravitational acceleration is also included, which accounts for the working pressure becoming the piezometric pressure. The body force due to gravity can be included in the momentum. The first-order wave is given by first-order approximation to Stokes' wave theory. It generates a wave that has a regular periodic sinusoidal profile with a specific wave height and wave period. The simulations are analyzed for a wide range of sea states: wave height (H) and wavelength (L) ratio ranging from 0.0051 to 0.1002. i) Inlet boundary is set as a velocity inlet; ii) outlet boundary is set as a pressure outlet; and iii) inner cylinder and outer perforated cylinder are set as wall boundaries. The time-step for the Implicit Unsteady solver is set to 0.001 s. The maximum physical time is set to 5 secs. The simulated models are subjected to unidirectional waves of (5, 10, 15 cm) wave heights while wave periods are varied from 0.8 s to 2.5 s. Fig. 2.5 shows the simulation of the outer perforated cylinder with a solid inner cylinder.

The Reynolds-averaged Navier–Stokes equation is solved, which is assumed to be converged when the residuals decrease by the multiple orders before settling around 0.001. Fig. 2.6 shows the variation of continuity throughout the simulation.

A similar procedure is followed for solid cylinders without perforation to investigate the effect of perforation. Fig. 2.7 shows the simulation of the inner, solid cylinder. Results of both the perforated and solid cylinders are compared. The sea states are grouped into three categories of steepness, namely steep, medium, and low wave, based on the steepness index. The steepness indices (H/L) range between (0.0051 to 0.0154), (0.0160 to 0.0321), and (0.0327 to 0.1002) for low, medium, and steep waves, respectively. Figs. 2.8 and 2.9 show the force variations in the inner and outer cylinder for a representative value of steep wave. There is a significant reduction in the force amplitude in the inner cylinder when provided with a perforated outer cover.

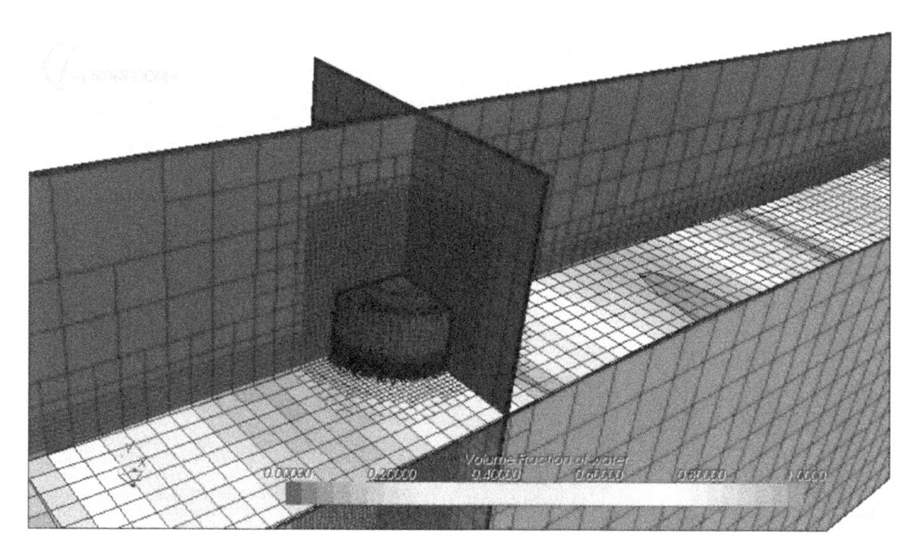

FIGURE 2.5 Simulation of inner solid cylinder with perforated outer cover

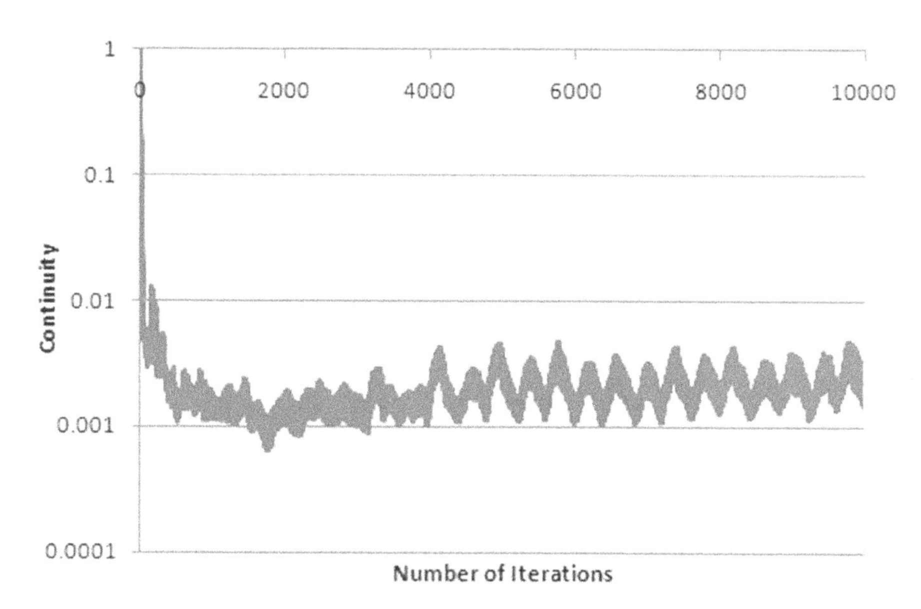

FIGURE 2.6 Residuals during numerical simulation

2.3 EFFECT OF ANNULAR SPACING AND PERFORATION RATIO ON FORCE REDUCTION

Annular spacing is the distance between the inner cylinder and the outer cylinder. The ratio of the radius of the inner cylinder (a) to that of the outer cylinder (b) is chosen between 0.35 to 0.60 to study its effect on force reduction (Chandrasekaran and Madhavi, 2016; 2015a; 2015b; 2015c; 2015d; 2016; Chandarsekaran et al., 2013c).

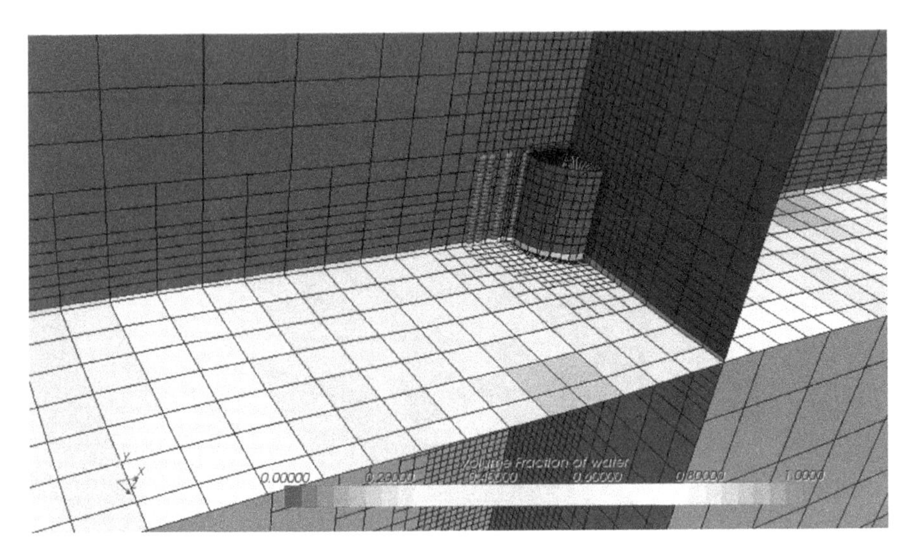

FIGURE 2.7 Numerical simulation of inner cylinder

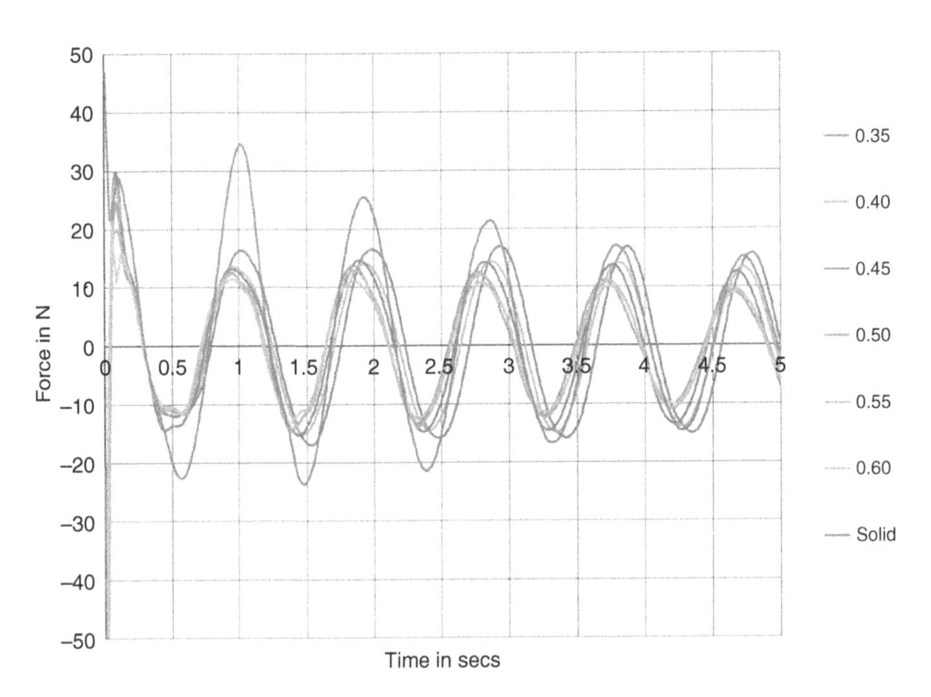

FIGURE 2.8 Variation of hydrodynamic force amplitude on inner cylinder ($H/L = 0.0962$)

The inner cylinder diameter is taken the same as that of the existing column member of the MARS TLP (22 m) and scaled as 0.22 m (1:100). Variation in the horizontal water particle velocity for different annular spacing ratios is plotted in Fig. 2.10. The percentage reduction in the hydrodynamic force on the inner cylinder, with and without perforated outer cylinder, is tabulated in Table 2.2.

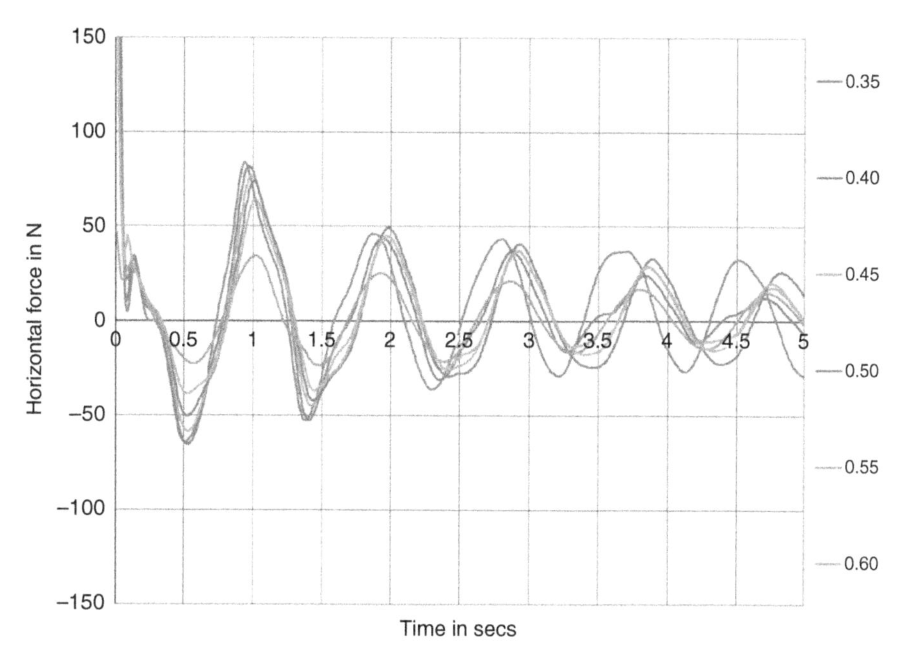

FIGURE 2.9 Variation of hydrodynamic force amplitude on outer cylinder ($H/L = 0.0962$)

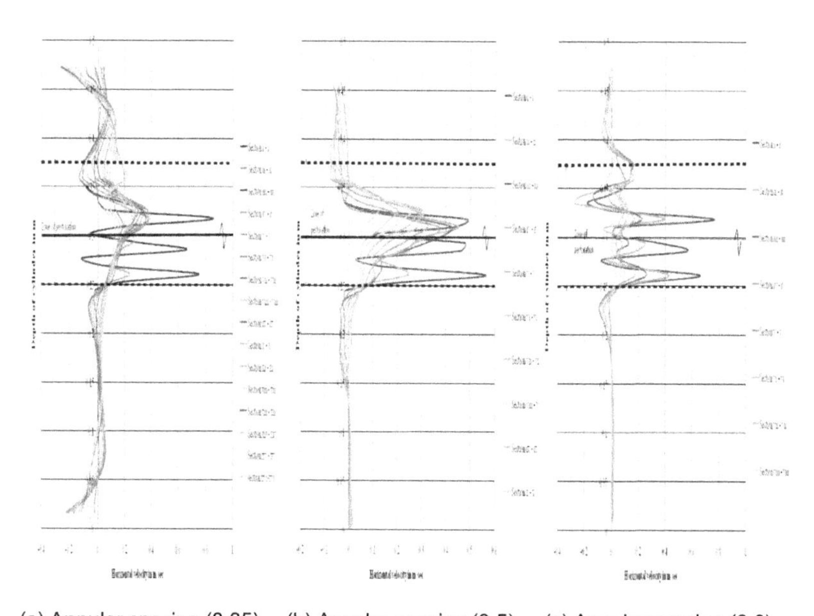

(a) Annular spacing (0.35) (b) Annular spacing (0.5) (c) Annular spacing (0.6)

FIGURE 2.10 Variation of horizontal velocity along with the depth of cylinder ($H/L = 0.0250$)

TABLE 2.2
Percentage Reduction in Hydrodynamic Force on Inner Cylinder

(*a*/*b*) Ratio	Horizontal Force	Percentage Reduction
Solid	13.5888	NA
0.35	10.8779	20%
0.40	9.4689	30%
0.45	9.3416	31%
0.50	8.0248	41%
0.55	8.2126	40%
0.60	8.9411	34%

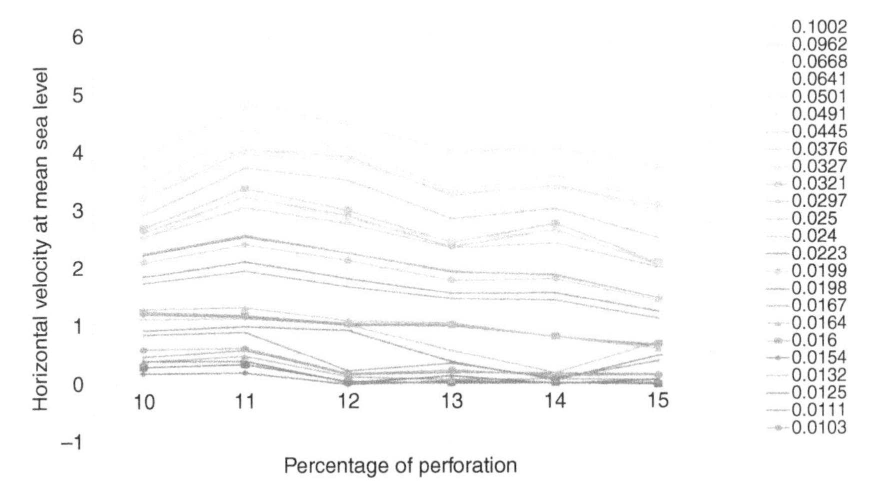

FIGURE 2.11 Variation of horizontal velocity at the MSL for different perforation ratio

The figure shows that the hydrodynamic performance of the perforated cylinders is better with an annular spacing ratio (*a*/*b*) of 0.5. There is a higher force reduction on the inner cylinder than without perforation; a similar pattern was observed for all sea states but not reported. The published results can be more detailed (Chandrasekaran and Madhavi, 2014a; 2014b; 2014c; 2014d). The horizontal water particle varies almost linear for lower (*a*/*b*) ratio values, indicating a tranquility zone. This is because the effect of the perforation becomes insignificant after a certain distance greater than the annular spacing between the cylinders.

The effect of the perforation ratio is further examined by varying the percentage of the perforation ratio in the range (10–15%). Studies reported a significant reduction in the horizontal water particle velocity for the chosen range of perforation ratio (Chandrasekaran and Madhavi, 2015c; 2015d; 2016). Fig. 2.11 shows the variations in the horizontal velocity at mean sea level for different sea states for various percentages of perforation. It is seen from the plots that there is a significant reduction in the horizontal velocity at mean sea level (MSL) for a range of 11–12% of perforation ratio for all the sea states. The main reason for the force reduction on the inner cylinder is a perforated outer cover. In the case of effective force reduction on column members of wave energy devices, this can be one of the effective ways that can be practiced.

2.4 EFFECT OF PERFORATION PARAMETERS ON FORCE REDUCTION

This section examines the influence of various parameters of perforation, namely the diameter of perforation, the spread of perforation, and spacing of perforation on the force reduction. Previous studies reported a significant influence of these parameters (Chandrasekaran et al., 2014a; 2014b; 2015a); however, a few results are highlighted for continuity of understanding. The perforation diameter is varied in the range (10, 12, 16 mm) on the scaled model while the perforation ratio is maintained as 11%; it was found to be more effective from the above section. The variation of horizontal velocity at the MSL for the various diameter of perforation and different sea states is shown in Fig. 2.12. It can be seen that the horizontal velocity reduces with the increase in the diameter. However, when the diameter of perforation is increased beyond 12 mm, the water beam entering through the perforation gets disturbed, resulting in loss of tranquility. It creates a disturbed zone between the cylinders, which can significantly influence the damping of members of wave energy devices. Hence, the diameter of perforation with 12 mm is more effective in force reduction on the inner cylinder for a wide range of sea states.

The spread of perforation also influences the force reduction (Chandrasekaran and Madhavi, 2016; 2015c; 2015d; 2015a). With the perforation diameter of 12 mm and an 11% perforation ratio, perforation is provided only on the splash zone to maximize

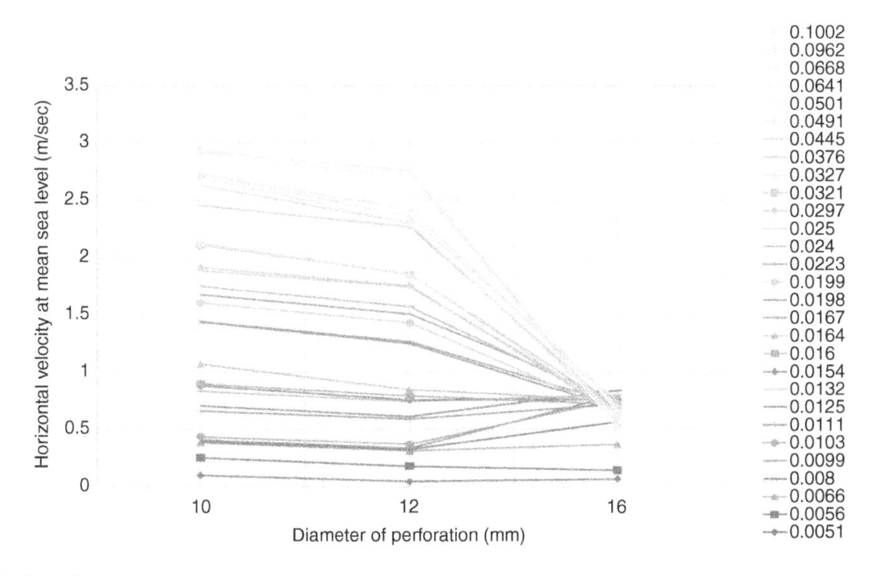

FIGURE 2.12 Variation of horizontal velocity at the MSL for varying diameter of perforation

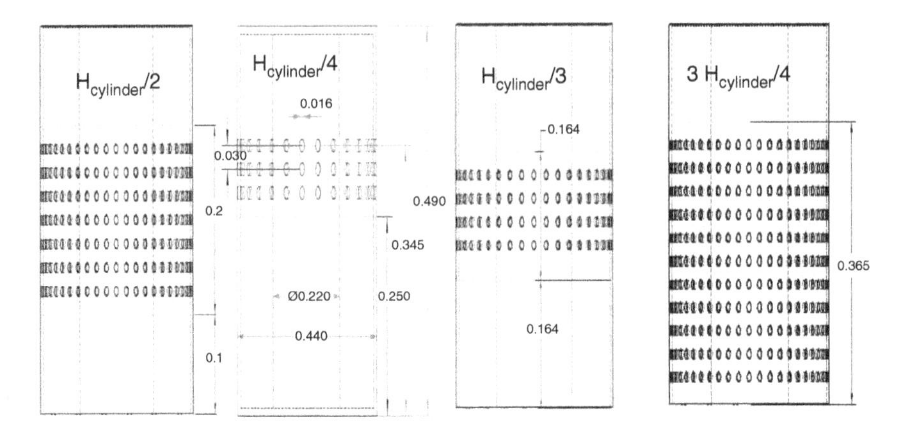

FIGURE 2.13 Spread of perforation along with the height

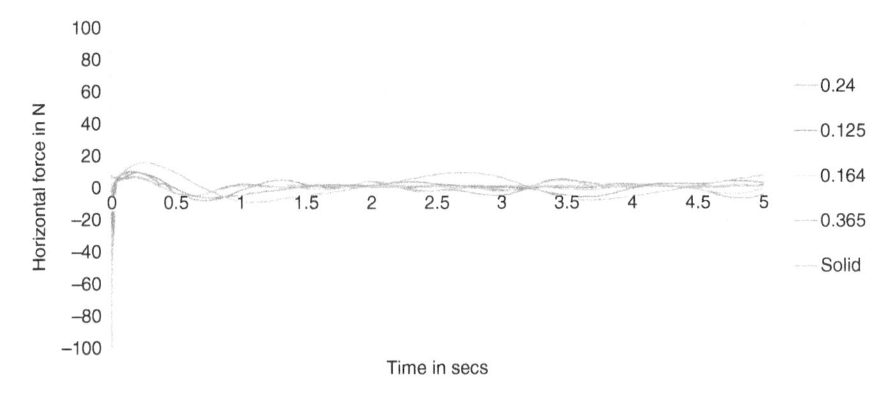

FIGURE 2.14 Variation of horizontal force on inner cylinder for spread of perforation ($H/L = 0.0154$)

force reduction. A different perforation spread of one-fourth, half, one-third, and three-fourths of the height of the cylinder are chosen to examine their influence on the force reduction on the inner cylinder, as shown in Fig. 2.13. The origin of position is chosen at the bottom of the cylinder. Figs. 2.14 to 2.16 show the horizontal force in the inner cylinder for various spread of perforation and sea states (H/L) ranging from low, medium and steep waves. Figs. 2.17 to 2.19 show the horizontal force variation of the outer cylinder for various spread of perforation and sea states (H/L) ranging from low, medium and steep waves. For all sea states ranging from low to steep, the hydrodynamic performance of the model with the spread of perforation 0.125 m ($H_{cylinder}/4$) is found to be more effective amongst the chosen cases. A maximum of 76% reduction in the hydrodynamic force is observed compared to the member without perforated cover. This geometrical configuration is found to be more effective as it takes full advantage of the free surface. Table 2.3 shows quantitatively the reduction of hydrodynamic force on the inner cylinder covered by the perforated outer

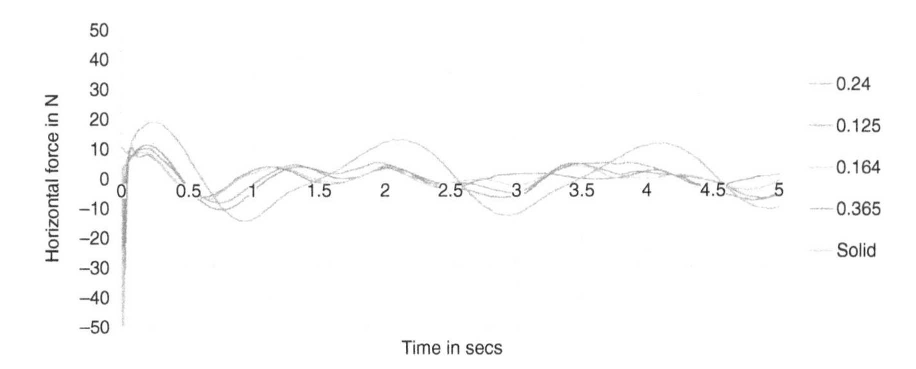

FIGURE 2.15 Variation of horizontal force on inner cylinder for spread of perforation ($H/L = 0.0240$)

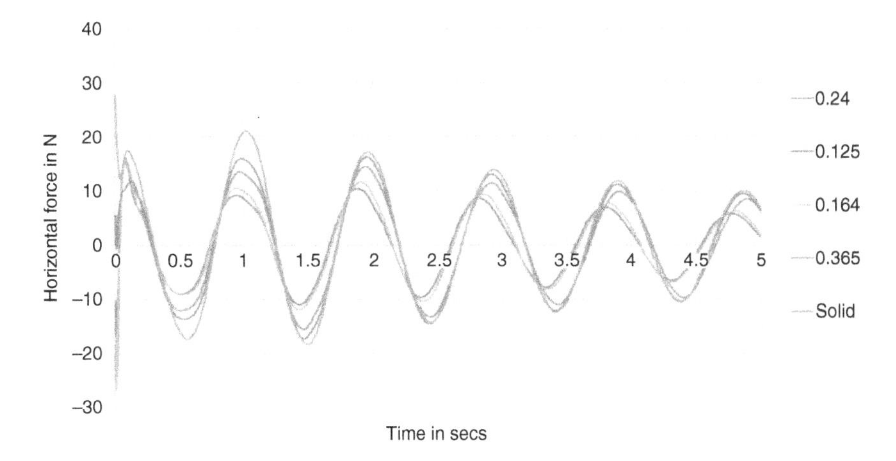

FIGURE 2.16 Variation of horizontal force on inner cylinder for spread of perforation ($H/L = 0.0641$)

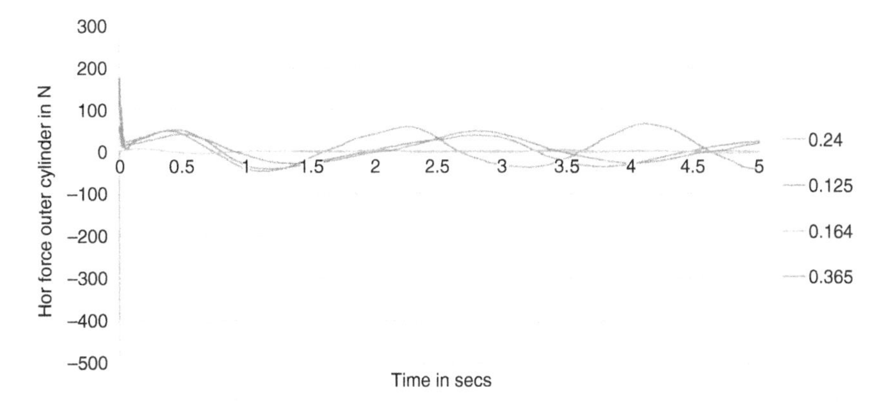

FIGURE 2.17 Variation of horizontal force on outer cylinder for spread of perforation ($H/L = 0.0154$)

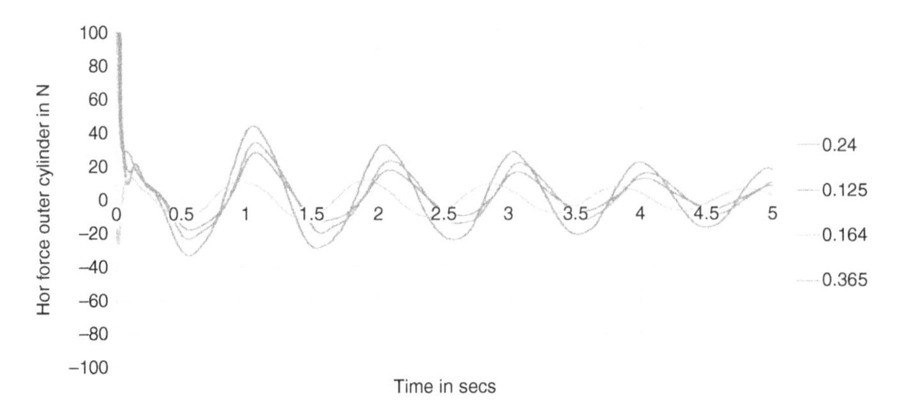

FIGURE 2.18 Variation of horizontal force on outer cylinder for spread of perforation ($H/L = 0.0240$)

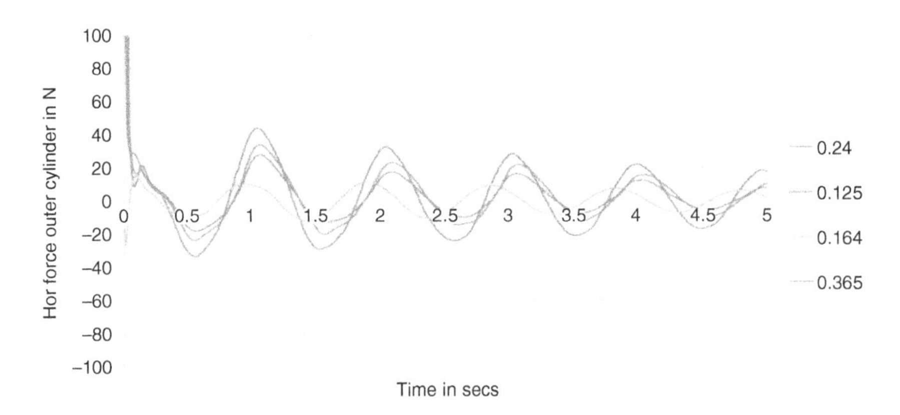

FIGURE 2.19 Variation of horizontal force on outer cylinder for spread of perforation ($H/L = 0.0641$)

cylinder with different proposed configurations for various sea states. The hydrodynamic force on the outer cylinder is also tabulated.

It is observed that the phenomenon is found to be reversing for the outer cylinder. The geometrical configuration of $H_{cylinder}$ /4 attracts more force than that of other chosen configurations, and $H_{cylinder}$ /3 attracts less force. To understand the mechanism of force reduction, it is necessary to understand the behavior of kinematics. The sections are taken at every 0.01 m interval from the outer cylinder towards the inner cylinder, as shown in Fig. 2.20. Figs. 2.21 to 2.24 show the variation of horizontal velocity along with the depth of cylinder for every section considered between the annular spacing of cylinder. The velocity profile is found highly non-proportional in the zones of perforation. Due to the disturbance caused by the perforation along the path of water particles, there is force reduction in the inner cylinder as the energy is dissipated along the advancing direction.

TABLE 2.3
Percentage Reduction in Horizontal Force on Inner Cylinder for Spread of Perforation

H/L = 0.0066				H/L = 0.0103			
	Inner Cylinder (N)	Percentage Reduction	Outer Cylinder (N)		Inner Cylinder (N)	Percentage Reduction	Outer Cylinder (N)
$H/2$	1.03	55	7.77	$H/2$	1.55	59	14.24
$H/4$	0.75	68	10.41	$H/4$	0.92	76	15.60
$H/3$	0.88	62	0.98	$H/3$	0.98	74	0.98
$3H/4$	1.58	32	5.90	$3H/4$	2.23	41	12.17
Solid	2.32			Solid	3.80		
H/L = 0.0164				H/L = 0.0240			
	Inner Cylinder (N)	Percentage Reduction	Outer Cylinder (N)		Inner Cylinder (N)	Percentage Reduction	Outer Cylinder (N)
$H/2$	2.39	41	9.93	$H/2$	3.41	58	31.71
$H/4$	1.34	67	13.80	$H/4$	2.34	71	33.47
$H/3$	1.58	61	1.58	$H/3$	7.12	70	2.44
$3H/4$	3.27	19	6.95	$3H/4$	4.61	43	28.10
Solid	4.02			Solid	8.12		
H/L = 0.0962				H/L = 0.0376			
	Inner Cylinder (N)	Percentage Reduction	Outer Cylinder (N)		Inner Cylinder (N)	Percentage Reduction	Outer Cylinder (N)
$H/2$	1.55	59	14.24	$H/2$	4.64	56	37.30
$H/4$	0.92	76	15.60	$H/4$	4.01	62	41.53
$H/3$	0.98	74	0.98	$H/3$	4.36	59	0.98
$3H/4$	2.23	41	12.17	$3H/4$	6.12	43	32.04
Solid	3.80			Solid	10.64		

Earlier studies further investigated the spacing between perforations (Srinivasan and Madhavi, 2015a; Srinivasan et al., 2015). Three models are examined for (20, 30, 40 mm) spacing between the perforation as shown in Fig. 2.24. Variation of hydrodynamic forces is compared with and without perforated outer cover for different sea states ranging from steep, medium, and low waves; results for different spacing are shown in Figs. 2.25 to 2.27, respectively. The figures show that the percentage reduction in the hydrodynamic force is higher with 30 mm spacing for all the sea states considered in the study. It is attributed to the uninterrupted beam of water entering the perforated cylinder as this spacing utilizes the free surface effect to its maximum. Table 2.4 summarizes force reduction on the inner cylinder for various perforation spacing and sea states considered.

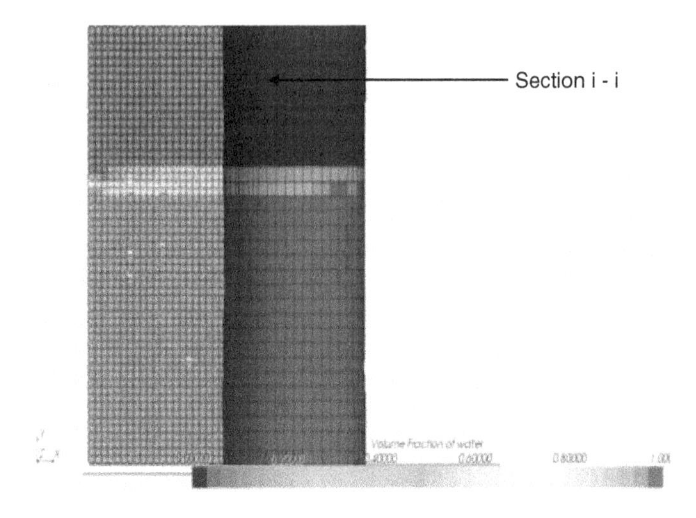

FIGURE 2.20 Sections considered for monitoring the water particle kinematics

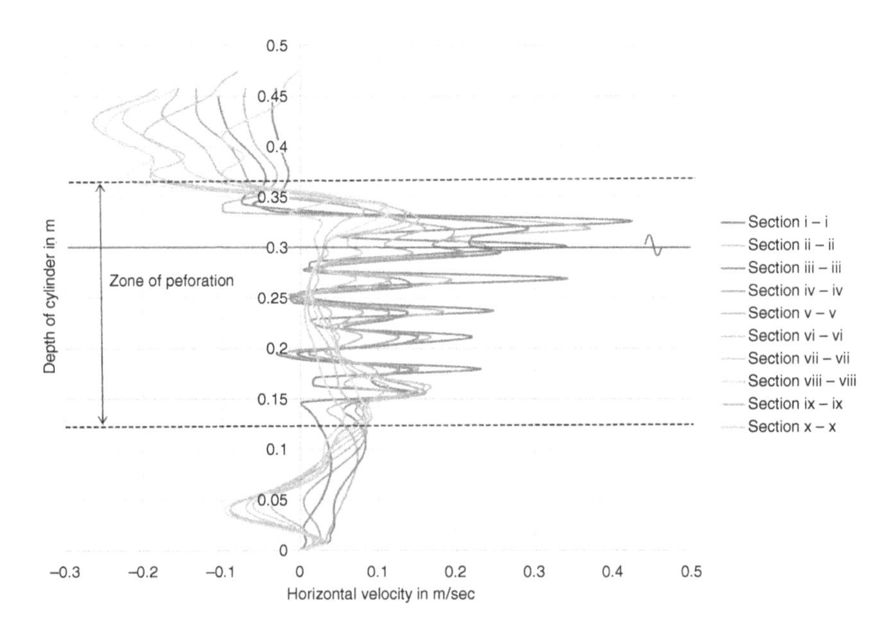

FIGURE 2.21 Variation of horizontal velocity in between the cylinder (spread of perforation, $H/2$; $H/L = 0.0962$)

Members of the floating wave energy converters are designed as positively buoyant. The reduction of lateral forces on these members improves the performance characteristics of the device. One such simple but effective method of providing a perforated outer cover is discussed. Based on the discussions, we can understand that the wave forces on the inner cylinder increase with the perforation ratio and

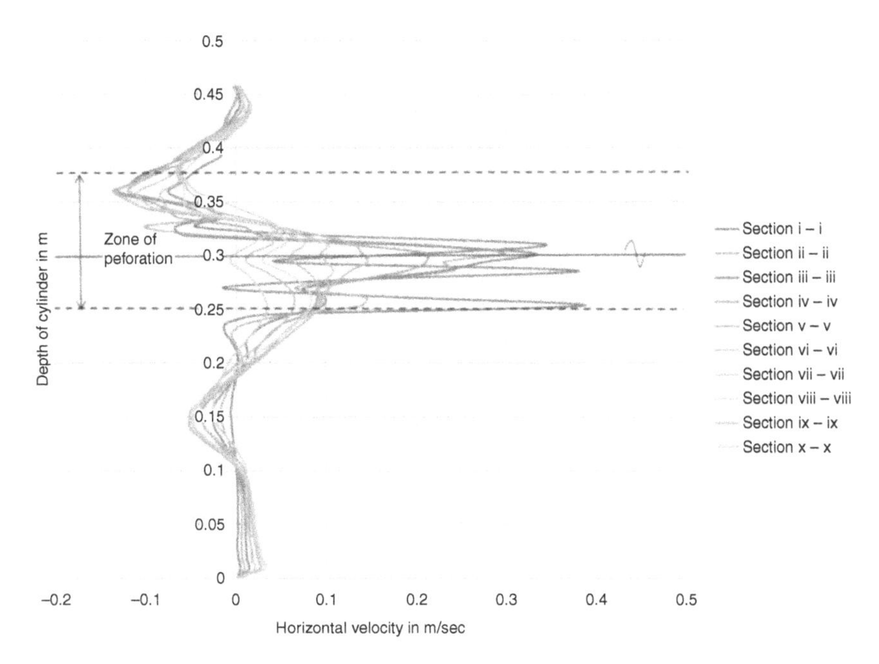

FIGURE 2.22 Variation of horizontal velocity in between the cylinder (spread of perforation, $H/4$; $H/L = 0.0962$)

decrease with the increase in the perforation size. It is also seen that the horizontal forces on the inner cylinder get reduced significantly in the presence of a perforated outer cover. When the annular spacing between the cylinder is reduced, the flow velocity is disturbed due to the turbulence effect. It can be attributed to the breaking down of the water particles, which significantly alters the cylinder's response. The (a/b) ratio of 0.5 showed a significant reduction of the horizontal forces on the solid inner cylinder, attached with a perforated outer cover. The perforation ratio of (11 to 12%) is more effective with a 12 mm diameter perforation. The spread of perforation corresponding to ($H_{cylinder}/4$) with 30 mm spacing is found to be performing well for all the sea states considered as it utilizes the free surface effect. It is important to note that the change in the velocity profile within the perforated zone is non-proportional. Hence, the recommended technique by covering the existing member with the perforated outer cover is more effective and economical by decreasing the device's response. Further, this technique does not demand replacement of the member as it just recommends a perforated outer cover, a feasible and practical approach for force reduction on members.

2.5 TWIN PERFORATED CYLINDERS

The presence of the outer perforated cylinder reduces the direct wave impact on the inner cylinder, which is established by many researchers (Chandrasekaran et al., 2014a; 2014b; Chandrasekaran and Madhavi, 2014a). In the case of wave energy

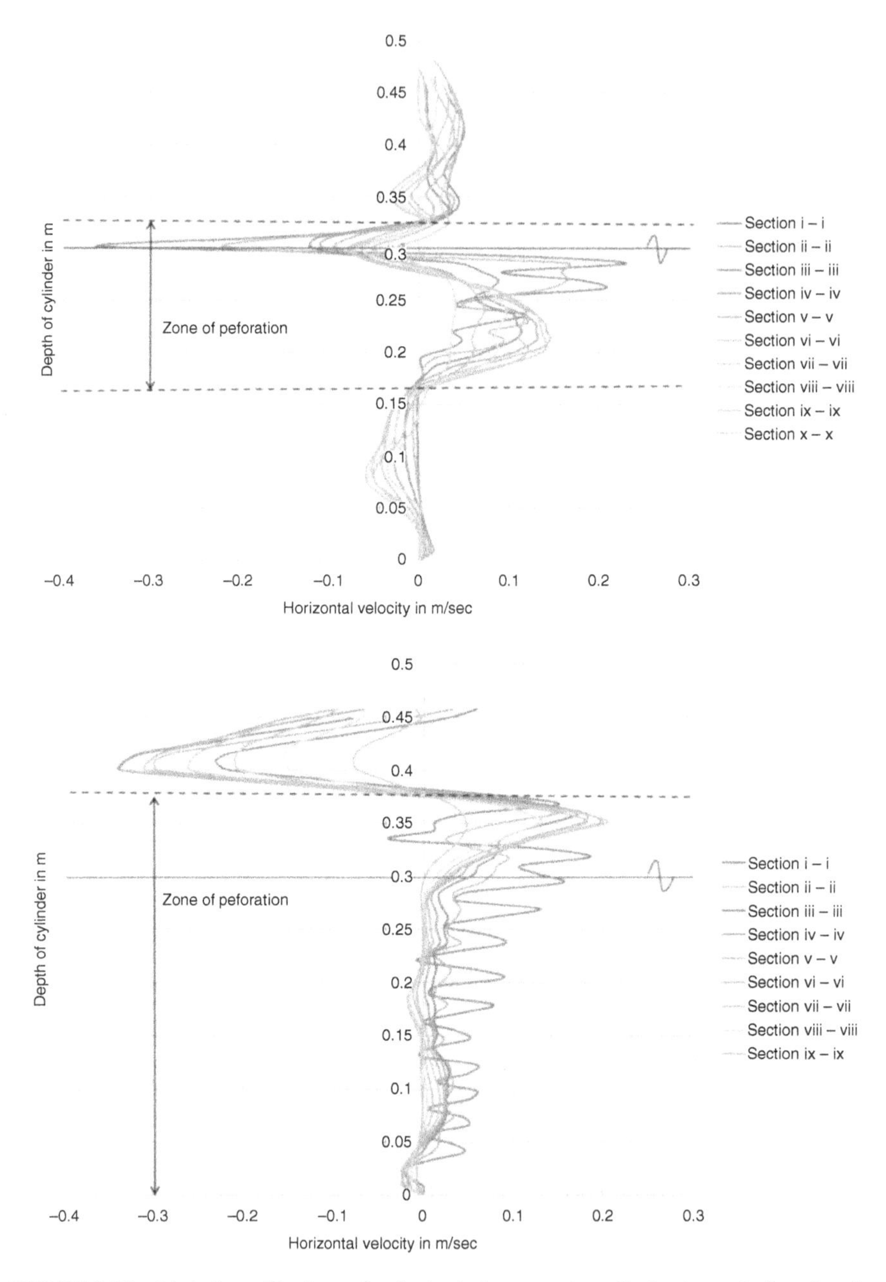

FIGURE 2.23 Variation of horizontal velocity in between the cylinder (spread of perforation $3H/4$; $H/L = 0.0962$)

devices, which comprise cylindrical members, it is interesting to understand the force reduction phenomenon with twin cylinders. The flow field around the twin cylinders with different orientations is examined with and without a perforated outer cover (Chandrasekaran and Madhavi, 2015b). Mechanisms contributing to the force

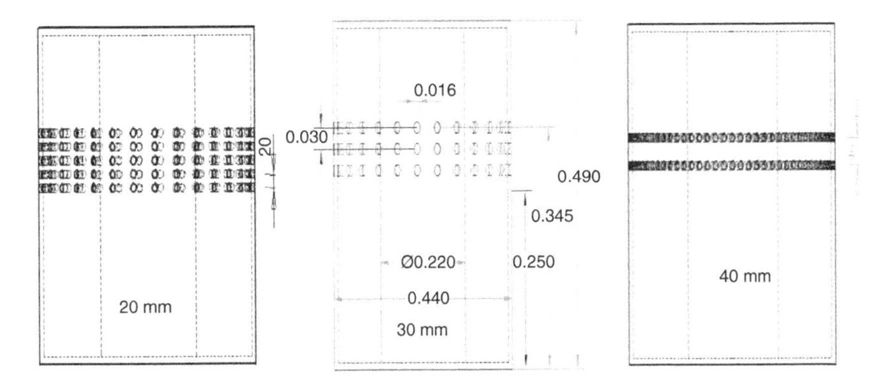

FIGURE 2.24 Spacing of perforation

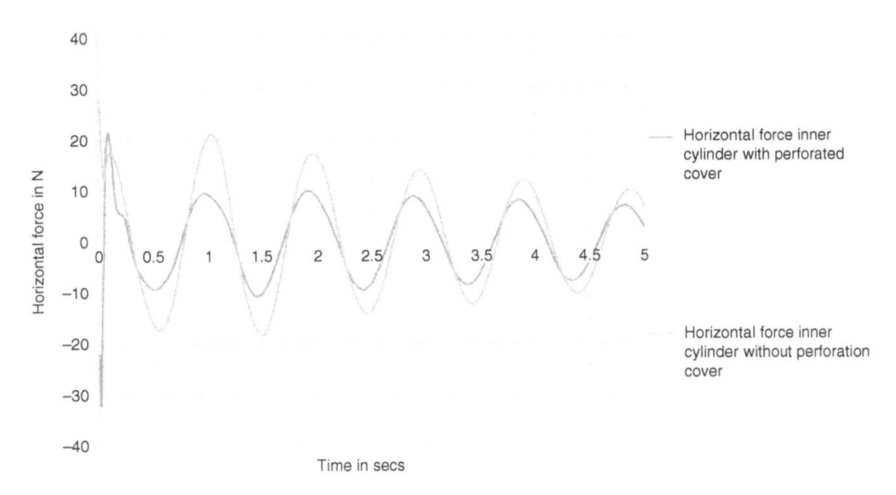

FIGURE 2.25 Variation of horizontal force for 20 mm spacing ($H/L = 0.0641$)

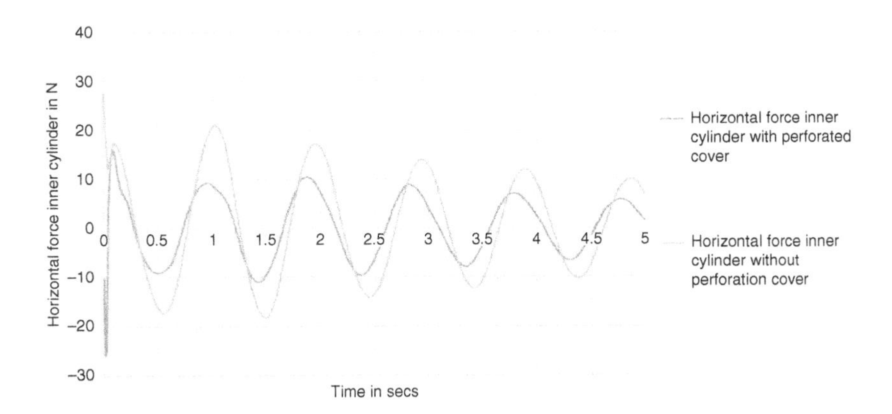

FIGURE 2.26 Variation of horizontal force for 30 mm spacing ($H/L = 0.0641$)

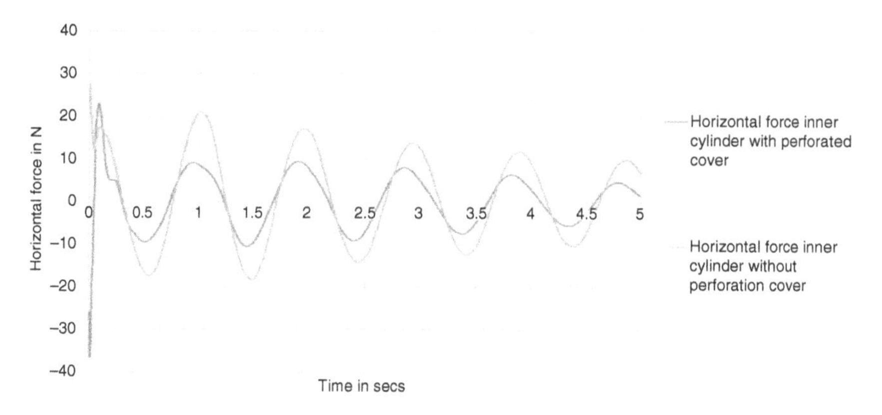

FIGURE 2.27 Variation of horizontal force for 40 mm spacing ($H/L = 0.0641$)

TABLE 2.4
Force Reduction on Inner Cylinder for Different Perforation Spacing

	20 mm		30 mm		40 mm		Solid
Steepness	**Horizontal Force (N)**	**Reduction (%)**	**Horizontal Force (N)**	**Reduction (%)**	**Horizontal Force (N)**	**Reduction (%)**	**Horizontal Force (N)**
Steep	6.0151	38	4.4377	54	5.5015	43	9.6472
Medium	2.577	75	2.3343	76	3.1901	74	7.0735
Low	0.9665	64	1.0044	67	0.9193	55	3.797

reduction are examined in physical terms and later quantified. Detailed discussions in this section of the chapter shall encourage readers to understand the behavior with a more practical approach.

Typically, wave energy devices comprise tubular members due to their geometrical advantages. Jarlan (1961) estimated the energy dissipation factor as a function of various parameters: wall porosity, wall thickness, and chamber width to wavelength ratio. Recent studies recommended the outer perforation cover as one of the passive techniques to control vortex-induced vibration due to wave damping effects (Chung et al., 1994; Terret et al., 1968; Jun Liu et al., 2012). A critical literature review emphasizes significant force reduction on the existing members when encompassed by the perforated outer cover. Still, the mechanism is highly complex due to wave-porous structure interaction. Factors contributing to this complexity include non-linearity of water particle kinematics, variable submergence effects on the floating bodies, free surface effects, turbulence, and dynamic effects arising from vortex shedding. Water particle kinematics is significantly altered in the presence of a perforated member. Variations of water particle kinematics around the perforation for twin cylinders that are oriented along and against the wave advancing direction are discussed in detail. A numerical model of a twin-cylinder, with and without a perforated outer cover, is shown in Fig. 2.28.

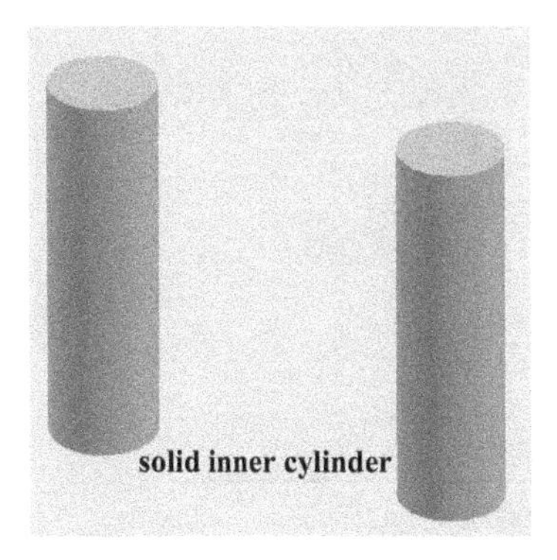

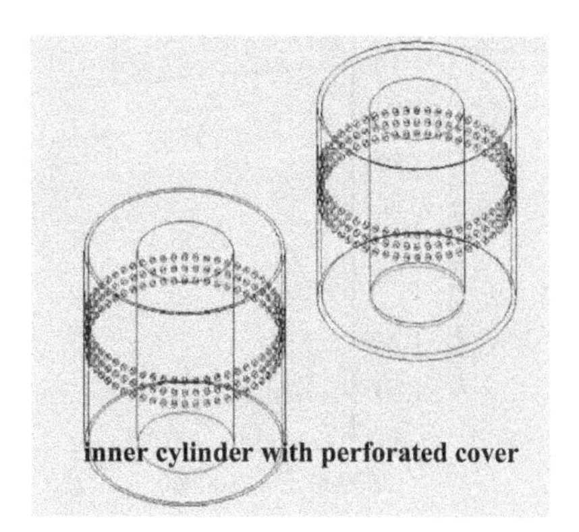

FIGURE 2.28 Twin cylinders with perforated cover

The perforation ratio expressed as a percentage is the ratio of the surface area of the perforation to the solid surface area of the cylinder; present discussions are limited to the ratio of 11%, while the area of each perforation is 0.59% of that of the total perforation area. For a more practical approach, perforation is assumed to be provided in the splash zone. To accommodate the study for all sea states, discussions include steepness indices ranging between (0.0051 to 0.0154), (0.0160 to 0.0321), and (0.0327 to 0.1002), which are categorized as low, medium, and steep waves, respectively (Chandrasekaran and Madhavi 2014a). In the numerical simulation, the block's width is equivalent to that of the wave flume. The height of the domain is taken as the characteristic length of the inner cylinder that is effective during wave-structure interaction. The numerical wave tank is thus capable of generating waves with a minimum frequency of 0.312 Hz, which corresponds to an incident wavelength of 638 m for the prototype. Numerical simulations are carried out for chosen sea states wherein a minimum of two wavelengths of waves are made to pass through the structure to illustrate the physics behind the force reduction mechanism. Full mesh diagnostics reports are applied to check the mesh for mesh validity, face validity, volume change statistics, maximum interior skewness angle, and maximum boundary skewness angle; unclosed cells and zero area cells are checked and found to be absent while the skewness angle is restricted to 85°.

The perforations are modeled physically such that the flow is allowed to pass through them so that real physics can be generated. Coarse mesh is generated near the domain and finer mesh near the free surface. As the mesh size is insensitive (z-direction) along the width of the domain, the number of cells along that direction of the mesh is kept constant for the study, as shown in Fig. 2.29.

Figs. 2.30 to 2.33 5–8 show different orientations of the twin cylinders with and without perforated outer cover. To simulate the real physics behind the problem assigning appropriate boundary conditions is vital. The domain is assigned with wall boundary conditions and no-slip boundaries. Inlet is taken as velocity inlet, and the outlet is assigned as the pressure outlet to mimic a free boundary. This is to ensure that the air can flow either into or out of the domain. Boundary face pressure is extrapolated from the adjacent cells using reconstruction gradients, while the pressure

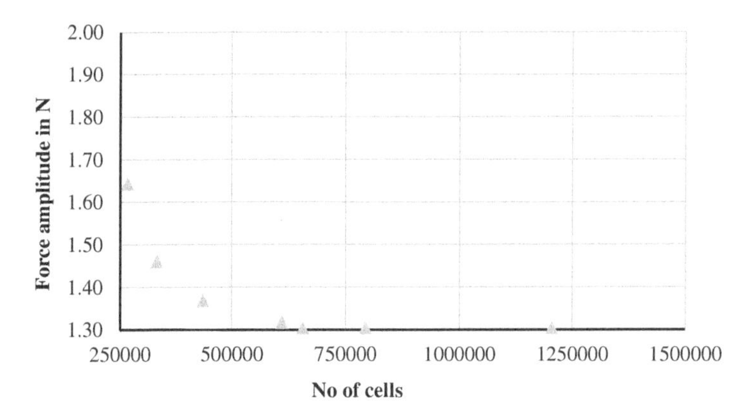

FIGURE 2.29 Mesh convergence study

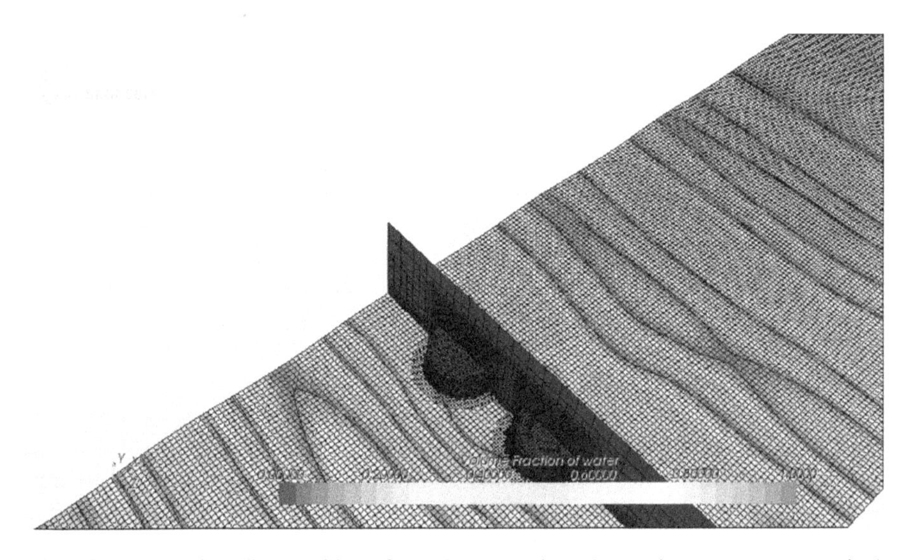

FIGURE 2.30 Twin cylinder with perforated cover (oriented opposite to wave propagation)

outlet specifies the pressure on the outlet boundary to be equal to zero. At the pressure outlet, boundary face velocity is extrapolated from the interior using reconstruction gradients. A momentum source term, which is a dissipation term, is added through the Navier–Stokes equations. The source term is composed of two parts, namely the viscous dissipation term and an inertia term. In the present simulation, wave energy could be dissipated effectively using the viscous dissipation term alone. Using this boundary condition, the tangential velocity at the boundary is set as zero resulting in no velocity through a boundary defined as 'wall'. The physics module defined above simulates the physical phenomenon in a continuum model. As the numerical simulation involves multiple phases like air and water, the definition and monitoring of this interface are of primary importance in this analysis. The volume of fluid (VOF) method is appropriate to capture the flow features around the free surface waves.

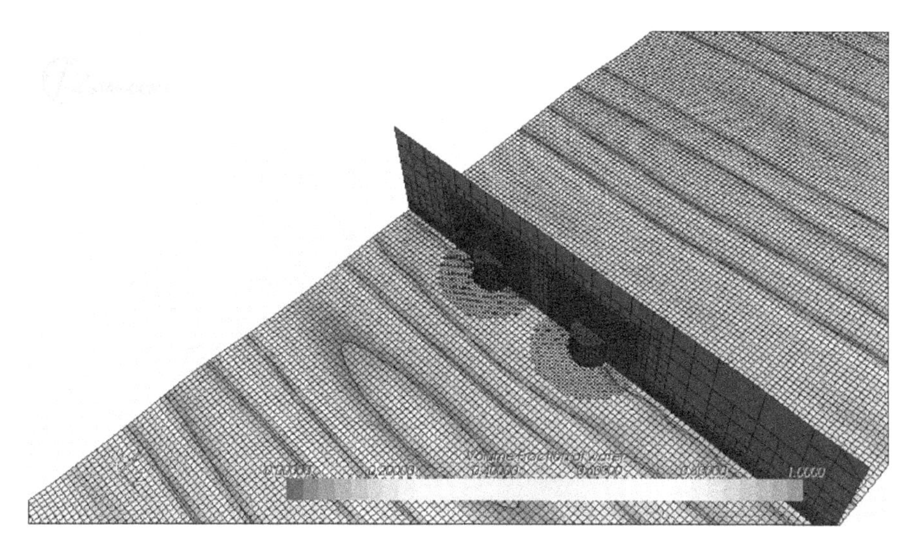

FIGURE 2.31 Twin-cylinder without perforated cover (oriented opposite to wave propagation)

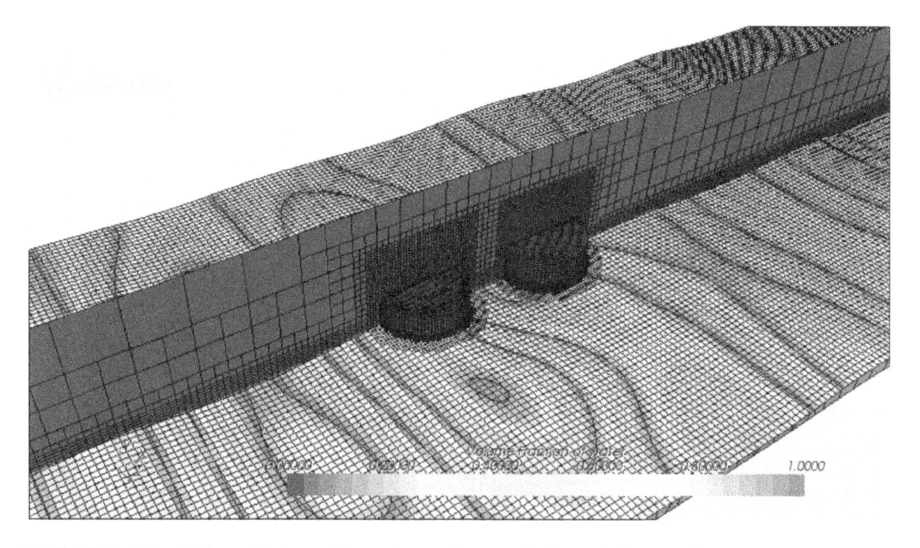

FIGURE 2.32 Twin cylinder with perforated cover (oriented along with wave propagation)

The method places the free surface in cells that are partially filled with water, and a volume fraction is calculated, representing the portion of the cell filled with a pre-determined fluid type. Once the volume fraction is known, the virtual interface can be subsequently solved.

Variation of the horizontal water particle velocity is examined around the twin cylinders. Figs. 2.34 and 2.35 shows the flow visualization around the twin cylinders oriented opposite and along the wave-advancing direction. Under unsteady flow

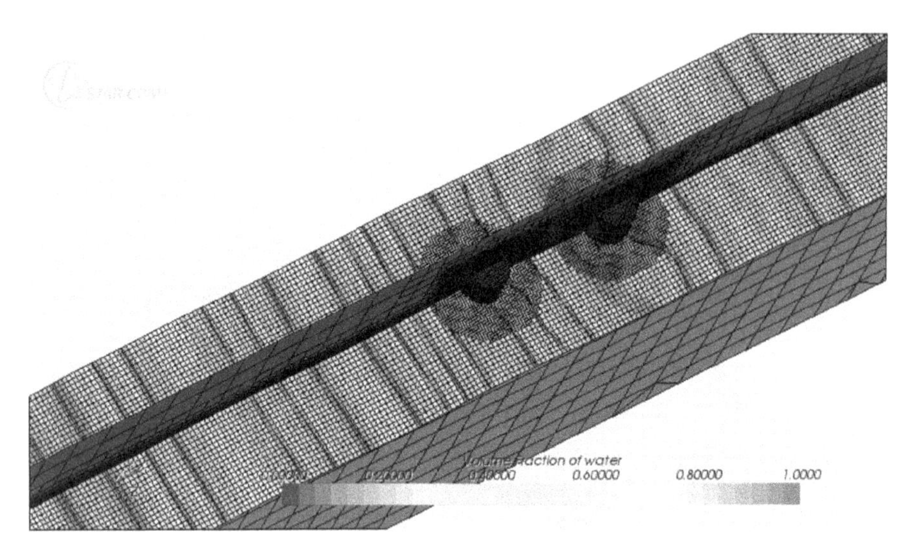

FIGURE 2.33 Twin-cylinder without perforated cover (oriented along with wave propagation)

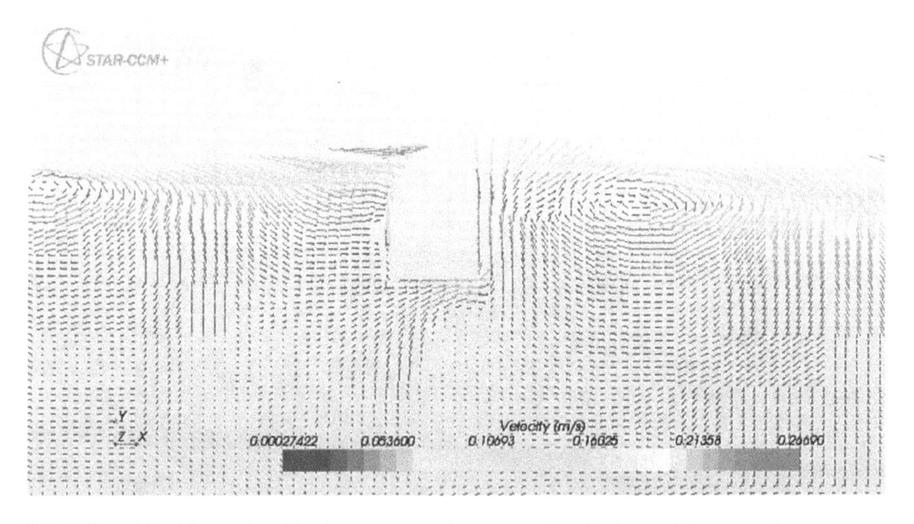

FIGURE 2.34 Flow visualization around the array of cylinders without perforated cover (oriented opposite to wave propagation)

conditions, it is seen from the flow visualization pattern that the fluid velocity at a given point changes with time. Streamline flow is symmetric about the direction of the flow while the velocity becomes zero at the stagnation point. The streamlines bend around the cylinder when placed in the wave advancing direction, as shown in Figs. 2.36 and 2.37. However, the flow field gets altered significantly in the presence of a perforated outer cover, as seen in Figs. 2.38 and 2.39. When the wave reaches the front side of the structure, the water particle is reflected partially through the

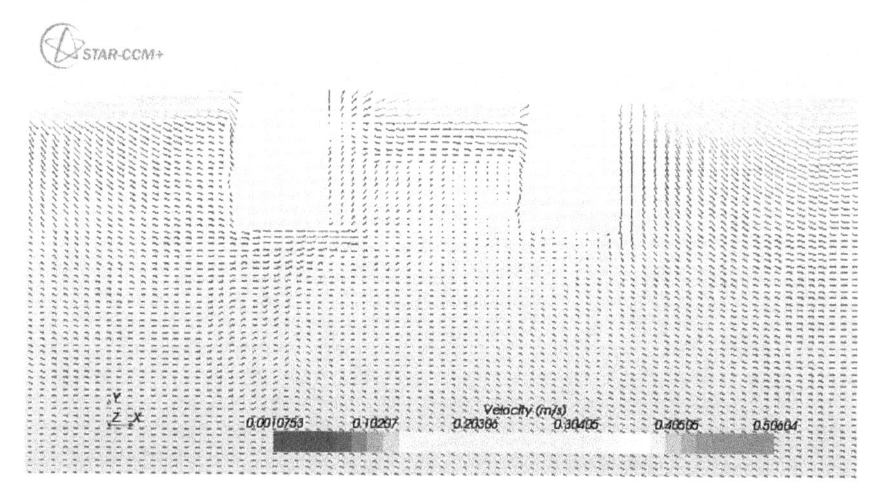

FIGURE 2.35 Flow visualization around the array of cylinders without perforated cover (oriented along with wave propagation)

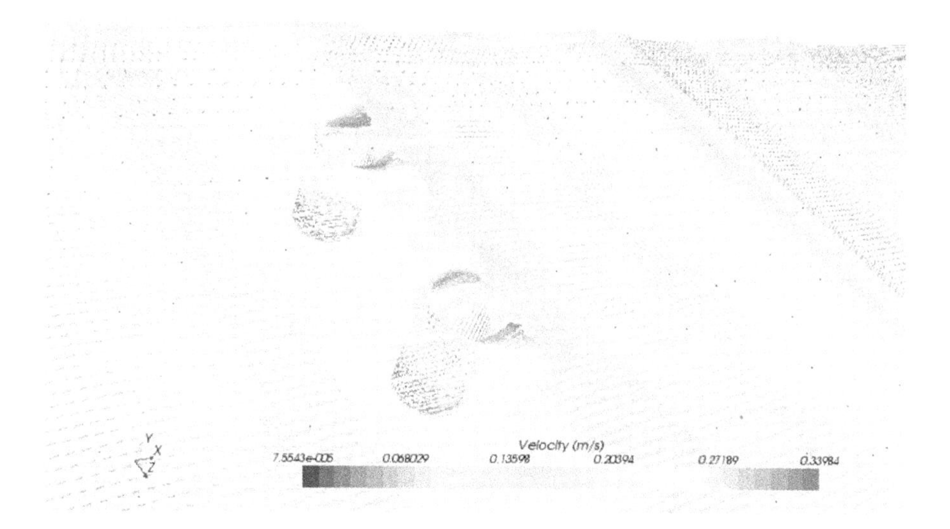

FIGURE 2.36 Velocity vector around the array of cylinders without perforated cover (oriented opposite to wave propagation)

perforated cover. It partly enters through the perforated zone and reaches the inner existing impermeable cylinder.

During this process, wave energy gets dissipated, reducing forces and the turbulence effect created within the annular spacing. When a wave train passes through the perforated cylinder, entrapped water particle blocks are restricted to enter the annular spacing. Hence, the force gets reduced on the inner existing cylindrical

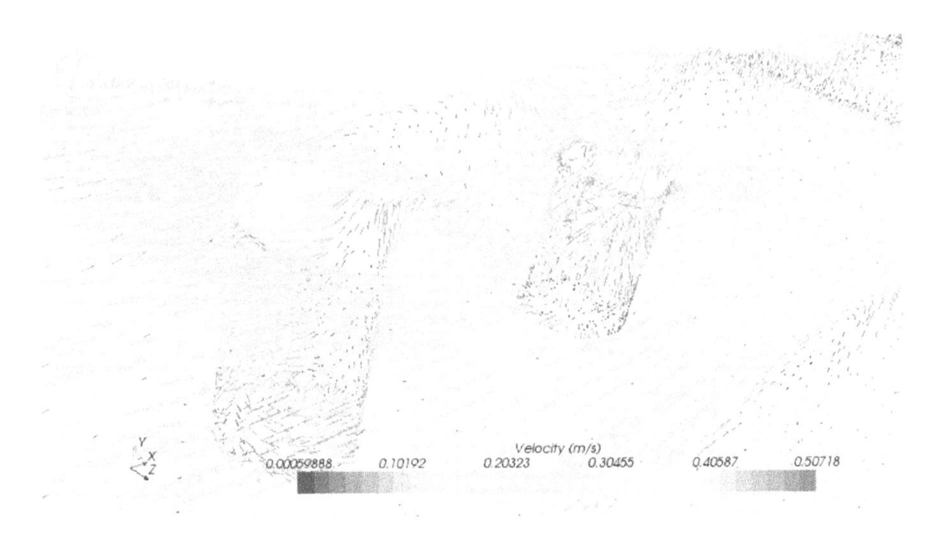

FIGURE 2.37 Velocity vector around the array of cylinders without perforated cover (oriented along with wave propagation)

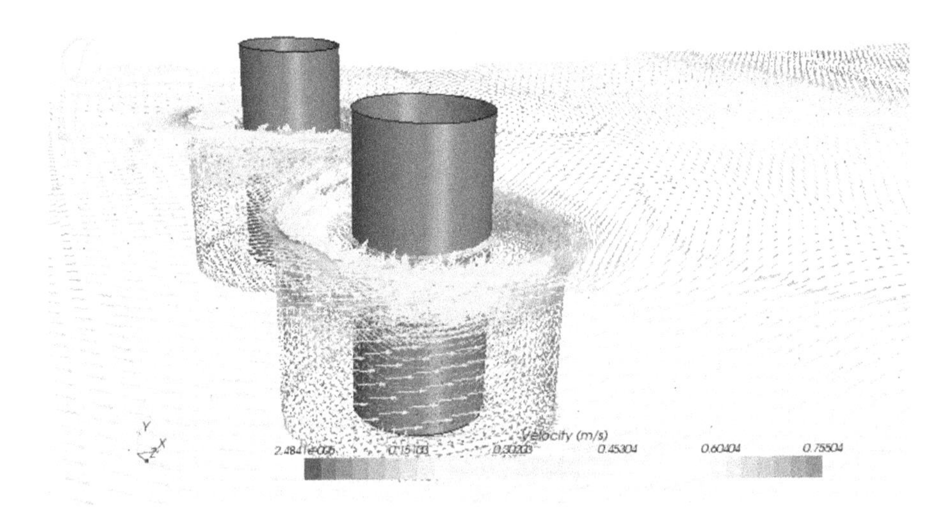

FIGURE 2.38 Velocity vector around the array of cylinders with perforated cover (oriented opposite to wave propagation)

structure significantly. This results in the turbulence effect within the annular spacing, causing a reduction of forces on the inner cylinder. For the inner cylinder without perforated outer cover, stagnation points lie almost in the middle. When the circulation increases, stagnation points move towards the lower half of the cylinder. The two companion fluid particles follow different routes to react with the downstream

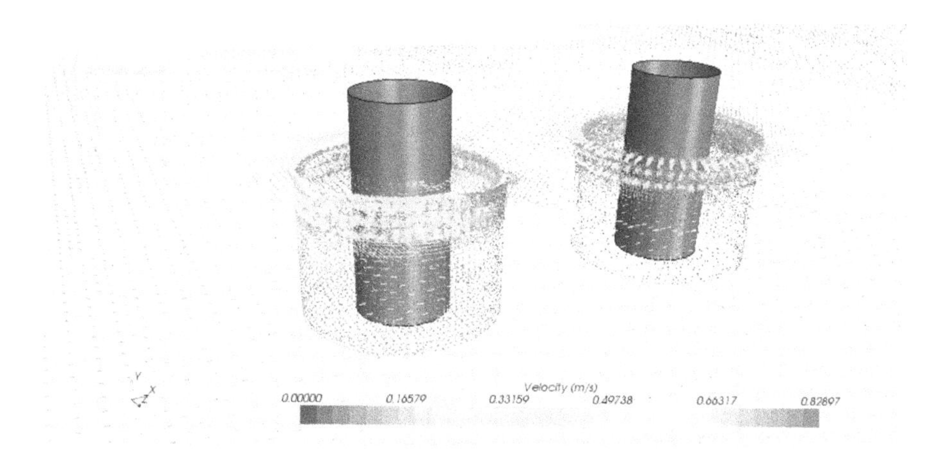

FIGURE 2.39 Velocity vector around the array of cylinders with perforated cover (oriented along with wave propagation)

stagnation point. In particular, fluid particles that travel above the cylinder make a longer route concerning their companion to reach them at the downstream point, which results in a higher velocity. The flow becomes non-linear in the zone of perforation. It results in the phase change of velocity vector due to the back-and-forth movement of the water particles. The flow is altered because of the additional resistance generated due to slotting and clogging of water particles in the perforation, as shown in the figures.

The complexity increases in the perforated zone with the increase in perforation ratio and area of perforation. Inflow through the perforation causes significant pressure loss and results in the dissipation of wave energy; a higher loss of wave energy will occur when the induced water particle velocity on the perforated structure is higher. Fig. 2.40 shows the flow velocity through the perforation zone, indicating significant variation in the water particle kinematics. To track these variations, different sections at closer intervals of 0.01 m are generated, as shown in Fig. 2.41

Fig. 2.42 shows the representative plot showing the variation of the water particle kinematics along with the depth of the cylinder within the zones of perforation for the chosen sections generated along the wave-advancing direction. It is seen that for steep waves, horizontal water particle velocity increases near the perforated outer cover and stabilizes after that. This results in a significant reduction near the vicinity of the inner cylinder, which can be attributed to the free surface effects. In medium steep waves, the velocity increases near the outer cover and drops continuously until it reaches the inner cylinder. There is a significant fluctuation of water particle velocity for low steep waves within the annular space due to the vertical water column accumulated within this space. The above comparisons clearly show that perforated outer cover results in significant force reduction on the inner cylinder. Due to the resistance developed by clogging and slotting of the water particles, there is a significant change of phase of the peaks for all the sea states considered in the study.

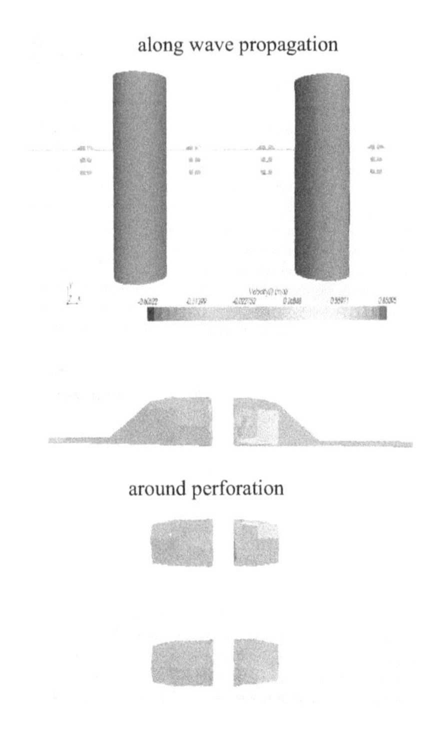

FIGURE 2.40 Velocity variation around perforation

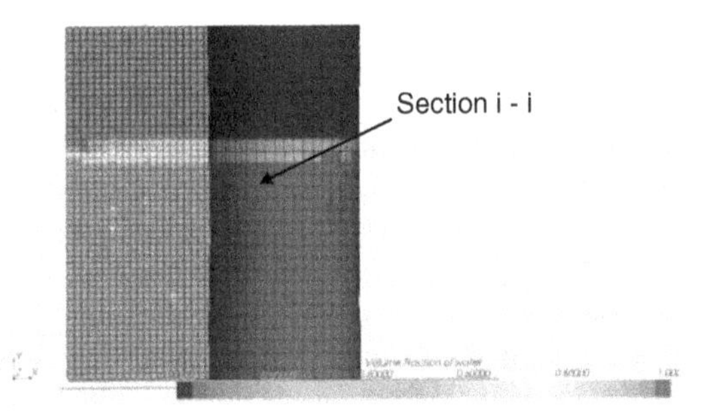

FIGURE 2.41 Representative section to show velocity variation in zone of perforation

Resistance developed near the perforations reduces the horizontal velocity along the wave-advancing direction at different sections within the annular spacing. This also validates the phenomenon of force reduction on the existing structure.

The variations of the flow field in the presence of the perforated cover of a twin-cylinder are non-linear in the perforation zones. The perforation makes the transmission scheme more complex, but the turbulence developed within the annular spacing significantly influences the water particle kinematics. It is also observed that the presence of a perforated cover reduces the horizontal water particle velocity

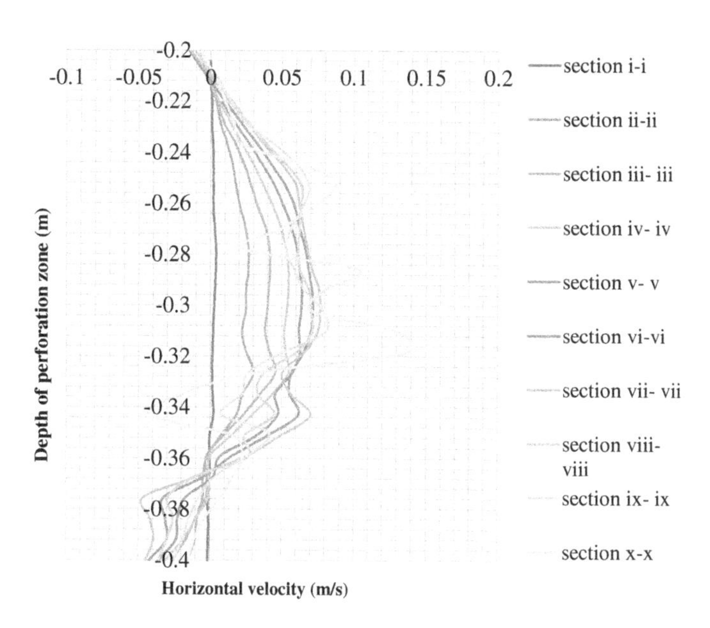

FIGURE 2.42 Velocity variation in zone of perforation along depth of cylinder ($H/L = 0.0064$)

on the inner cylinder significantly in the wave-advancing direction. For low steep waves, variations are due to the vertical water column accumulated within the annular spacing.

EXERCISES

1. Compare the various methods of reducing wave forces on offshore members.
2. What factors govern the reduction in wave forces on the inner solid cylinder when circumscribed by a perforated outer cover?
3. Explain the physics behind force reduction on the inner cylinder when circumscribed by a perforated outer cover.
4. Plot the typical force variation on the inner cylinder near the MSL.
5. How do twin cylinders behave when covered using a perforated cover?
6. What is the effect of annular spacing and perforation ratio on force reduction?
7. Discuss the effect of various parameters of perforation on force reduction on the inner cylinder when covered using a perforated cover.
8. How does spacing of perforation affect force reduction?

3 Floating Wave Energy Converter

3.1 INTRODUCTION

While unleashing the joy of an ocean drive, riding the waves, underwater diving, or any such interactions with the ocean, one would experience a greater magnitude of force when exposed to sea water. It is because seawater is roughly 1000 times denser than air. The solar power arriving on the Earth, besides the Sun–Earth–Moon interactions, instigates the ocean's movement; it generates the surface waves, tides, and currents. It has been a century since the research on harnessing ocean energy started worldwide. Interestingly, the 1979 oil crisis, the 2008 world financial crash, and the COVID-19 pandemic are the path-breaking reality checks that have shaped ocean energy and, as a matter of fact, the entire renewable energy sector.

Ocean energy is one of the alternatives, which is widely being developed. Compared to the massive generation of renewable energy from solar and wind, the contribution from ocean energy remains insignificant at present. But, as per statistics, out of all forms of ocean energy in India, waves outrun the others with a potential estimate of about 40 GW. The sea-surface elevation, resulting in the waves' undulations, is stochastic; yet it is predictable with seasonal variations and defined using statistical parameters such as significant wave height and peak wave period. The wave energy density along the Indian Coast varies between 5 and 30 kW per unit meter of wave front exposure, leaving many hot spots.

A device that harnesses wave energy and converts it into a useful form of energy (mechanical, hydraulic, or electrical) is termed a floating wave energy converter (FWEC). The basic design concept is that the FWECs are ideally designed to perform in the resonance band with the dominant wave period of its deployment site; it must harness the maximum power (Chandrasekaran and Sricharan, 2020a; 2020b). Therefore, unlike other ocean and land-based structures, the design of a FWEC has to deal with the critical performance of the structure (the device) as it should not fail to operate even under the resonance band. The literature refers to many technologies useful in harnessing wave energy: point-absorbers, oscillating water columns (OWC), and overtopping devices (Chandrasekaran and Harender, 2011; 2014; 2015; 2012; Chandrasekaran et al., 2013d). FWEC technologies in India are in the preliminary research stage and require huge funding to enter the prototype stage (Amarkarthik et al., 2012; Chandrasekaran and Harender, 2012).

As the geographic location and the ocean climate influence the design of a wave energy converter, the geometric design cannot be developed for general conditions; it has to be site-specific (Chandrasekaran and Sricharan, 2020a; 2020b; 2020c; 2021; 2019). The design must be based on the local deployment site's characteristics

DOI: 10.1201/9781003281429-3

and estimated sea state to maximize the power outputs (Sheng and Lewis, 2012; Chandrasekaran and Sricharan, 2019). Various complexities involved in the operation, capital investment, and technological challenges to progress from a lab-scale model to the prototype of an FWEC govern the impediment to progress (Babarit et al., 2004; Clément *et al.*, 2002). There are three stages in the wave energy conversion: i) transferring energy from the sea to the oscillating system, comprising the hydrodynamic and reaction subsystems; ii) converting the mechanical energy into either hydraulic or pneumatic power using a suitable PTO system; and iii) electricity generation (Sricharan and Chandrasekaran, 2021; Chandrasekaran et al., 2014c; Chandrasekaran and Raghavi, 2015).

A cam-shaped FWEC, Salter's duck, is popular, and first-generation FWECs (Salter, 1974). Wehausen (1971) and Newman (1962; 1979) developed the analytical methods for marine systems and derived hydrodynamic coefficients from the body oscillations in waves. Falcão (2010) and Mei (2012) classified wave energy converters based on wave energy absorption and wave energy extraction theories. A complete mathematical system for wave energy converters of single oscillating bodies was developed by Falnes (2001; 2002). A floating body active in heave motion (say, for example, a point absorber) radiates circular waves, which interfere destructively with the incident wave to convert the energy (Clément *et al.*, 2002). It is conceptually shown in Fig. 3.1: case (a) represents the calm incoming waves; cases (b, c) represent an oscillating body in the heave and pitch/roll motion, respectively, such that they radiate waves with amplitude half that of the incoming waves. As shown in case (d), the sum of the radiated waves cancels out the incoming waves, thereby absorbing the total incoming energy exerted onto it. The wave pattern of the two interfering waves, namely the incoming and radiated waves, is shown in Fig. 3.2 for a symmetric heaving body.

Based on the geometric shape, size, and positioning of the wave energy extraction devices, they are classified as point absorbers (PA), attenuators, and terminators. The schematic diagrams of these are shown in Fig. 3.3. Attenuators and terminators are

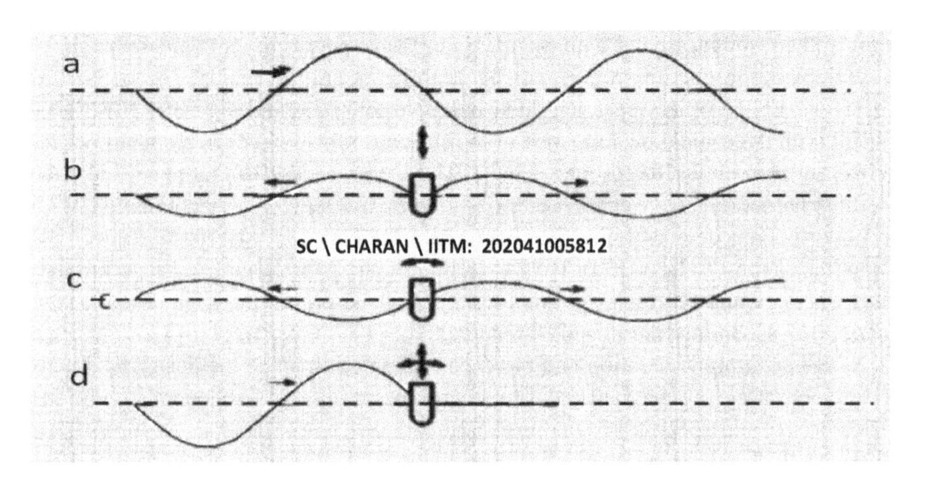

FIGURE 3.1 Wave pattern and body interference

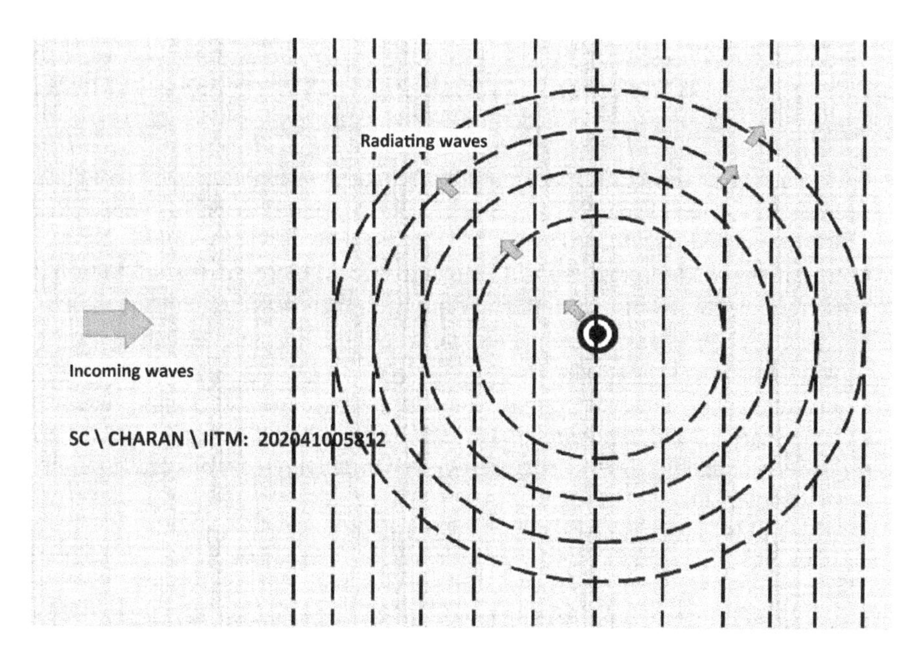

FIGURE 3.2 Plan view of the wave interference

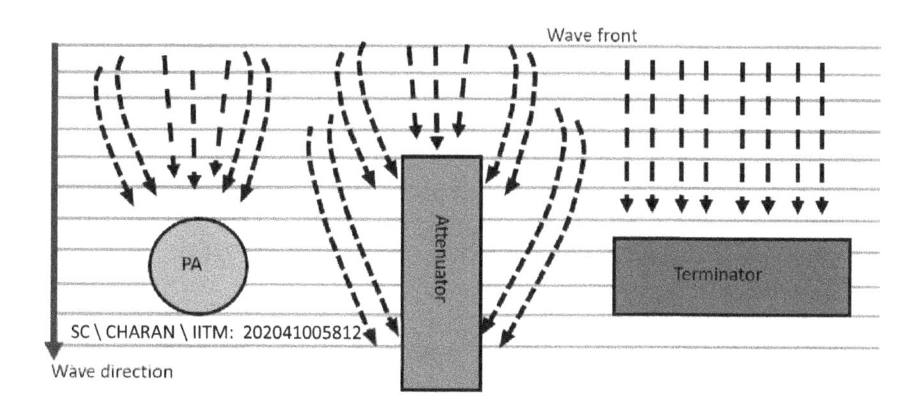

FIGURE 3.3 Classification of wave-energy devices

larger than the incoming wavelengths and are installed parallel and perpendicular to the wave direction. Point absorbers are smaller in size compared to incoming wavelengths and are directionally insensitive.

Based on their operations, wave energy devices can be further sub-classified as follows: i) submerged pressure differentials like CETO-FWEC (Sergiienko *et al.*, 2020); ii) oscillating water column (OWC) (Sharmila *et al.*, 2004); iii) surge or oscillating pitch converters (Oyster); and iv) overtopping devices such as wave dragon and Tapchan (Christensen et al., 2005; Diaconu and Rusu, 2013). Considering the advantages of floating-point absorbers in comparison with the fixed or near-shore devices, the following factors are important (Eidsmoen, 1998; Babarit, 2015; Babarit et al., 2004; 2012):

- A multi-body device invokes multiple frequencies, enabling it to stay closer to the resonance band (Yemm *et al.*, 2012; Meriguad and Ringwood, 2018; Ringwood and Bacelli, 2014).
- Multi-degree absorption devices, active in two or more modes, are proven to perform better (Evans, 1976; French and Bracewell, 1987; Santo et al., 2020).
- The device should be insensitive to the wave direction and the seasonal wave-wind directional variations.
- Floating positive buoyant systems are convenient to maintain under survivable conditions (Chandrasekaran, 2016a, 2016b, 2019b).
- The device should be versatile in design and installation in terms of the location of deployment.
- It should be integrated with other existing structures.
- It should possess the capability to serve as a stand-alone or an auxiliary system.
- Convenience in modeling and fabrication makes them cost-competitive.

For the point-absorbers, the following features are complemented with the desired characteristics, as discussed above:

- wave-wind-PV integrated systems and possible synergic combinations
- wave-powered freshwater production units, like desalination plants integrated with FWECs
- coastal protection, like Breakwater, integrated with FWECs
- FWEC integrated with the oil and gas production platforms in deeper waters
- meteorological measuring devices, integrated with FWECs
- wave-powered, reef generating devices, integrated with FWEC
- potential FWEC, commissioned in arrays.

Various studies conducted on the influence of the geometric shape of the floats on the device's performance are quite interesting. Different shapes such as flat cylinders, hemispherical cylinders, spherical buoys, asymmetric floats, and their size showed different performance (Clauss and Birk, 1996; Birk, 2009). By varying the draft, and other geometric properties, the hydrodynamic parameters of the device are highly influenced, which either improves or worsens the performance of the FWECs (Vantorre et al., 2004; Goggins and Finnegan, 2014). Rounded bottom floats possess a higher oscillation (about 64%) than other shapes due to the reduced viscous drag (Tom and Yeung, 2014). The performance of wave energy devices is compared using the absorbed power and capture width ratio (CWR). The CWR does not reflect the device's efficiency. They attract more energy from the undisturbed incoming waves under the resonance band; in fact, a few studies recorded devices with CWR greater than 100% (Babarit, 2015; Stansby et al., 2017).

3.2 WAVE-TO-WIRE TRANSFER

A suitable PTO is essential for the wave-to-wire transfer (Rodriguez et al., 2019). An overview of PTO systems is given in Fig. 3.4. The literature has widely contributed to various PTO systems regarding their influence on the overall performance and suitability with various FWEC concepts (António, 2007; López et al., 2017; Jin et al., 2019). A recent review by Ahamed et al. (2020) gives comprehensive information for FWEC developers. Mueller (2002), Ivanova et al. (2005), and Rhinefrank et al. (2012), through numerical simulations, explained different ways to tune the damping and stiffness coefficients of a PTO system to improve the power absorption. Hydraulic PTO is one of the most commonly deployed FWEC as it can convert the low-frequency input wave motion to a high-frequency motor response (Gaspar et al., 2016; Jusoh et al., 2019). Cargo et al. (2012; 2014) discussed the effects of sizing of the components of a PTO, accounting for compressibility and pressure loss effects under random waves. Different successful prototype devices of wave energy converters, namely Wavestar (Hansen and Kramer, 2011; Kim et al., 2019), the Scottish Pelamis (Yemm et al., 2012), are good examples of the hydraulic PTOs. However, the other alternatives are not popular due to their larger initial investments.

The preliminary design of a wave-energy converter requires a combined understanding of wave characteristics, wave-FWEC interactions, and aspects of the mechanical design of components of the FWEC. Exposure to a couple of analysis tools and a good experience in conducting experiments are also important to fabricate a workable model of an FWEC. Fig. 3.5 describes the flowchart, highlighting the necessary steps required in the analysis and design phases of the FWEC. The procedure, conventionally practiced in the industry and the research laboratories is an iterative process, comprising of three distinct stages: The first stage is to the wave characteristics at the location of interest; the second stage is to prepare a CAD model to assess the device's parameters, and the last stage is to improve its absorbed

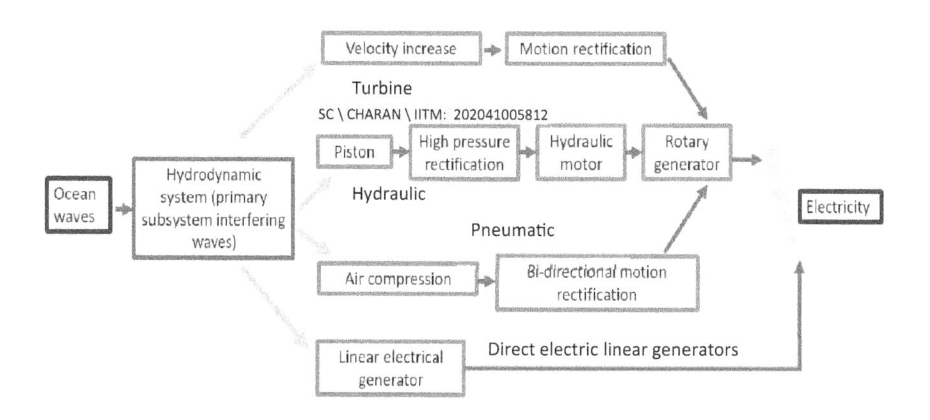

FIGURE 3.4 Power take-off systems: Classification

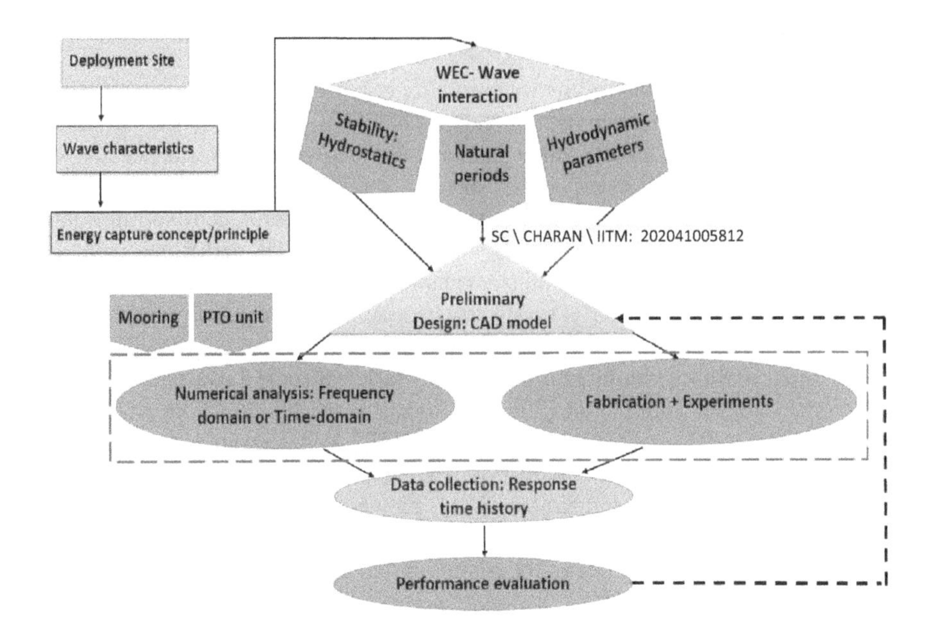

FIGURE 3.5 Overview of modeling and analyses of FWEC

power and overall performance, based on experimental tests on scaled models. Please note that a wave-to-wire transfer does not deal with the design of the electrical grid, compatible with the hydraulic output of the FWEC. Based on the observations from the numerical analysis, a scaled model shall be fabricated and tested for showcasing the proof-of-concept in the early stages of development. This is termed as technology-readiness.

The hydrodynamic interaction of the FWEC and ocean waves is a complex problem to solve as they are highly nonlinear. However, simplified versions of such interactions are possible under certain accepted assumptions, making it a linear problem (Falnes, 2002; 2001). In the preliminary design stages, it is a widely accepted practice to model the reactive subsystems of the FWEC, such as mooring and PTO, as a linear spring and linear damper, respectively. They are conveniently modeled in the frequency domain to represent the FWEC dynamics.

3.3 NUMERICAL MODELING

A schematic illustration modeling FWEC as the spring-mass-damper system is shown in Fig. 3.6. Consider a hollow, floating buoy being pushed into the ocean. The volume of water surrounding it will displace, developing an opposite force resulting in the bobbing of the buoy. It is similar to the oscillations of a spring-supported mass. In the presence of viscosity and body-generated waves, part of its energy is dissipated into the medium. A damping effect is induced externally onto the bobbing buoy by a PTO system. It absorbs energy from the harmonic motion of the buoy, which is analogous to the dashpot-damping of the mass. The motion

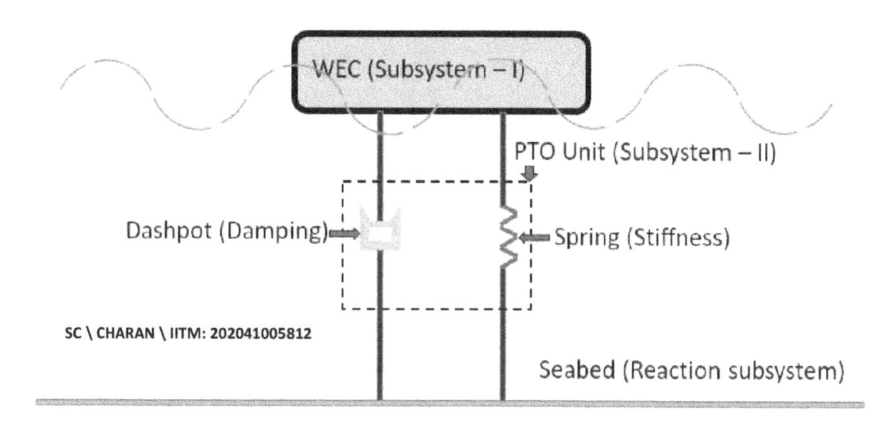

FIGURE 3.6 Schematic representation of FWEC-PTO system

of the waves is described using velocity potentials; it is a widely accepted practice to represent these as first-order waves using Airy wave theory while assuming the flow as a potential flow. The theory uses the linearized Bernoulli equation, assuming waves with small amplitude and large wavelengths under uniform water depth; bathymetry effects are usually neglected. In the current scenario, the fluid is assumed to be inviscid, irrotational, and incompressible, enabling the solution of the linear potentials of diffracted and radiated waves. It is often referred to as the potential wave theory approach.

Bernoulli's equation on the free surface, conserving the mass and the momentum at every point of the fluid domain, is written as follows:

$$-\frac{\partial\phi}{\partial t}+\frac{p_a}{\rho}+\frac{1}{2}(\nabla\phi)^2+g\eta=C \text{ on free surface } z=\eta(x,y,t) \tag{3.1}$$

However, the linearized version implies the boundary conditions on the FWEC's surface at its equilibrium position and thus neglects the instantaneous variation of sea surface elevation. Mathematically, instead of defining the equation on the free surface, $z=\eta(x,y,t)$, it is defined on $z=0$.

$$-\frac{\partial\phi}{\partial t}+\frac{p_a}{\rho}+g\eta=0 \text{ on } z=0 \tag{3.2}$$

ϕ is the velocity potential, and the atmospheric pressure, p_a, is taken as a reference for the pressure variation and is therefore set to be zero on the free surface elevation, η. Therefore, substituting the value of p in the above equation at the free surface, we get:

$$\eta=\frac{1}{g}\left(\frac{\partial\phi}{\partial t}\right)_{z=\eta} \tag{3.3}$$

Since the wave amplitude is assumed to be small, it is rewritten as below:

$$\eta = -\frac{1}{g}\left(\frac{\partial \phi}{\partial t}\right)_{z=0} \tag{3.4}$$

A linearized, kinematic boundary condition, which describes that the component of the fluid velocity normal to a structure must be equal to the free-surface velocity and is written as:

$$\frac{\partial \eta}{\partial t} + \frac{\partial \phi}{\partial z} = 0 \text{ on } z = 0 \tag{3.5}$$

For the linearized version of the governing flow equation, the free-surface boundary condition can be written as follows, after due substitution:

$$\frac{\partial^2 \phi}{\partial t^2} + \frac{1}{g}\frac{\partial \phi}{\partial z} = 0 \text{ on } z = 0 \tag{3.6}$$

The potential flow theory is solved by developing the Laplace equation with the boundary conditions: free-surface, body-surface, far-field, and the seabed. The first three are given above. The far-field boundary condition states that the waves generated by the FWEC should die down far away from the FWEC's position; only the incident waves must be present. The literature suggests this decay as in the form of radiation condition, whose magnitude is inversely proportional to the square of the radial distance (r) and the wave number (k) from the structure and is given as follows:

$$\phi \propto \frac{1}{\sqrt{kr}} e^{-ikr} \text{ as } r \to \infty \tag{3.7}$$

The total velocity potential of a typical wave-floating FWEC's interactions is decomposed as:

$$\phi_{tot} = \phi_I + \phi_S + \phi_R \tag{3.8}$$

ϕ_{tot} is the total wave potential, decomposed into incident wave (ϕ_I), scattered wave (ϕ_S) and radiated wave (ϕ_R). The diffracted wave is composed of both the incident and the scattered waves ($\phi_D = \phi_I + \phi_S$). The incident waves are calm waves of the ocean without any interference effect from the structure, while the diffraction field is the disturbance created in the incident wave field. At the same time, the radiated field is created due to the structure's oscillations in the calm water. The boundary conditions associated with each potential are not discussed here; further details can be seen in recent studies (Chandrasekaran and Sricharan, 2019; 2020c; 2021).

Since the waves are sinusoidal, their decomposition is based on the linear superposition of incident, diffracted, and radiated waves. Numerical tools are used to solve the velocity potentials. As the dynamic response of the FWEC is also combined, solving them through analytical methods is difficult. Alternatively, they can be solved using the appropriate numerical tools in any of these domains, namely frequency domain, time domain, or spectral domain (Chandrasekaran and Sricharan, 2020c).

3.4 FREQUENCY-DOMAIN MODELING

For the known shape and the overall geometry of the FWEC, frequency-domain analysis is convenient, estimating the absorbed power output and performance characteristics under a virtual-linear PTO system. As the frequency domain is a linear model, nonlinearities generated during wave-structure interactions are ignored; hence, a near-resonance performance cannot be captured accurately. However, the preliminary results obtained from this analysis are helpful to modify the wave loads, transfer functions and optimize the shape and size of the FWEC. The hydrostatic parameters are estimated first and followed by the estimation of frequency-dependent, hydrodynamic parameters. Hydrostatics provide the device's stability characteristics in terms of center of gravity, the center of buoyancy, transverse and longitudinal metacentric height, water-plane area, submerged volume at the designed draft, overturning, and restoring moments. They add appreciable perception to the preliminary design of the FWEC.

The hydrodynamic parameters such as the added mass, radiation damping, and excitation force are frequency-dependent. Hence, in this approach, the response amplitude operators (RAO) of the device's motion are estimated assuming linear potential wave theory (PWT). The power absorption capacity and the performance of the FWEC in the active degrees of freedom are estimated by providing appropriate PTO properties. Subsequently, the output is improved by altering the shape, size, and other design parameters of the chosen FWEC configuration. A frequency-domain approach using the commercial solvers yields a reasonably accurate estimate of the response of FWEC in all the active degrees-of-freedom. Still, the responses are generally limited to the heave motion of the point absorber, in general. Fig. 3.7 shows a block diagram of the frequency-domain model with user inputs and the relevant outputs yielded from the numerical tools. A frequency domain-specific workflow to Ansys AQWA is

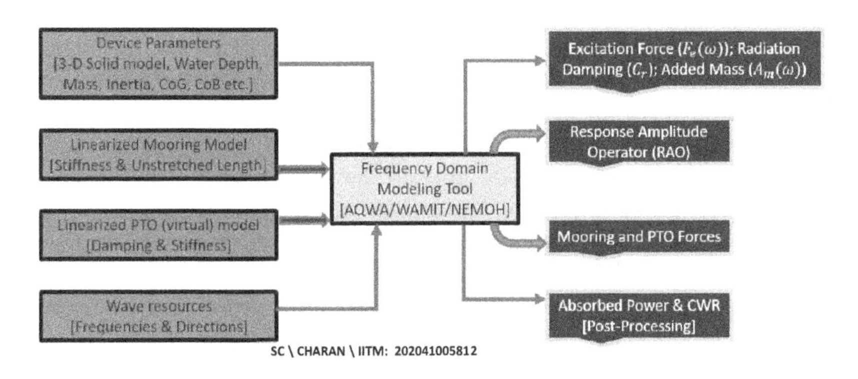

FIGURE 3.7 High-end block diagram of the frequency-domain model

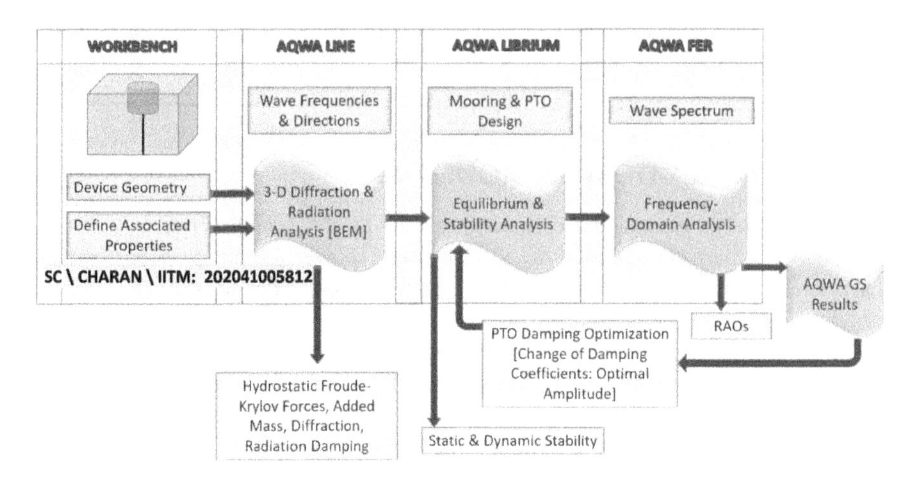

FIGURE 3.8 Frequency domain-specific workflow to Ansys AQWA

given in Fig. 3.8. However, one of the major tasks in such analysis is the meshing of the submerged surface. In the linear analysis, it is required to calculate the pressure and the resultant forces on each element of the mesh; the accuracy of the simulation time depends on the mesh's number of elements. In the case of a multi-body FWEC, the number of elements further increases. Hence, one has to carefully decide the number of mesh elements, which shall govern the accuracy of the response. In general, mesh size is reduced within the proximity of two bodies and at the edges in the case of a multi-body analysis; hence, a coarse mesh can be chosen for the remaining part of the geometry.

Software tools like WAMIT and NEMOH can solve the boundary value problem using the boundary element method (BEM) or the panel method. These help estimate the diffraction and radiation velocity potentials by applying the source functions and solving their complementary strengths on the submerged surfaces. Once the velocity potentials are solved, the excitation force added mass, damping, pressure fields, and the response amplitude operators (RAOs) can be subsequently estimated. However, one of the main advantages of using Ansys AQWA is the graphical user interface for modeling and simulation. The software has limited post-processing tools built-in, making the analysis more convenient.

The environmental forces acting on the FWEC are balanced by their inertial force based on Newton's second law. These forces include non-viscous and viscous forces (form drag and friction drag are part of the viscous forces). The boundary-layer frictional effects are represented by the friction drag terms, which are usually small in magnitude compared to the form drag; they can be neglected, but the form drag increases with the size (cross-section) of the structure. Further, in the potential wave theory approach, the viscous effects are ignored. Non-viscous forces include hydrodynamic forces (other than viscous force) acting on the FWEC's surface, including mooring and the PTO mechanism.

The wave potentials and the FWEC's response are represented in the complex terms in the frequency domain. The wave potential ϕ for a monochromatic wave can thus be represented as a combination of the temporal and spatial functions as shown below:

$$\phi(x,y,z,t) = Re\{\hat{\phi}(x,y,z,t)e^{i\omega t} \tag{3.9}$$

$\hat{\phi}$ are the complex wave potential function, ω and t are wave frequency and time, respectively. All the boundary conditions are thus represented in the complex form. For simplicity, let us assume FWEC as a single degree of freedom system. The forces and the response are thus represented in terms of a harmonic function (time-dependent parameter) and complex amplitude. Thus, the FWEC displacement is given by the following relationship:

$$x(t) = Re\{\hat{x}(\omega)e^{i\omega t}\} \tag{3.10}$$

Further, the velocity and the acceleration are given as below:

$$\dot{x}(t) = Re\{i\omega\hat{x}(\omega)e^{i\omega t}\} \tag{3.11}$$

$$\ddot{x}(t) = Re\{-\omega^2\hat{x}(\omega)e^{i\omega t}\} \tag{3.12}$$

From Newton's law, the complex equation of motion for an FWEC of mass 'M' can be written as below:

$$-M\omega^2\hat{x}(\omega)e^{i\omega t} = F_{tot}(t)e^{i\omega t} \tag{3.13}$$

The total force comprises the hydrodynamic forces acting on the device, excluding the viscous forces; however, external forces induced by the PTO system and mooring lines are included. Hence, the right-hand side of Eq. (3.13) is of a complex form. The hydrodynamic forces are classified into excitation and radiation forces, as shown in Fig. 3.9. They are further classified into the diffraction force and the Froude-Krylov

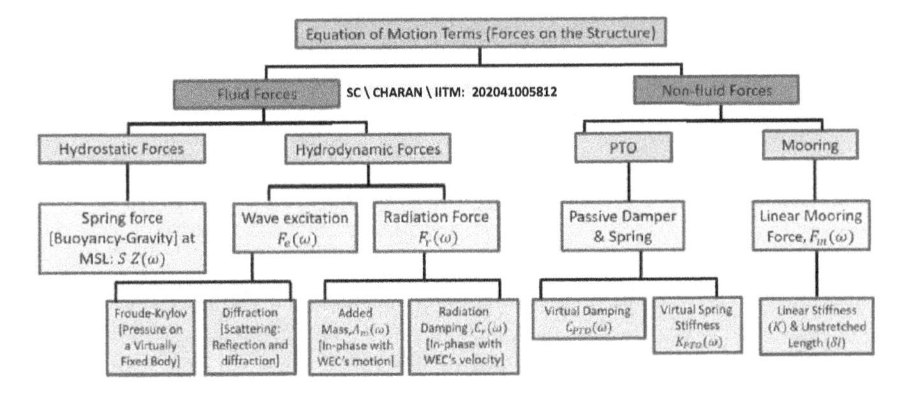

FIGURE 3.9 Non-viscous forces acting on the device

force. While the former occurs due to the scattering of waves in the device's presence, the latter is due to the pressure acting on the virtual FWEC (assumed to be fixed) in the absence of waves. The diffraction theory presumes that the device's dimensions are larger than the incident wavelength; the presence of the structure is felt, and waves get diffracted.

The radiation forces are those generated due to the oscillations of the device in the calm water; these waves are generated outwards. The net effect of the radiated and incident waves produces a complex wave-FWEC interaction, affecting the device's performance. The radiation force comprises the added mass and the radiation damping force. Added mass is the additional inertial force generated by the accelerated fluid around the device due to its oscillations in the water; it will be in phase with its motion. The radiation damping is the energy dissipation with the radiated (out-going) waves, which shall be out-of-phase with the device's motion, but shall be in-phase with its velocity. Thus, it is proportional to the velocity of the device and termed as the radiation-damping coefficient. The hydrostatic force (or restoring force) is the equilibrium with the gravity and buoyancy proportional to the wetted surface area at the mean sea level (MSL). The following equation holds:

$$\hat{F}_{tot}(\omega) = \hat{F}_{FK}(\omega) + \hat{F}_{D}(\omega) + \hat{F}_{R}(\omega) + \hat{F}_{H}(\omega) + \hat{F}_{PTO}(\omega) + \hat{F}_{M}(\omega) \tag{3.14}$$

The force is estimated by integrating the corresponding pressure over the surface, which will be a surface integral, $\int_S (P)\vec{n}\,dS$). Pressure can be determined from the first order Bernoulli equation ($P = -\rho\frac{\partial \phi}{\partial t}$). The radiation potential is assumed to be active in all degrees of freedom unless constrained, irrespective of the active degree of freedom of the device. Thus, Eq. (3.14) can be written in terms of potential as follows:

$$\begin{aligned}\hat{F}_{tot}(\omega) = i\omega\rho\int_S \left(\hat{\phi}_I + \hat{\phi}_S\right)\vec{n}\,dS + i\omega\rho\int_S \hat{\phi}_R\,\vec{n}\,dS - \rho g S\hat{x}(\omega)\\ - i\omega C_{PTO}\,\hat{x}(\omega) - K_{PTO}\,\hat{x}(\omega) - K_M\,\hat{x}(\omega)\end{aligned} \tag{3.15}$$

First-term in the above equation is the complex excitation force, $\hat{F}_E(\omega) = \hat{F}_{FK}(\omega) + \hat{F}_D(\omega)$. The second term is the complex radiation force; (j, N) represent the mode and the number of free modes (oscillating/unconstrained modes), respectively. It is expanded in the following form:

$$-\omega^2\rho\int_S \sum_{j=1}^{N} \hat{x}_j(\omega)\,\phi_j\,\vec{n}\,dS \tag{3.16}$$

where $\hat{\phi}_R = i\omega\sum_{j=1}^{N}\hat{x}_j(\omega)\,\phi_j$. It is further split into radiation damping and the added mass. The complex radiation force is represented as follows:

$$\hat{F}_R = -i\omega\hat{x}(\omega)Z_R \tag{3.17}$$

Comparing the expanded form of the second term of Eq. (3.15) for any one degree of radiation with this term, radiation impedance will be given as follows:

$$Z_R = -i\omega\rho\int_S \phi\vec{n}\, dS \tag{3.18}$$

Impedance is a complex function, with the real part representing the resistance and the imaginary part representing the reactance. Hence, Eq. (3.18) can be rewritten as follows:

$$Z_R = -i\omega\rho\int_S \phi\vec{n}\, dS = R(\omega) + i\,I(\omega) \tag{3.19}$$

where the real part (R) is the radiation damping coefficient, and the imaginary part ($I(\omega)$) is the added mass. Z_R is also represented in a different form:

$$Z_R = C_{rd}(\omega) + i\,\omega A_m(\omega) \tag{3.20}$$

The real and imaginary parts are derived using the Kramers-Kronig and Haskind relations (Haskind, 1957; Newman, 1962; 1979; Folley, 2016).

The third term in Eq. (3.15) represents the complex hydrostatic force, while the fourth and fifth terms represent the damping and spring stiffness of the PTO system, respectively. The sixth term represents the complex mooring force. Summing them, one can write the equation of motion in the frequency-domain as follows:

$$\begin{aligned} &-\omega^2\left(M + A_m(\omega)\right)\hat{x}(\omega) - i\,\omega C_{rd}(\omega)\hat{x}(\omega) + \rho g S\hat{x}(\omega) + i\omega C_{PTO}\hat{x}(\omega) \\ &+ K_{PTO}\,\hat{x}(\omega) + K_M\,\hat{x}(\omega) = \hat{F}_E(\omega) \end{aligned} \tag{3.21}$$

The following relationship gives the complex amplitude of the device:

$$\hat{x}(\omega) = \frac{\hat{F}_E(\omega)}{\left[-\omega^2\left(M + A_m(\omega) + K_M + K_H + K_{PTO}\right] + i\omega\left[C_{rd}(\omega) + C_{PTO}\right]} \tag{3.22}$$

If the diameter of a circular point absorber is greater than 15% of the incoming wavelength, then Morison's equation is no longer valid. In such cases, forces are estimated using linear diffraction and Froude-Krylov methods (Chakrabarthi, 1982; Chakrabarti and Tam, 1975; Chakrabarti et al., 1976). Scattering of the incoming waves increases if the effective diameter is significantly larger than the incoming wavelengths, signifying the importance of the diffraction force. Further, for (*D*/*L*) greater than unity, it is a scenario of a pure reflection. In the design of FWEC, the majority of the incoming wave force is exerted due to the dominance of Froude-Krylov forces; the remaining is from diffraction.

Response amplitude operator (RAO) is the transfer function, which is a system property. It represents the motion characteristics of the device and is useful in estimating the dominant frequency band within which the device shall absorb the maximum power. RAO is expressed as a ratio of the response amplitude to the wave amplitude. It is quite easy to infer that the peak RAO amplitude is seen at the resonance band. However, in a multi-body FWEC, the peaks can occur at various frequencies, activating the device to absorb energy over a wider range of frequencies. The phase term of the complex RAO gives the phase difference between the motion of the device and the incident waves. To optimize an FWEC, it is necessary to optimize the response amplitude while keeping the device's velocity in phase with that of the incident waves; it is done by passive controlling. However, the external actuating system needs to actuating a zero-phase difference, termed active controlling. Usually, this is modeled and assessed in the time domain due to the virtue of nonlinearities present in the scheme.

The PTO force is calculated using the damping value. The stiffness of the PTO is used to tune the device with the incoming wave periods, unlike hydrostatic stiffness. The latter alters the natural period by changing the wetted surface area. Commercial frequency-domain packages like Ansys AQWA do not have a provision to incorporate the virtual PTO stiffness values; hence, damping coefficients can be used to calculate PTO force and the absorbed power. The damper is the sole contributor in absorbing the mechanical oscillations of the device. The average absorbed power over a range of wave period is given by:

$$P_a = \frac{1}{T}\int_0^T C_{PTO}\dot{x}^2 dt = \frac{1}{2}C_{PTO}\,\omega^2 \mid \hat{x}(\omega) \mid^2 \tag{3.23}$$

A wide range of PTO damping values can be considered to estimate the corresponding absorbed power. An envelope can be generated highlighting the power absorption characteristics at various frequencies and damping values. The output will be a spectral density function in irregular waves, referred to as a response spectrum; the area under the curve gives the absorbed power. Thus, the frequency-domain analysis provides a rough estimate of the optimal PTO damping values required for a device. This analysis method has limitations in terms of incompetence in representing the nonlinearities that arise from various terms in the equation of motion. In addition, real PTO systems and dedicated active control systems cannot be incorporated.

3.5 TIME-DOMAIN MODELING

Nonlinear forces arise from various sources, namely PTO system, control system, viscous quadratic forces, and wave potential. They can be handled in the time-domain approach but at the expense of higher computational time and improved accuracy. Alternate methods, namely spectral-domain analysis (Folley, 2016) and nonlinear frequency domain analysis (Merigaud and Ringwood, 2018), were formulated to provide the trade-off between accuracy and speed. In the time domain approach, governing equations are described as a function of time. The frequency-dependent hydrodynamic coefficients, RAO, and hydrostatics obtained from the frequency domain are the building blocks for a time-domain model. The wave elevation and the velocity potential of the incident waves, assuming a sinusoidal motion, are represented as follows:

$$\eta(t) = \eta_a \sin(\omega t - kx cos\beta - ky sin\beta + \varphi) \tag{3.24}$$

$$\phi(x,y,t) = \frac{\eta_a g}{\omega} e^{kz} \cos(\omega t - kx sin\beta - ky cos\beta + \varphi) \tag{3.25}$$

β is the direction of propagation of the incident waves, and φ is the wave phase. For unidirectional polychromatic waves, wave elevation is represented as a summation of monochromatic wave elevation with a difference of frequencies and amplitudes, as given below:

$$\eta(t) = \sum_{k=1}^{N_w} \eta_{ak} \sin(k_k x - \omega_k t(k) + \varphi_k) \tag{3.26}$$

$$\eta_{aw,k} = \sqrt{2 S_{pdf-w}(f_k) \Delta f_w} \tag{3.27}$$

where N_w is the number of frequency components, k_k is the wave number, φ_w is the phase difference $\{\in [0\ 2\pi]\}$, S_{pdf-w} is the Wave spectral density function, f_k is the frequency in Hz and Δf_w is the frequency step. Based on Newton's law, the governing equation for the floating FWEC is formulated using the Cummins relationship (Cummins, 1962); FWEC is considered a single degree of freedom system.

3.5.1 Linear Time-Domain Model (LTD)

By assuming a linear system, the Cummins equation defines the parameters in terms of impulse responses. According to Cummins, a body from rest is excited for a short duration and exhibits an impulsive oscillation at a constant velocity. Due to this, the water particles also get displaced and are represented by the corresponding velocity potential; this is proportional to the velocity of the device within the range (t_0, $t_0 + \delta t$). Even after the short duration, an energy transfer occurs from the device to the medium, oscillating the water particles. It is well established that an impulse displacement at any time instant shall influence the surrounding medium during that time instant and the time afterward. Hence, water particle motion shall influence the past response, which is termed the memory effect. Thus, the total time response is calculated by accumulating the successive impulse displacements. The velocity potential can be expressed as follows:

$$\phi(t) = \dot{x}(t)\psi + \int_{-\infty}^{t} \chi(t-\tau) . \dot{x}(\tau) d\tau \tag{3.28}$$

The first term in the above equation represents the potential during δt, while the second term is the successive impulse response (memory effect). (ψ, χ) are the normalized velocity potentials caused by the displacement during δt and the subsequent time steps, respectively. Using Eq. (3.28), the hydrodynamic force can be obtained as follows:

$$F_R(t) = A_m\,\ddot{x}(t) + \int_{-\infty}^{t} K_r(t-\tau)\,.\dot{x}(\tau)d = -\iint_S P_R\,\vec{n}\,ds \tag{3.29}$$

Now, the equation of motion reduces to the following form:

$$(M + A_m)\ddot{x}(t) + \int_0^{\infty} K_r(\tau)\dot{x}(t-\tau)d\tau + K_H\,x(t) = F_E(t) \tag{3.30}$$

In Eq. (3.30), it is necessary to estimate $K_r(\tau)$ and A_m Which can be done using the frequency domain approach (Ogilvie, 1964). The following equation holds good:

$$A_m(\omega) = A_m - \frac{1}{\omega}\int_0^{\infty} K_r(\tau)\,\sin(\omega\tau)\,d\tau \tag{3.31}$$

$$C_r(\omega) = \int_0^{\infty} K_r(\tau)\,\cos(\omega\tau)\,d\tau \tag{3.32}$$

By an inverse Fourier transform, one can obtain the time-domain retardation coefficient, $K_r(\tau)$. By substituting in Eq. (3.31), one can derive the added mass, which is now valid across all the frequencies. The added mass is measured at an infinite frequency and thus represented as infinite added mass (A_∞). Numerical techniques can be used to estimate these parameters while solving the equation of motion. The hydrodynamic force under wave excitation can be written as:

$$F_E(t) = F_{FK}(t) + F_D(t) = -\iint_S P_I\,\vec{n}\,ds - \iint_S P_D\,\vec{n}\,ds \tag{3.33}$$

Based on the linear time-domain model, the pressure estimates as $P_I = -\rho\frac{\partial\phi_I}{\partial t}$ and $P_D = -\rho\frac{\partial\phi_D}{\partial t}$, respectively. The total radiation force is obtained by convoluting the radiation retardation term, also known as impulse response function (IRF), with FWEC's velocity for the desired degree of freedom. The excitation force in the time domain is often expressed in terms of the impulse response function and is given by:

$$F_E(t) = \int_{-\infty}^{\infty} \eta(\tau)K_E(t-\tau)d\tau \tag{3.34}$$

The above is obtained by convoluting the excitation IRF with the wave elevation. Hence, the equation of motion takes the following form:

$$(M - A_\infty)\ddot{x}(t) + \int_{-\infty}^{t} K_r(t-\tau)\dot{x}(\tau)d\tau\ddot{x} + C_{PTO}\dot{x}(t) + K_M x(t) = \int_{-\infty}^{\infty} \eta(\tau)K_E(t-\tau)d\tau \tag{3.35}$$

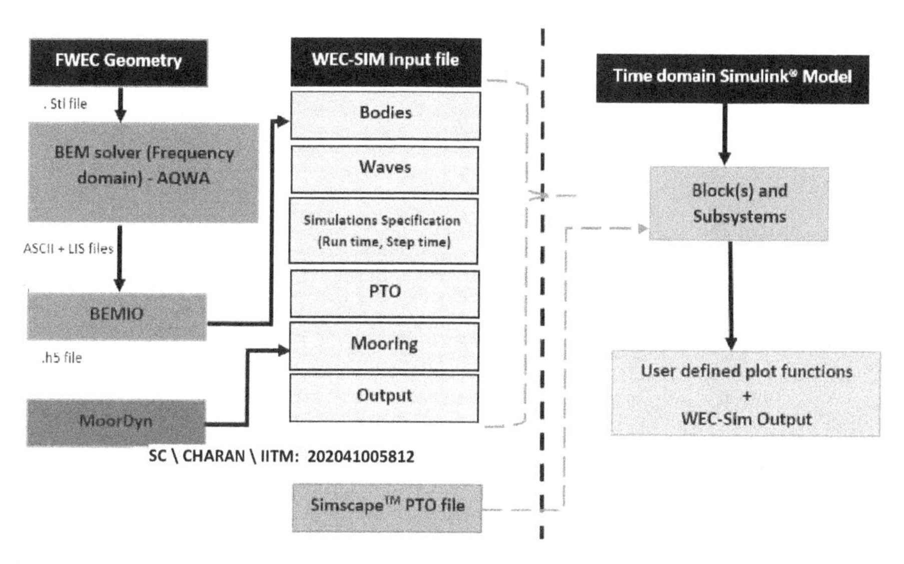

FIGURE 3.10 Workflow of WEC-Sim

3.6 COMPUTATIONAL TOOLS

The object-oriented numerical tool developed by Sandia National Laboratories (SNL) and the National Renewable Energy Laboratory (NREL), called WEC-Sim (Wave Energy Converter Simulator), is useful for modeling FWEC in the linear time-domain (FWEC-Sim, 2020). This tool makes the modeling convenient to customize with the open-source code, and reduces the simulation processing time and cost with an acceptance of accuracy. The workflow of WEC-Sim is given in Fig. 3.10.

The input parameters such as time step, simulation time, wave conditions, geometric and material properties of the mechanical components of the FWEC are defined in the MATLAB® input file. A set of ordinary differential equations representing the dynamics of FWEC are solved at each time step to obtain the next set of values at the next time step. Before starting the simulation, excitation impulse response functions, Froude-Krylov force, and the radiation term (to account for the memory effect) are evaluated at every time step. The extent of the memory effect from a given time instant can be fixed to reduce computational time. One can ensure that both IRFs estimated before the simulation should decay after some time instant, but this depends on the interference effects. WEC-Sim is also capable of handling a multi-body FWEC. The interference effects generated due to the multi-body oscillations are accounted for in the frequency-domain AQWA solver, which is subsequently handled using WEC-Sim in the time domain. Fig. 3.11 illustrates the dynamic model of FWEC in the Simulink platform.

3.7 MULTIBODY FLOATING WAVE ENERGY CONVERTER

To complement the efforts and a motivation to contribute to Indian renewable energy, since 2017, a novel concept known as the 'Bean Floating Wave Energy Converter'

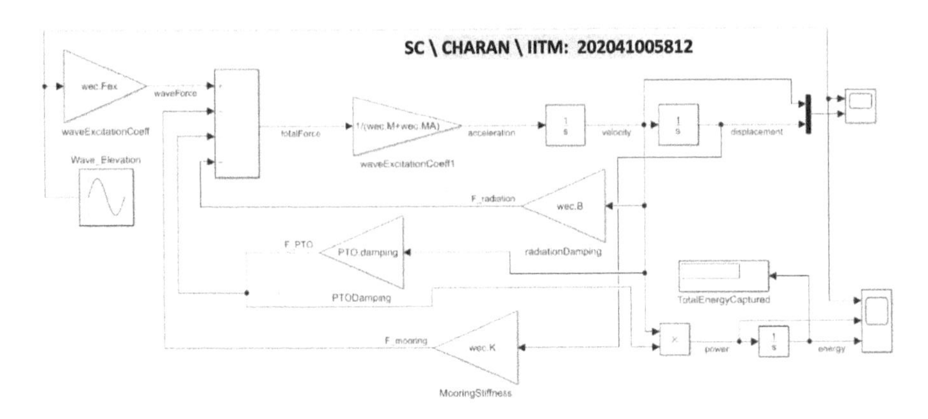

FIGURE 3.11 Dynamic model of FWEC in the Simulink platform

(BFWEC) is being rigorously designed and developed indigenously for harnessing wave energy from Indian waters at the Department of Ocean Engineering, IIT Madras. The authors acknowledge the administrative and technical support extended by the Indian Institute of Technology Madras in conducting the studies and acknowledge the support extended by the Prime Minister's Research Fellowship (PMRF), Ministry of Education, Govt. of India. A brief discussion is presented to strengthen the Indian renewable energy sector's capacity development for knowledge dissemination.

The proposed device consists of a central reference unit (CB), a set of bean-shaped floats and supporting frames, mooring, and a hydraulic power take-off unit, as shown in Fig. 3.12.

All the floats and the central buoy are designed as positively buoyant to stay afloat without capsizing. The central buoy is anchored to the seabed using a set of moorings lines. The sizing of the device members is smaller compared to the dominant wave lengths of typical Indian waters. A typical FWEC shall comprise about three to ten bean floats placed around the central buoy; it depends upon the application and the characteristics of the deployment site. Each float is connected to the central buoy using a supporting frame; the connecting end of the float is rigid while the end at the central buoy is hinged. The central buoy integrates the float-frame units and also houses a PTO unit. To achieve this function, the motion of the central buoy should be the minimum in all possible active degrees of freedom. The bean floats are responsible for harnessing the wave energy; when exposed to ocean waves, their pitching motion harnesses the energy collectively.

The central buoy is a hollow, cylindrical member integrating all the components of the device. It is a deep-draft cylinder with about three-fourths of submerged depth, inspired by a single point anchor reservoir (SPAR) buoy. The floating central buoy is position-restrained using a set of taut-moored tethers, mimicking a TLP. The taut-moored buoy remains stiff in the vertical plane but compliant in the horizontal plane. Heave, pitch, and roll motion are restrained, but free to oscillate in the surge and sway. Yaw motion is restrained as the buoy is connected to the floats circumferentially. The natural period in the stiff degrees is about 0.5 s, and that in the compliant motion is

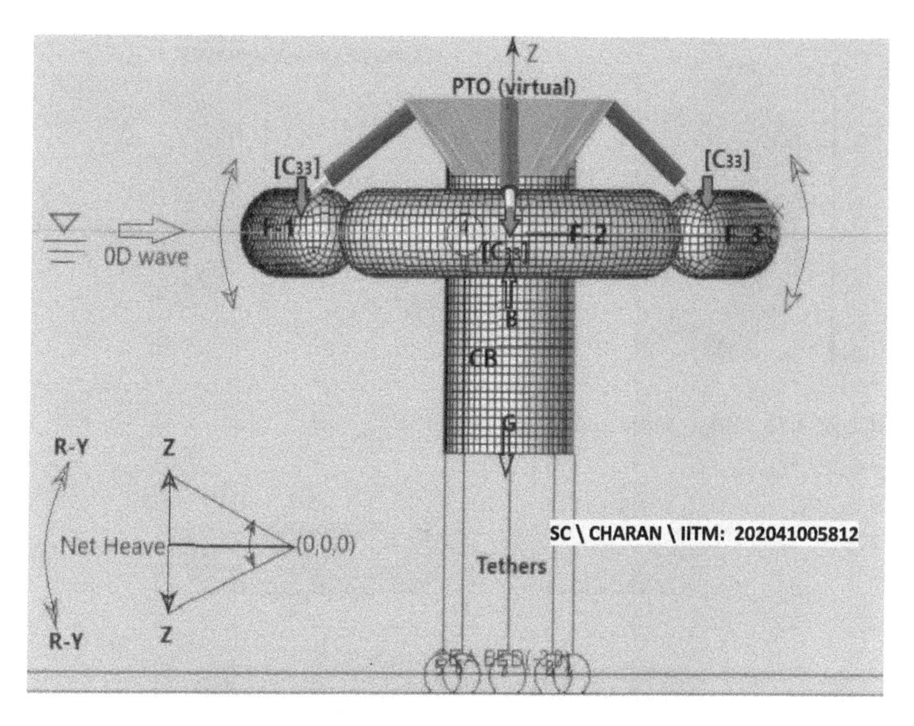

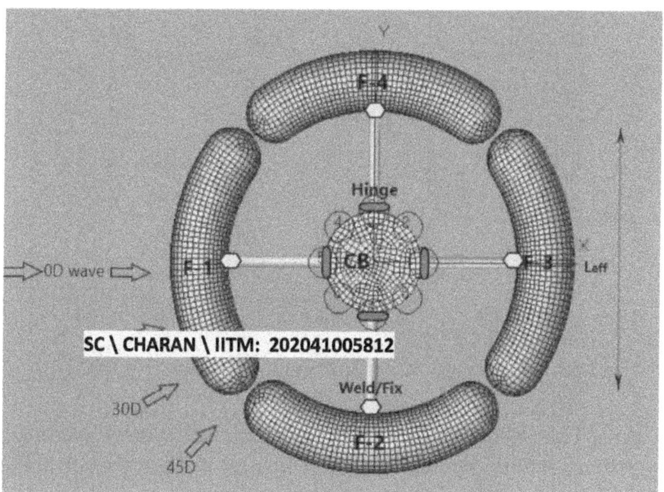

FIGURE 3.12 Multi-body, FWEC

about 1.2 s. The central buoy is self-floating, positively buoyant, and will float even under the tether plug-out conditions. The center of gravity of the FWEC is maintained well below the center of buoyancy to ensure dynamic stability by adding sufficient ballast at the bottom of the central buoy. Floats are bean-shaped, smoothly finished, and rounded hollow floating members are connected to the central buoy using the

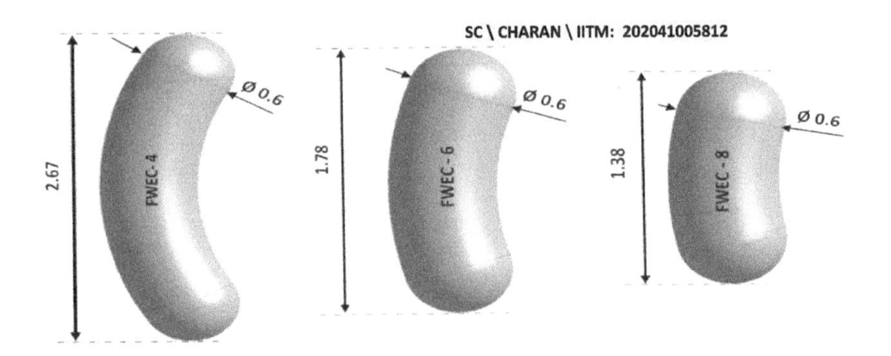

FIGURE 3.13 Bean-shaped floats of the device

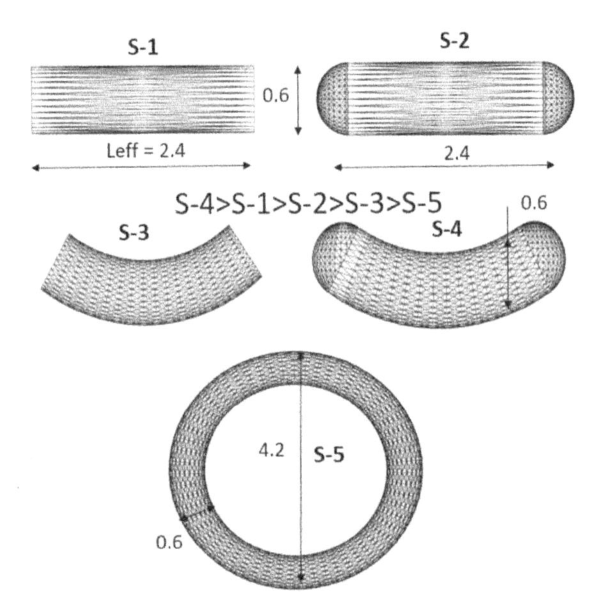

FIGURE 3.14 Various float shapes (Chandrasekaran and Sricharan, 2020c)

connecting arms. This arrangement adds stability to the device while making it directionally insensitive, enabling power harvest under all wave approach angles. Relative motion between the central buoy and the floats is harnessed. As the wave approaches the device, floats rotate about the hinge, harnessing their relative motion. Fig. 3.13 shows a typical bean-shaped float with different configuration numbers, namely four, six and eight.

As shown in Fig. 3.14, floats of five different configurations are investigated in the frequency domain to compare the capture width ratio (CWR) and absorbed power (Chandrasekaran and Sricharan, 2020c). Three design layouts of the device with four, six, and eight bean floats are further to assess their performance in different sea states (Chandrasekaran and Sricharan, 2020c).

The relative motion between the floats and the central buoy is enabled by the appropriate positioning of the hinges on the central buoy. The hinge is free to rotate about the y-axis while the other degrees of freedom are restrained. The schematic diagram of the hinge shown in Fig. 3.15 illustrates the working principle. The mechanical properties of the hinge are defined by rotational stiffness and rotational damping. The rotational stiffness arises due to the static and dynamic friction, which is neglected by assuming optimum lubrication and smooth motion. The rotational damping governs the dynamic response and the hydrodynamic efficiency of the device. Rotational damping is analogous to viscous damping, which can be considered a virtual PTO in the analysis.

The floats rotate about hinges due to wave-device interactions, absorbing energy from the waves. The device's oscillations result in diffraction and radiation interferences. The power absorbed by the device is termed hydrodynamic absorbed power. Considering the device as analogous to a spring-mass-dashpot system, one can consider two cases: the mass is connected only with the spring, and the other is only with the damper. In the former, the displaced mass will exhibit a continuous energy exchange between the mass and the spring in kinetic and potential energy. Neglecting the resistance offered by the medium will result in no absorption of energy in the system. In the latter, the damper absorbs energy, which is required in the case of an FWEC. As an analogy, mass is derived from the components of the FWEC, while hydrostatic stiffness arises from waves; the difference in the weight and buoyancy of the system is correlated to the stiffness. The damper is analogous to the viscosity, and the external damping exerted on the floats; this is termed PTO damping. In simple terms, PTO provides external damping to the floats and enables capture of the mechanical motion of the floats; it is subsequently converted into the desired form of energy.

In the first part of the numerical analysis, viscous dampers (or hinges) are used as PTO, providing the necessary external damping to the floats. The damping coefficient of the viscous damper is an important parameter as it directly influences the device's overall performance. Therefore, it is necessary to estimate the optimal damping coefficient to maximize the energy absorption. Subsequently, a hydraulic PTO will be introduced in the analysis where the hydraulic motor indirectly provides the required damping. The device is designed to harness energy collectively by the relative heave motion between the floats and the central buoy. Fig. 3.16 shows the arrangement of six-float FWEC. An additional damping force is assumed in the heave motion on each float in the absence of a physical PTO. Common terminology is as follows: float facing the incident wave heading is designated as F-1 while others are numbered

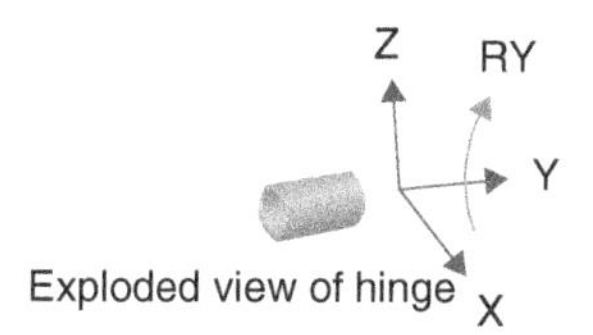

FIGURE 3.15 Hinge connecting the floats to central buoy

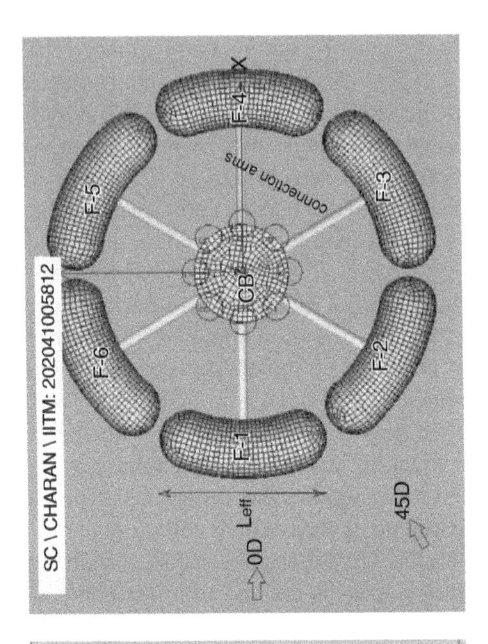

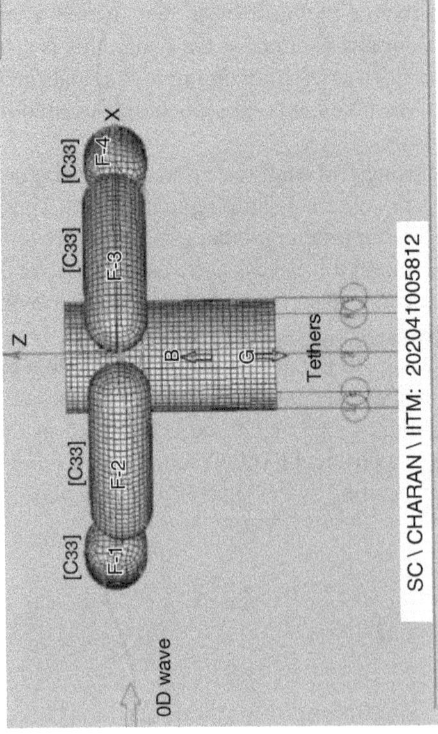

FIGURE 3.16 Six-float FWEC configuration

TABLE 3.1
Geometrical Details of the Device

Description	Values	Units
Water	3.0	m
Weight of the central buoy	10668.37	N
Weight of each float	1039.9	N
Buoyancy force of the central buoy	11845.57	N
Buoyancy force each float	1137.22	N
Stiffness of mooring lines	13541.35	N/m
Freeboard of the central buoy	0.5	m
Freeboard of the float	0.5	m
Draft of the central buoy	1.5	m
Draft of each float	0.1	m
Length of the connecting arm	1.0	m
Central buoy (height, diameter)	2, 1	m
Float effective length (diameter)	2.1, 0.6	m

in the anti-clockwise direction. As the device is symmetric about the z-axis, one quadrant of varying the wave approach angle will suffice to examine the directional behavior of the device. The design parameters of the device are given in Table 3.1.

Frequency-independent damping constants of magnitude 500, 1000, 1500, 2000, 2500, and 3000 kg/s are used to assess the performance of the FWEC. By ensuring the same total mass in all the layouts, the natural heave period is maintained. A massless, virtual PTO is considered to induce damping, and the equation of motion for heave degree of freedom is written as follows:

$$\begin{aligned}(M+A_\infty)\ddot{z}(t)+K_H\,z(t)+\int_{-\infty}^{t}K_r(t-\tau)\dot{z}(\tau)d\tau \\ +C_{LPTO}\left(\Delta\dot{z}(t)\right)-F_m=\int_{-\infty}^{\infty}\eta(\tau)K_E(t-\tau)d\tau\end{aligned} \tag{3.36}$$

where $\ddot{z}(t)$, and $\dot{z}(t)$ are time-dependent acceleration and velocity of bodies in the device, η is the free-surface elevation induced by the impulse response function (IRF), C_{LPTO} is a linear damping coefficient.

The mass matrix includes the dry and the added mass of a typical float coupled with the adjacent floats and the central buoy in the matrix form. In a more generalized, form it is given by:

$$\boldsymbol{M}_f = \begin{bmatrix} m_{f_j} + a_{11}^{cf_i} + \sum_{i=1}^{n} a_{11}^{f_j f_i} & 0 & a_{13}^{cf_j} + \sum_{i=1}^{n} a_{13}^{f_j f_i} & 0 & a_{15}^{cf_j} + \sum_{i=1}^{n} a_{15}^{f_j f_i} & 0 \\ 0 & m_{f_j} + a_{22}^{cf_i} + \sum_{i=1}^{n} a_{22}^{f_j f_i} & a_{23}^{cf_j} + \sum_{i=1}^{n} a_{23}^{f_j f_i} & a_{24}^{cf_j} + \sum_{i=1}^{n} a_{24}^{f_j f_i} & 0 & 0 \\ a_{31}^{cf_j} + \sum_{i=1}^{n} a_{31}^{f_j f_i} & a_{32}^{cf_j} + \sum_{i=1}^{n} a_{32}^{f_j f_i} & m_{f_j} + a_{33}^{cf_i} + \sum_{i=1}^{n} a_{33}^{f_j f_i} & a_{34}^{cf_j} + \sum_{i=1}^{n} a_{34}^{f_j f_i} & a_{35}^{cf_j} + \sum_{i=1}^{n} a_{35}^{f_j f_i} & 0 \\ 0 & a_{42}^{cf_j} + \sum_{i=1}^{n} a_{42}^{f_j f_i} & a_{43}^{cf_j} + \sum_{i=1}^{n} a_{43}^{f_j f_i} & I_{xx-f_j} + a_{44}^{cf_j} + \sum_{i=1}^{n} a_{44}^{f_j f_i} & 0 & 0 \\ a_{51}^{cf_j} + \sum_{i=1}^{n} a_{51}^{f_j f_i} & 0 & a_{53}^{cf_j} + \sum_{i=1}^{n} a_{53}^{f_j f_i} & 0 & I_{YY-f_j} + a_{55}^{cf_j} + \sum_{i=1}^{n} a_{55}^{f_j f_i} & 0 \\ 0 & 0 & 0 & 0 & 0 & I_{zz-f_j} \end{bmatrix} \quad (3.37a)$$

Mass matrix ($\boldsymbol{M}_c$) including dry mass and the added mass of the central buoy is coupled with the floats and can be written as follows:

$$\boldsymbol{M}_c = \begin{bmatrix} m_c + a_{11}^{cc} + \sum_{i=1}^{n} a_{11}^{f_i c} & 0 & a_{13}^{cc} + \sum_{i=1}^{n} a_{13}^{f_i c} & 0 & a_{15}^{cc} + \sum_{i=1}^{n} a_{15}^{f_i c} & 0 \\ 0 & m_c + a_{22}^{cc} + \sum_{i=1}^{n} a_{22}^{f_i c} & a_{23}^{cc} + \sum_{i=1}^{n} a_{23}^{f_i c} & a_{24}^{cc} + \sum_{i=1}^{n} a_{24}^{f_i c} & 0 & 0 \\ a_{31}^{cc} + \sum_{i=1}^{n} a_{31}^{f_i c} & a_{32}^{cc} + \sum_{i=1}^{n} a_{32}^{f_i c} & m_c + a_{33}^{cc} + \sum_{i=1}^{n} a_{33}^{f_i c} & a_{34}^{cc} + \sum_{i=1}^{n} a_{34}^{f_i c} & a_{35}^{cc} + \sum_{i=1}^{n} a_{35}^{f_i c} & 0 \\ 0 & a_{42}^{cc} + \sum_{i=1}^{n} a_{42}^{f_i c} & a_{43}^{cc} + \sum_{i=1}^{n} a_{43}^{f_i c} & I_{xx-c} + a_{44}^{cc} + \sum_{i=1}^{n} a_{44}^{f_i c} & 0 & 0 \\ a_{51}^{cc} + \sum_{i=1}^{n} a_{51}^{f_i c} & 0 & a_{53}^{cc} + \sum_{i=1}^{n} a_{53}^{f_i c} & 0 & I_{YY-c} + a_{55}^{cc} + \sum_{i=1}^{n} a_{55}^{f_i c} & 0 \\ 0 & 0 & 0 & 0 & 0 & I_{zz-f_j} \end{bmatrix} \quad (3.37b)$$

Similarly, the radiation damping matrix of a typical float, $\boldsymbol{C}_{rf}$ and CB, $\boldsymbol{C}_{rB}$ are written as:

$$\left[\boldsymbol{C}_{rf}\right] = \begin{bmatrix} b_{11}^{cf_j} + \sum_{i=1}^{n} b_{11}^{f_j f_i} & 0 & b_{13}^{cf_j} + \sum_{i=1}^{n} b_{13}^{f_j f_i} & 0 & b_{15}^{cf_j} + \sum_{i=1}^{n} b_{15}^{f_j f_i} & 0 \\ 0 & b_{22}^{cf_j} + \sum_{i=1}^{n} b_{22}^{f_j f_i} & b_{23}^{cf_j} + \sum_{i=1}^{n} b_{23}^{f_j f_i} & b_{24}^{cf_j} + \sum_{i=1}^{n} b_{24}^{f_j f_i} & 0 & 0 \\ b_{31}^{cf_j} + \sum_{i=1}^{n} b_{31}^{f_j f_i} & b_{32}^{cf_j} + \sum_{i=1}^{n} b_{32}^{f_j f_i} & b_{33}^{cf_j} + \sum_{i=1}^{n} b_{33}^{f_j f_i} & b_{34}^{cf_j} + \sum_{i=1}^{n} b_{34}^{f_j f_i} & b_{35}^{cf_j} + \sum_{i=1}^{n} b_{35}^{f_j f_i} & 0 \\ 0 & b_{42}^{cf_j} + \sum_{i=1}^{n} b_{42}^{f_j f_i} & b_{43}^{cf_j} + \sum_{i=1}^{n} b_{43}^{f_j f_i} & b_{44}^{cf_j} + \sum_{i=1}^{n} b_{44}^{f_j f_i} & 0 & 0 \\ b_{51}^{cf_j} + \sum_{i=1}^{n} b_{51}^{f_j f_i} & 0 & b_{53}^{cf_j} + \sum_{i=1}^{n} b_{53}^{f_j f_i} & 0 & b_{55}^{cf_j} + \sum_{i=1}^{n} b_{55}^{f_j f_i} & 0 \\ 0 & 0 & 0 & 0 & 0 & 0 \end{bmatrix}$$

$$\left[\boldsymbol{C}_{rf}\right] = \begin{bmatrix} b_{11}^{cf_j} + \sum_{i=1}^{n} b_{11}^{f_j f_i} & 0 & b_{13}^{cf_j} + \sum_{i=1}^{n} b_{13}^{f_j f_i} & 0 & b_{15}^{cf_j} + \sum_{i=1}^{n} b_{15}^{f_j f_i} & 0 \\ 0 & b_{22}^{cf_j} + \sum_{i=1}^{n} b_{22}^{f_j f_i} & b_{23}^{cf_j} + \sum_{i=1}^{n} b_{23}^{f_j f_i} & b_{24}^{cf_j} + \sum_{i=1}^{n} b_{24}^{f_j f_i} & 0 & 0 \\ b_{31}^{cf_j} + \sum_{i=1}^{n} b_{31}^{f_j f_i} & b_{32}^{cf_j} + \sum_{i=1}^{n} b_{32}^{f_j f_i} & b_{33}^{cf_j} + \sum_{i=1}^{n} b_{33}^{f_j f_i} & b_{34}^{cf_j} + \sum_{i=1}^{n} b_{34}^{f_j f_i} & b_{35}^{cf_j} + \sum_{i=1}^{n} b_{35}^{f_j f_i} & 0 \\ 0 & b_{42}^{cf_j} + \sum_{i=1}^{n} b_{42}^{f_j f_i} & b_{43}^{cf_j} + \sum_{i=1}^{n} b_{43}^{f_j f_i} & b_{44}^{cf_j} + \sum_{i=1}^{n} b_{44}^{f_j f_i} & 0 & 0 \\ b_{51}^{cf_j} + \sum_{i=1}^{n} b_{51}^{f_j f_i} & 0 & b_{53}^{cf_j} + \sum_{i=1}^{n} b_{53}^{f_j f_i} & 0 & b_{55}^{cf_j} + \sum_{i=1}^{n} b_{55}^{f_j f_i} & 0 \\ 0 & 0 & 0 & 0 & 0 & 0 \end{bmatrix} \tag{3.38}$$

$$[c_{rc}] = \begin{bmatrix} b_{11}^{cc}+\sum_{i=1}^{n} b_{11}^{fc_i} & 0 & b_{13}^{cc}+\sum_{i=1}^{n} b_{13}^{fc_i} & 0 & b_{15}^{cc}+\sum_{i=1}^{n} b_{15}^{fc_i} & 0 \\ 0 & b_{22}^{cc}+\sum_{i=1}^{n} b_{22}^{fc_i} & b_{23}^{cc}+\sum_{i=1}^{n} b_{23}^{fc_i} & b_{24}^{cc}+\sum_{i=1}^{n} b_{24}^{fc_i} & 0 & 0 \\ b_{31}^{cc}+\sum_{i=1}^{n} b_{31}^{fc_i} & b_{32}^{cc}+\sum_{i=1}^{n} b_{32}^{fc_i} & b_{33}^{cc}+\sum_{i=1}^{n} b_{33}^{fc_i} & b_{34}^{cc}+\sum_{i=1}^{n} b_{34}^{fc_i} & b_{35}^{cc}+\sum_{i=1}^{n} b_{35}^{fc_i} & 0 \\ 0 & b_{42}^{cc}+\sum_{i=1}^{n} b_{42}^{fc_i} & b_{43}^{cc}+\sum_{i=1}^{n} b_{43}^{fc_i} & b_{44}^{cc}+\sum_{i=1}^{n} b_{44}^{fc_i} & 0 & 0 \\ b_{51}^{cc}+\sum_{i=1}^{n} b_{51}^{fc_i} & 0 & b_{53}^{cc}+\sum_{i=1}^{n} b_{53}^{fc_i} & 0 & b_{55}^{cc}+\sum_{i=1}^{n} b_{55}^{fc_i} & 0 \\ 0 & 0 & 0 & 0 & 0 & 0 \end{bmatrix} \quad (3.39)$$

where m_{f_j} and m_c are the dry mass of the float and the central buoy, respectively. I_{f_j}, and I_c are the solid mass moment of inertia of the float and the central buoy, respectively. In the above equations, (j) is a particular float number for which the matrix is defined in a typical FWEC configuration, (n) is the total number of floats per configuration, and (i) is the float number. As all the floats are identical for a given configuration, their mass is the same (m_f). It is important to note that (p, q) represent the respective modes of motion.

The above sets of matrix equations are coupled in all the degrees of freedom except in yaw. The mooring lines restrain the motion of the central buoy; further, the device is symmetric in the x-y plane. In linear models, if surge or pitch is activated in the absence of transverse instabilities, only the corresponding degrees are influenced; the rest degrees-of-freedom, such as sway and roll, are considered inactive. The same holds if the response is induced in sway or roll. Heave is coupled with all the degrees of freedom. There will be a set-down effect when the device offsets, causing heave motion, but it is uncoupled with yaw motion.

The number of floats participating in the wave extraction, the spatial distance between them for each layout, and the wave directionality influence the device's overall performance. The existing numerical tools based on the boundary element method use the linear potential theory and address the interactions of such multiple-body floats layouts. The oscillation of the device in a calm wave causes additional inertial effects and hydrodynamic damping. Since the viscous effects are neglected, only radiation contributes to the hydrodynamic damping. The total radiation force in heave, in turn, the kernel K_r arises from the following:

- radiation force from the motion of the central buoy and its influence on the floats and vice versa, represented by b^{cc}_{pq}, $b^{f_jc}_{pq}$ and $b^{cf_j}_{pq}$, respectively
- influence of radiation force of one float on and on the central buoy, expressed as $\sum_{i=1}^{n} b^{f_jf_i}_{pq}$, which is the impulse response function.

The terms in the radiation damping matrix are represented in the form of kernel (K_r) in the Cummins equation. Further, the stiffness matrix consists of the hydrostatic stiffness components offered by the mooring lines connected to the central buoy. The hydrostatic stiffness matrix is also coupled to the mass and damping matrices. These coupled terms in heave, roll, and pitch are as below:

$$K_{H33} = \rho g A_w \tag{3.40a}$$

$$K_{H44} = \rho g V_w \overline{GM}_{xx} \tag{3.40b}$$

$$K_{H55} = \rho g V_w \overline{GM}_{yy} \tag{3.40c}$$

where $\overline{GM}$ is the meta-centric height. There is no hydrostatic stiffness in the surge, sway, and yaw for the floats and the central buoy; hence, $(K_{H11}, K_{H22}, K_{H66})$ will be zero. In addition, the cross-coupling terms to these degrees will also be negligible. The heave motion due to relative displacement of the floats and the central buoy is given as below:

a) For the float: $i = 1\,to\,n$,

$$\left(m_{f_j}\right)\ddot{z}_{f_j}(t) + a_{33}^{f_j c}\ddot{z}_c(t) + \sum_{i=1}^{n} a_{33}^{f_j f_i}\ddot{z}_{f_i}(t) + \int_0^t \sum_{i=1}^{N} K_{r33}^{f_j f_i}(\tau)\dot{z}_{f_i}(t-\tau)d\tau$$
$$+\int_0^t K_{r33}^{f_j c}(\tau)\dot{z}_c(t-\tau)d\tau + K_{H33}^{f_j} z_{f_j}(t) + C_{LPTO}\left(\dot{z}_{f_j}(t) - \dot{z}_c(t)\right)$$
$$= \int_{-\infty}^{+\infty} \eta(\tau) K_{E33}^{f_j}(t-\tau)d\tau \tag{3.41}$$

b) For the central buoy:

$$\left(m_c + a_{33}^{cc}\right)\ddot{z}_c(t) + \sum_{i=1}^{n} a_{33}^{cf_i}\ddot{z}_{f_i}(t) + \int_0^t K_{r33}^{cc}(\tau)\dot{z}_c(t-\tau)d\tau$$
$$+\int_0^t \sum_{i=1}^{N} K_{r33}^{cf_i}(\tau)\dot{z}_{f_i}(t-\tau)d\tau + K_{H33} z_c(t) + K_{M33}\dot{z}_c(t)$$
$$= \int_{-\infty}^{+\infty} \eta(\tau) K_{E33}^{c}(t-\tau)d\tau \tag{3.42}$$

where n = 4, 6 or 8, depending on the number of floats in the FWEC layout. The mooring stiffness matrix is based on a taut-mooring system, which can be seen from the established results in the literature (Chandrasekaran and Jain, 2001; 2002; Chandrasekaran et al., 2004; 2007a; 2007b; 2007c).

Mooring stiffness matrix (taut-mooring),

$$\boldsymbol{K}_M = \begin{bmatrix} \frac{4T}{L} & 0 & 0 & 0 & -\frac{4Th}{L} & 0 \\ 0 & \frac{4T}{L} & 0 & \frac{4Th}{L} & 0 & 0 \\ 0 & 0 & \frac{4AE}{L_o} & 0 & 0 & 0 \\ 0 & \frac{4Th}{L} & 0 & 4[Th + \frac{2Th^2}{L} + \frac{AEd_c^2}{8L_o}] & 0 & 0 \\ -\frac{4Th}{L} & 0 & 0 & 0 & 4[Th + \frac{2Th^2}{L} + \frac{AEd_c^2}{8L_o}] & 0 \\ 0 & 0 & 0 & 0 & 0 & \frac{Td_c^2}{L} \end{bmatrix} \tag{3.43}$$

where T, L, L_0, A and E are the pre-tension, length, unstretched length, cross-sectional area (8 mm diameter), and Young's modulus (Steel, 250 MPa) of a mooring line; h is the distance between the center of gravity of the central buoy and the fair-lead point of the mooring line (0.2m); and d_c is the diameter of the central buoy (1 m). The proposed FWEC has four mooring lines, and hence, the coefficients of Eq. (4.34) have a prefix of 4. The relation between the frequency and time-domain for excitation and radiation kernel is given by:

$$K_{E33}^{c\ or\ f_j}(t) = \frac{1}{2\pi}\int_{-\infty}^{+\infty} e^{i\omega t} F_{E33}^{c\ or\ f_j}(i\omega)\,d\omega \tag{3.44}$$

$$K_{r33}^{c\ or\ f_j}(t) = \frac{2}{\pi}\int_{0}^{+\infty} C_{r33}^{c\ or\ f_j}(\omega)\cos(\omega t)\,d\omega \tag{3.45}$$

where K_E is the excitation impulse response function (kernel), $F_E(i\omega)$ is the frequency-dependent complex excitation force, and $C_r(\omega)$ is the frequency-domain radiating force. The total wave force due to external pressure on the device comprises hydro-dynamic and hydrostatic forces. Multiple bodies will significantly affect the overall performance compared to the isolated single float; combination with the central buoy increases the net force acting on the individual float.

3.8 HYDRODYNAMIC COEFFICIENTS OF FLOATING WAVE ENERGY CONVERTERS

The hydrodynamic coefficients (added mass, radiation damping, excitation force) of the floats in the heave degree on a frequency band obtained from the AQWA modeling are shown in Figs. 3.17 to 3.19. Figs. 3.20 to 3.21 show the free-floating RAO

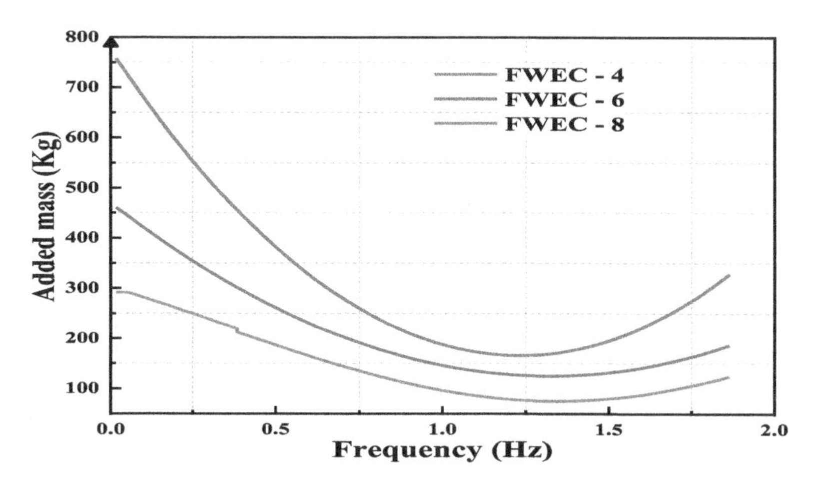

FIGURE 3.17 Added mass of floats in heave

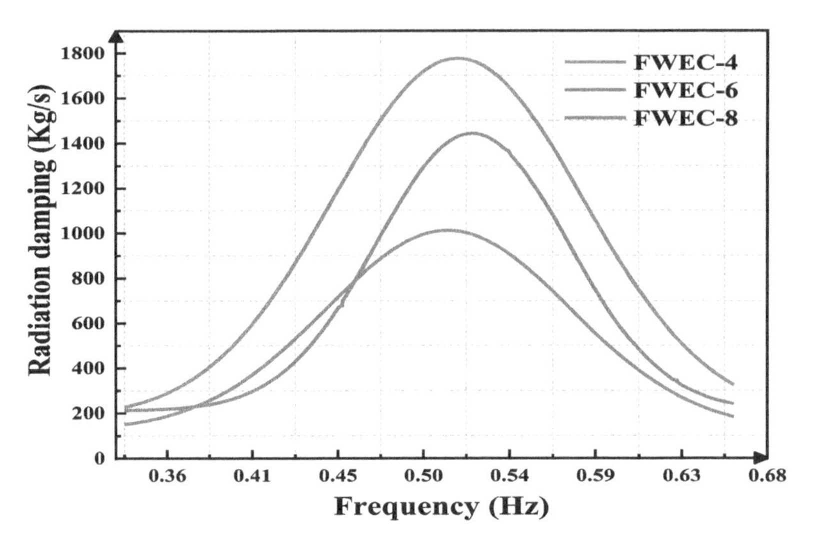

FIGURE 3.18 Radiation damping of floats in heave

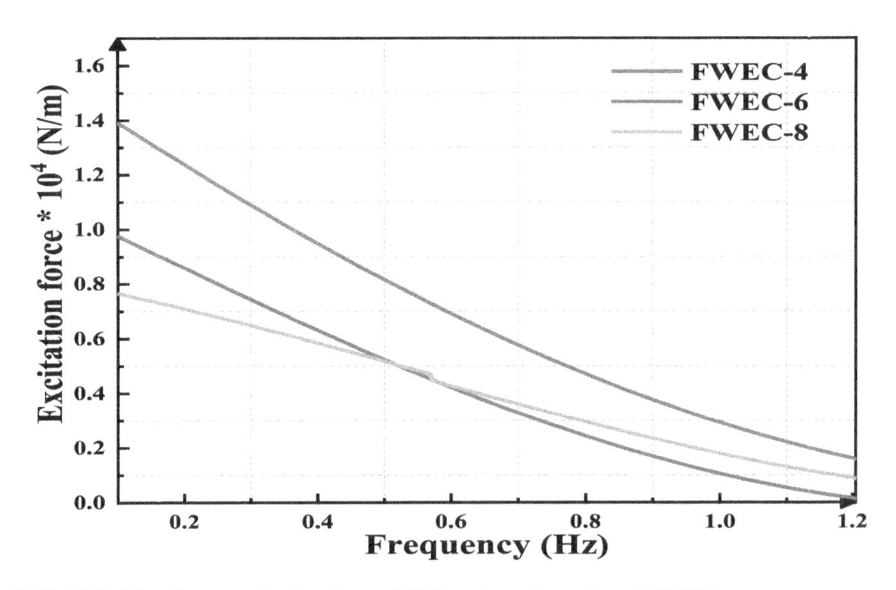

FIGURE 3.19 Heave excitation force of different configuration of FWEC

of the floats in the absence and presence of PTO, respectively. It can be seen from the figures that the influence of the added mass and the excitation force are dominant near the wave peak frequency (0.36 Hz). In comparison, the system-generated properties such as radiation damping and PTO damping are dominant at the natural heave period of the floats (= 0.56 Hz). However, as the damping is imposed on the floats by

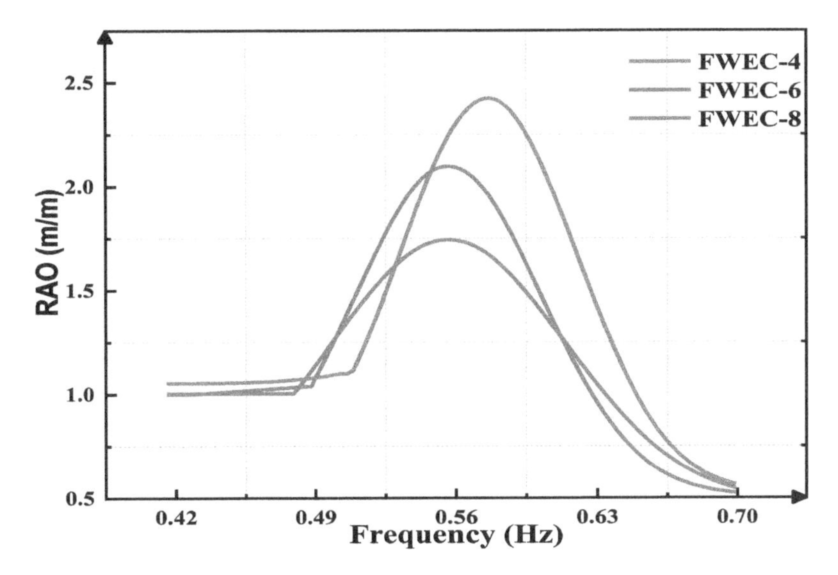

FIGURE 3.20 Free-floating heave RAOs of different configuration of FWEC (C_{33} =0)

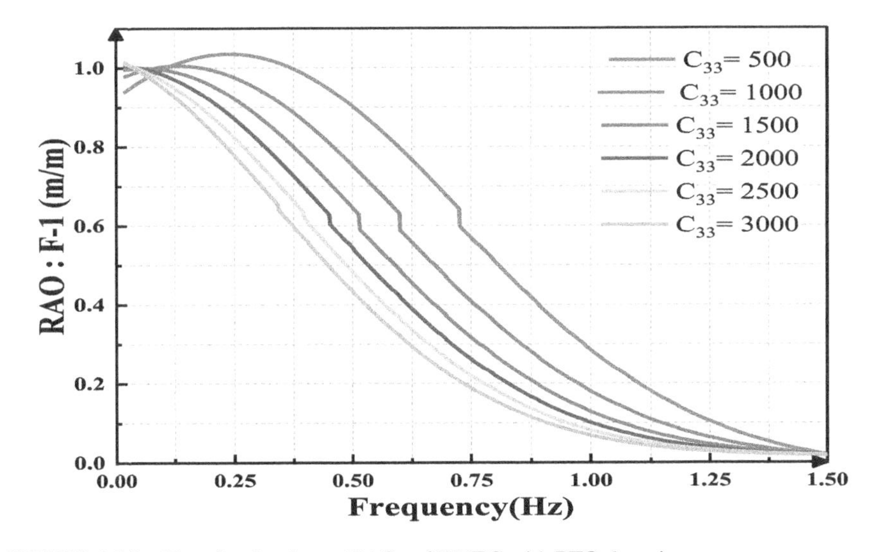

FIGURE 3.21 Free-floating heave RAOs of FWEC with PTO damping

the PTO, the magnitude of heave motion decreases with the increase in the damping coefficient. No dominant peaks are observed at the resonant frequency in RAO of the floats. The value of heave excitation at zero frequency signifies the maximum force limit on the floats in the heave motion. Typical values are (14, 9.9, 7.98 kN/m) for the four, six and eight float configurations, respectively.

3.9 WAVE POWER

3.9.1 Regular Waves

Regular waves are considered ideal, monochromatic waves to highlight the response behavior of the device. Kinetic and potential energies constitute the total ocean energy and are represented in terms of the total average energy per unit surface area:

$$\bar{E}_R = \frac{\rho_w g H^2}{8} \tag{3.46}$$

The rate at which energy is transferred, known as energy flux or wave power per unit length of the wave front, is given as below:

$$P_{Rwave} = \int_0^T \int_{-d}^{\eta} \frac{P_D u \, dt \, dz}{T} \tag{3.47}$$

$$P_{Rwave} = \bar{E}_R C_g \tag{3.48}$$

Total wave power can be expressed as below:

$$TP_R = P_{Rwave} L_{eff} \tag{3.49}$$

where L_{eff} is the effective length of the floats. It is equal to the length of the wave-front approaching the float at any instant of time. As floats are positioned around the central buoy, the effective length of each float will be different. In such cases, group celerity, which is the rate at which a train of propagating waves travel, is important and is given as below:

$$C_g = \frac{C}{2}\left(1 + \frac{2kd}{\sinh(2kd)}\right) \tag{3.50}$$

P_D is the dynamic pressure of a wave, u is horizontal water particle velocity, H is the wave height, k is the wave number, C is wave celerity, and d is the water depth. For shallow water depths ($d/L \leq 0.05$), $C_g = C$ and the wave celerity can be expressed as the function of water depth as follows:

$$C = \sqrt{gd} \tag{3.51}$$

For deep waters ($d/L \geq 0.5$), $C_g = 0.5C$. Thus, wave power available in shallow waters and deep waters can be computed using the following expressions:

$$P_{R\_shallow} = \frac{\rho_w g^2 H^2 T}{16\pi} \tag{3.52}$$

$$P_{R\_deep} = \frac{\rho_w g H^2 T}{32\pi} \tag{3.53}$$

Power absorbed by the proposed device is expected to reduce the energy content present in the ocean. The oscillatory motion of the device shall generate waves and radiate waves in the opposite direction to that of the approaching waves; it shall lead to destructive interference. Thus, the transferred energy can be captured by the damping offered by PTO. Average power absorbed under regular waves can be calculated as follows:

$$\overline{P_{R\_BFWEC}} = \overline{F_{PTO}(t)\dot{z}_2(t)} = \overline{C_{PTO}\dot{z}_2(t)\dot{z}_2(t)} = \frac{1}{2}C_{PTO}\omega^2 z_2^2 \tag{3.54}$$

The capture width (CW) or absorption length, L_b (Budar and Falnes, 1975), which is also referred to as the capture width of the device, is defined as the ratio of $\overline{P_{FWEC}}$ and P_{Rwave} is expressed as follows:

$$L_b = \frac{\overline{P_{R\_BFWEC}}}{P_{Rwave}} \tag{3.55}$$

The following relationship gives the capture width ratio (CWR) or hydrodynamic efficiency of the proposed device under regular waves:

$$CWR_{R\_BFWEC}\ or\ \eta_{R\_BFWEC} = \frac{L_b}{L_{eff}} \tag{3.56}$$

The maximum ideal capture width or absorption length is given by $\frac{L}{2\pi}$, where L is the wavelength. The average and maximum absorption length for the float F-1 is 0.19 and 0.50 m for the device with a four-float configuration. This value is comparable with that of the maximum possible width of 0.6 m. It is computed using the statistical mean of width values of F-1 under (0.5 m, 1.56 s) for all the damping constants.

3.9.2 Irregular Waves

Wave elevation for irregular waves is defined as follows:

$$\eta(t) = \sum_{i=1}^{n} a_i \cos(\omega_i t + \phi_i) \tag{3.57}$$

Wave characteristics of irregular waves are represented in terms of spectral parameters. The chosen wave spectrum can be expressed in terms of the Nth spectral moment as follows:

$$m_N = \int_0^\infty \omega^N S(\omega) d\omega \quad (3.58)$$

A unidirectional wave spectrum represents the spread of energy content over a frequency band varying from zero to positive infinity. The wave energy per unit surface area ($\bar{E}_{IR}$) is defined as the zeroth spectral moment of a particular wave spectrum or as the total variance of the wave elevation and is given by the following relationship:

$$m_0 = \sigma^2 = \bar{E}_R = \int_0^\infty S(\omega)\, d\omega \quad (3.59)$$

The wave amplitude can be calculated using the following relationship:

$$A_i = \sqrt{2S(\omega)\Delta\omega} \quad (3.60)$$

Generic spectral parameters, namely the significant wave height and mean wave period, are given as follows:

$$H_S = 4\sqrt{m_0} \quad (3.61)$$

$$T_M = 2\pi\frac{m_0}{m_1} \quad (3.62)$$

When the spectral shape is not known, then the period of the irregular waves can be represented using the energy period as below:

$$T_E = \frac{m_{-1}}{m_0} \quad (3.63)$$

In such cases, the zero-crossing period is given by the following relationship:

$$T_Z = \sqrt{\frac{m_0}{m_2}} \quad (3.64)$$

Maximum energy content, occurring at the peak period, is given by:

$$T_P = 1.42T_Z \quad (3.65)$$

The following relationship gives the total average energy per unit surface area for irregular waves:

$$\bar{E}_{IR} = \int_0^\infty \rho_w\, g\, S(\omega) d\omega = \rho_w\, g\frac{H_S^2}{16} \quad (3.66)$$

Wave power per unit wave front is given as:

$$P_{IRwave} = \bar{E}_{IR} C_g \tag{3.67}$$

$$P_{IR\_shallow} = \frac{\rho_w g^2 H_S^2 T_E}{32\pi} \tag{3.68}$$

$$P_{IR\_deep} = \frac{\rho_w g^2 H_S^2 T_E}{64\pi} \tag{3.69}$$

A two-parameter Pierson-Moskowitz spectrum, which is used for numerical studies on FWEC, has the following relationship of the spectral energy and total power:

$$S_{PM}(\omega) = (4\pi^3)\frac{H_S^2}{T_Z^4}\frac{1}{\omega^5} exp\left[-\frac{16\,\pi^3}{T_Z^4}\frac{1}{\omega^4}\right] \tag{3.70}$$

$$TP_{PM} = \int_0^{\infty} L_{eff}\,\rho_w\, g\, C_g\; S_{PM}(\omega)\,d\omega \tag{3.71}$$

Total power absorbed by the device, P_{IR_BFWEC} under irregular waves in the presence of virtual PTO is given by:

$$TP_{IR\_BFWEC} = \int_0^{\infty} C_{PTO}\,\omega^2 \left(\frac{z_n}{A_i}\right)^2 S_{PM}(\omega)\,d\omega \tag{3.72}$$

Alternatively, the device's efficiency can also be determined using the power spectral density (PSD) functions. This is the energy content for a particular wave set, computed by integrating the area under the PSD. Similarly, the total energy extracted by each float can be calculated by computing the area under the PSD in heave motion. Thus, total power captured by the device in heave motion under irregular waves is expressed as: $S_{FLOAT}(\omega)$

$$TP_{IR\_FLOAT} = \rho_w\, g\, \mathrm{V}\, L_{eff}\, S_{FLOAT}(\omega) \tag{3.73}$$

The following relationship gives the CWR or hydrodynamic efficiency of the device under irregular waves:

$$CWR_{IR\_BFWEC}\ or\ \eta_{IR\_BFWEC} = \frac{TP_{IR_{BFWEC}}\ or\ TP_{IR\_FLOAT}}{TP_{PM}} \tag{3.74}$$

Numerical studies are carried out on the FWEC model under regular and irregular waves. The wave parameters considered for the study are summarized in Tables 3.2 and 3.3.

TABLE 3.2
Regular Wave Parameters

Case	PTO Damping $[C_{33}]$ (kg/s)	Wave Height-H (m)	Wave Period-T (s)	Direction (°)
1	500	0.3	1.2	0
2	500	0.5	1.56	0
3	500	0.5	1.56	15
4	500	0.5	1.56	30
5	500	0.5	1.56	45
6	1000	0.3	1.2	0
7	1000	0.5	1.56	0
8	1000	0.5	1.56	15
9	1000	0.5	1.56	30
10	1000	0.5	1.56	45
11	1500	0.3	1.2	0
12	1500	0.5	1.56	0
13	1500	0.5	1.56	15
14	1500	0.5	1.56	30
15	1500	0.5	1.56	45
16	2000	0.3	1.2	0
17	2000	0.5	1.56	0
18	2000	0.5	1.56	15
19	2000	0.5	1.56	30
20	2000	0.5	1.56	45
21	2500	0.3	1.2	0
22	2500	0.5	1.56	0
23	2500	0.5	1.56	15
24	2500	0.5	1.56	30
25	2500	0.5	1.56	45
26	3000	0.3	1.2	0
27	3000	0.5	1.56	0
28	3000	0.5	1.56	15
29	3000	0.5	1.56	30
30	3000	0.5	1.56	45

The power and efficiency of the device are analyzed under two regular wave conditions: (0.3 m, 1.2 s) and (0.5 m, 1.56 s) with a set of wave directions (0, 15, 30, and 45°). The analysis is also carried out under one set of irregular wave conditions (0.3 m and 2 s). Wave power for shallow waters is found to be 212.91 W/m and 765.25 W/m for sets of regular wave sets, respectively. Table 3.4 summarizes the effective length of the floats for all the geometric layouts of the device. Table 3.5 shows the power extraction of the six-float configuration under irregular waves.

TABLE 3.3
Irregular Wave Parameters

Case Direction	PTO Damping $[C_{33}]$ (kg/s)	Significant Wave Height H_S (m)	(s)	Zero Crossing Period-T_Z (°)
31	500	0.3	2	0
32	1000	0.3	2	0
33	1500	0.3	2	0
34	2000	0.3	2	0
35	2500	0.3	2	0
36	3000	0.3	2	0

TABLE 3.4
Effective Length of the Floats

	Effective Length of the Floats L_{eff} (m)							
Layout	F-1	F-2	F-3	F-4	F-5	F-6	F-7	F-8
FWEC-4	1.735	1.735	1.735	1.735	-	-	-	-
FWEC-6	1.54	1.54	1.54	1.54	1.54	1.54	-	-
FWEC-8	1.29	1.29	1.29	1.29	1.29	1.29	1.29	1.29

TABLE 3.5
Parameters for Power Extraction Six-Float Configuration under Irregular Waves

	Area under Floats PSD $(m^2 * 10^{-4})$						Max Velocity of Float $\left(\frac{m}{s} * 10^{-2}\right)$					
Case	F-1	F-2	F-3	F-4	F-5	F-6	F-1	F-2	F-3	F-4	F-5	F-6
31	69	51	41	54	41	51	56	56	48	55	48	56
32	55	42	32	37	32	42	49	48	40	42	40	48
33	44	34	26	44	26	35	46	42	36	37	36	42
34	36	29	21	20	21	29	42	37	32	34	32	37
35	30	24	17	16	17	24	38	33	29	30	29	33
36	25	20	14	13	14	20	34	30	26	34	26	29

3.10 PERFORMANCE CURVES WITH VIRTUAL POWER TAKE-OFF SYSTEM

The aim is to evaluate the device's performance with three layouts under the influence of the external damping force generated from the PTO. Heave amplitude variation of the float F-1 of eight-float configuration is shown in Fig. 3.22. It is seen from the figure that under the influence of external damping, there is a marginal phase shift between the float motion and that of the wave elevation. The maximum heave is about 0.19 m, with a velocity of about 0.45 m/s. Irregular waves are represented by a unidirectional 2-D spectrum (PM) for the analysis. The energy content remains unchanged even under different wave approach angles varying from 0 to 45°. The total energy content of the input irregular wave over a band of chosen frequencies is 0.0056 m^2. The PM spectrum peak occurs at the wave's peak frequency (0.36 Hz), and the wave power per unit front is 305.19 W/m.

Fig. 3.23 shows the PSD plot of the F-1 in the eight-float configuration under the influence of PTO damping. The total hydrodynamic power extracted by the floats is

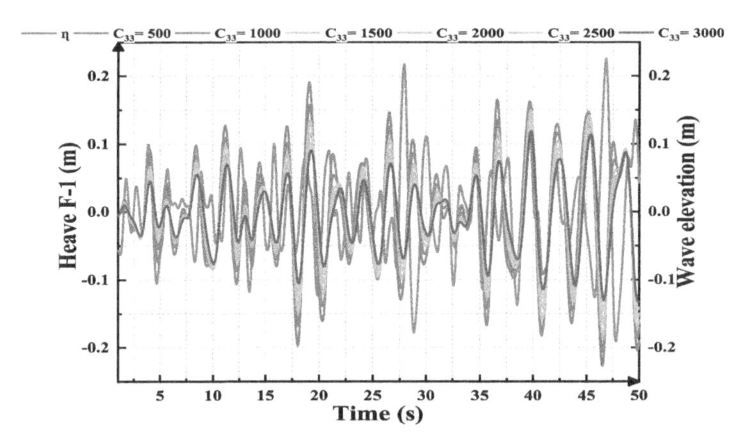

FIGURE 3.22 Heave amplitude of float F-1 for eight-float configuration under irregular waves

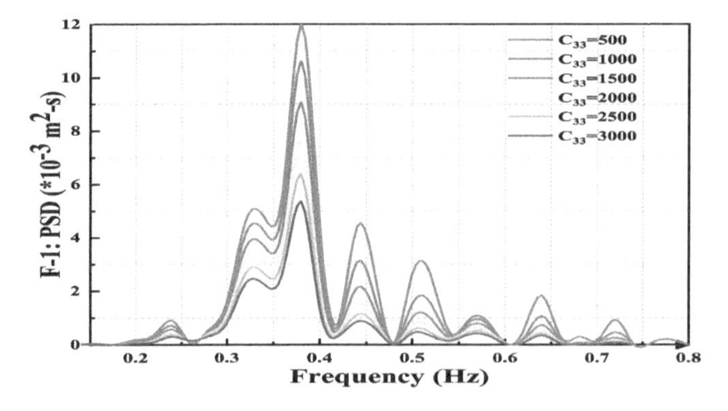

FIGURE 3.23 PSD of F-1 in eight-float configuration with PTO damping

calculated from the PTO damping. It is seen from the figure that the major peak is occurring at 0.36 Hz with a marginal shift of the peak wave frequency. The occurrence of multiple minor peaks, including a peak closer to the natural heave frequency, indicates the spread over a wide range of incoming frequencies. This is an indication of the ability of the proposed device to absorb energy at multiple incoming frequencies. Power captured by each float under irregular waves is calculated by converting the heave time series into frequency domain PSD curves using a fast Fourier transform (FFT) tool in MATLAB®. Using the computed area under each curve and the velocity of each float (please refer to Table 3.5), the total power extracted and the CWR (also referred to as hydraulic efficiency) under irregular waves can be computed using the above set of equations.

Fig. 3.24 shows the influence of PTO damping on the power extraction of float F-1 of the device under the eight-float configuration. It is seen from the figure that the wave power is maximum for damping of 500 kg/s; it reduces further with the increase in the damping constant. Fig. 3.25 shows the total power for all the floats under regular waves at a damping value of 500 kg/s over a window of the simulated time where the maximum and average power of F-1 is about 0.35 kW and 0.15 kW, respectively. Figs. 3.26 and 3.27 show the mean CWR of the device under regular and irregular waves, respectively. The plots are based on the statistical mean of efficiency of each float under the influence of the additional damping. It is seen from the figures that the efficiency of all the geometric layouts of the device increases with the increase in external damping, beyond which it decreases with the further increase in damping. The average and maximum efficiency of BFWEC-8 for regular waves are 41.32% and 53.47%, respectively, at a damping value of 1000 kg/s.

The average and maximum efficiency of the device under irregular waves are computed for all damping constants and found to be 19.49% and 48.97%, respectively,

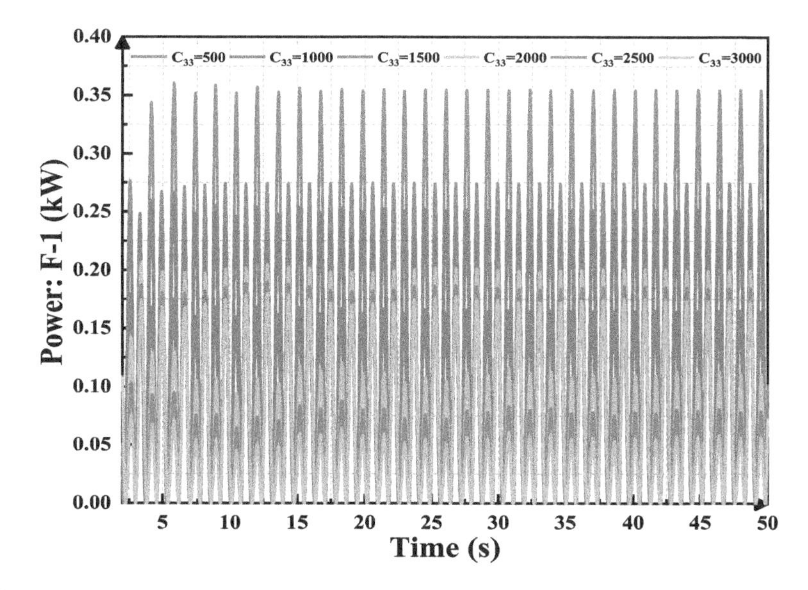

FIGURE 3.24 Total power captured by F-1 for eight-float configuration with PTO damping

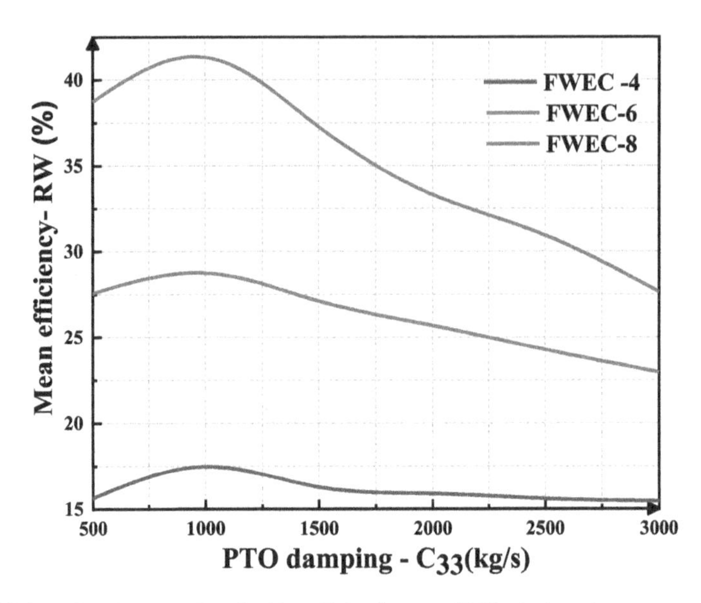

FIGURE 3.25 Total power absorbed by all the floats at 500 kg/s damping

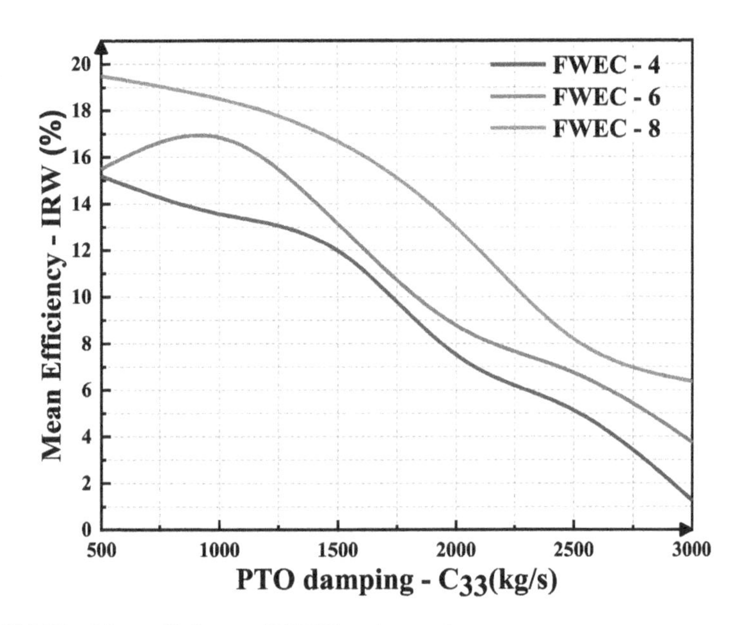

FIGURE 3.26 Mean efficiency of FWEC under regular waves

for the eight-float configuration at a damping constant of 500 kg/s. It is observed that the inclusion of external damping has a significant effect on the device's performance. Maximum efficiency occurs when the external PTO damping is equal to that of the hydrodynamic damping. Further, studies are also carried out to assess the directional insensitivity of the device. The average CWR (hydrodynamic efficiency) of all the floats in each geometric layout under different wave directions are plotted, as shown in Fig. 3.28.

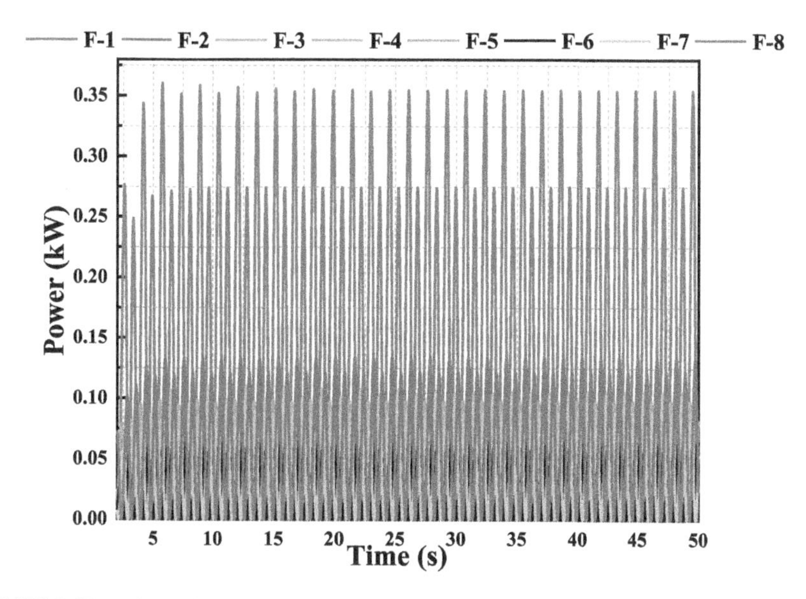

FIGURE 3.27 Mean efficiency of FWEC under irregular waves

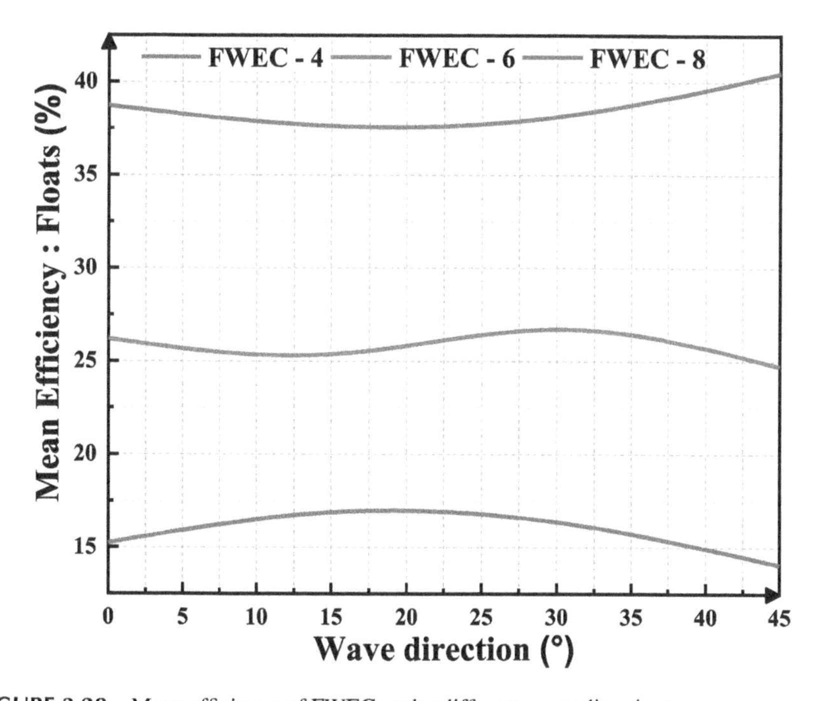

FIGURE 3.28 Mean efficiency of FWEC under different wave directions

It can be seen that the overall efficiency remains constant, confirming the fact that the proposed device is insensitive to the wave approach angle. For a constant PTO damping acting on the floats, floats placed at the wavefront are more efficient than all other floats, as they radiate and absorb the maximum energy.

It is necessary to deploy variable damping to improve the performance of the other floats. On the other hand, the damping constants acting on each float must be adjusted depending on its position in the layout and the wave directions. As observed from the study, the CWR of the device is increasing with the increase in the number of floats; however, choosing an optimum number of floats is crucial. It is important to note that the number of rotating components of the PTO increases with the increase in the number of floats; it reduces the CWR. A compromise is expected between them for better reliability.

3.11 ANALYSIS WITH HYDRAULIC POWER TAKE-OFF SYSTEM

FWEC with a four-float configuration is integrated with a hydraulic PTO unit and analyzed for its performance. One of the most dedicated tools to model the hydraulic PTO is WEC-Sim. With its integration to MATLAB®/Simulink and various domains of Simscape, it is a convenient tool to carry out such analyses. The total circuit is designed, including the hydraulic PTO's governing parameters, whose optimal values will be estimated numerically.

3.11.1 Governing Parameters

The Hydraulic PTO is well-known for its low, fluctuating input speed and higher torque output application. It is also interesting to note that these characteristics are typically seen in most FWECs, specific to the point absorbers. In addition to being adaptable to the revisable wave loads, hydraulic power take-off (HPTO) systems can have a smoothening effect on the output electric power by choosing the parameters appropriate to the chosen window of the wave frequencies and the wave heights; numerical estimations give a plausible output to move forward. The governing design parameters of the HPTO for variation in the output power and overall performance are identified for the circuit shown in Fig. 3.29. Each float is directly connected by one side of the double-acting hydraulic cylinder and a set of four directional check valves. The input velocity of each float at every time step pressurizes the oil in the cylinders. In other words, the net heave from the floats is directly linked to the stroke range of the hydraulic cylinders, with its velocity equal to the relative velocity. The floats oscillation under the encountered wave initiates a phase difference depending upon the incoming wave periods. All floats are connected to a single generator, a hydraulic motor, and a set of high and low-pressure gas-charged accumulators in each configuration.

In the first half cycle of wave input, when the float is lifted, the oil flow rushes from one of the ports of the hydraulic cylinders. Subsequently, oil enters the line of the high-pressure, gas-charged accumulator. Depending upon the severity and randomness of exerted forces from the waves in the real sea, oil is pressurized randomly in the circuit. The pressurized oil thus rotates the motor, whose motion must be smoothened; gas accumulators are commonly used in the circuit as they are cheap and reliable.

However, an improper selection of PTO parameters in regular waves can increase the fluctuations in the output. While flowing through the high-pressure hose, the oil

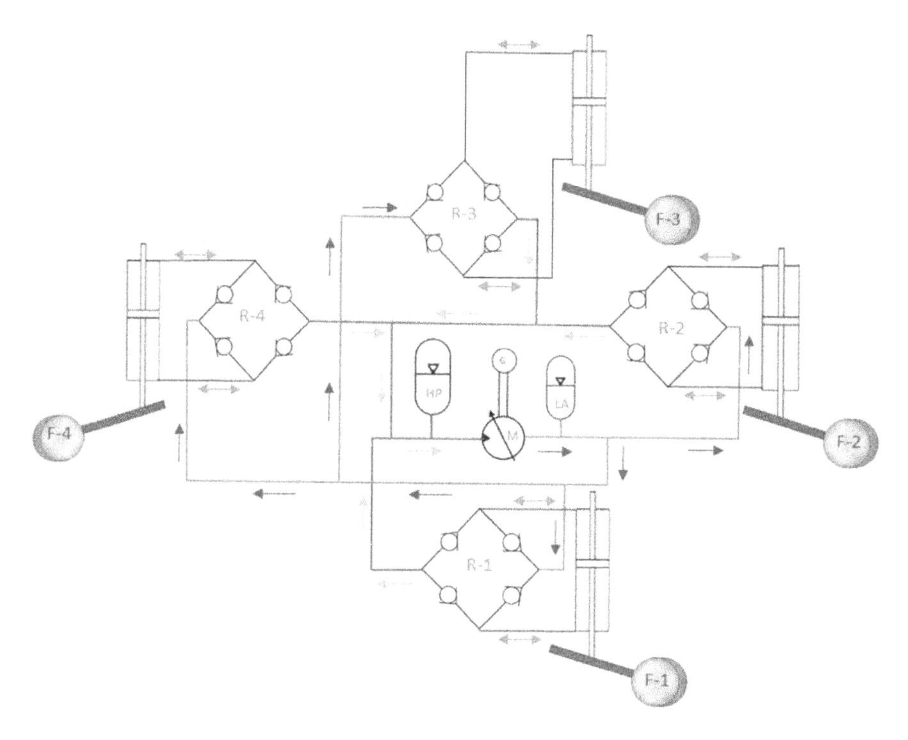

FIGURE 3.29 Hydraulic PTO circuit for four-float FWEC (Sricharan and Chandrasekaran, 2021)

is partially absorbed by the accumulator for a short period; the remaining flow enters the hydraulic motor. In the second half of the cycle, when the float moves down, oil is directed to the high-pressure line with the help of check valves though the oil is pumped through the lower end of the cylinder. Therefore, the oil first passes through the high-pressure line and enters the hydraulic motor irrespective of the piston movement. The pressure of the exit flow from the motor is reduced while the inlet flow is rotating the motor shaft. In turn, it rotates the electric motor. The low-pressure oil is again distributed to all the appropriate ports (opposite of where they entered the high-pressure hose line). A low-pressure gas accumulator reduces the flow fluctuations in the low-pressure line, maintaining a constant pressure difference across the hydraulic motor. The gas-charged accumulators are governed by pre-charge pressure and initial volume, whose values must be chosen appropriately for better results. Fig. 3.30 shows the FWEC with HPTO set up for a four-float configuration.

Choosing the parameters for HPTO is similar to the back-propagation process; it starts from the rated characteristics of the generator (last component of the PTO) to the hydraulic piston on the float (first component of the PTO). The influence of HPTO parameters on the absorbed power is now examined for a four-float configuration under regular waves. Generator damping, piston area, and motor displacement are the critical parameters whose values are adjusted (by trial and error) to change the power absorption and conversion characteristics of the FWEC. The problem is

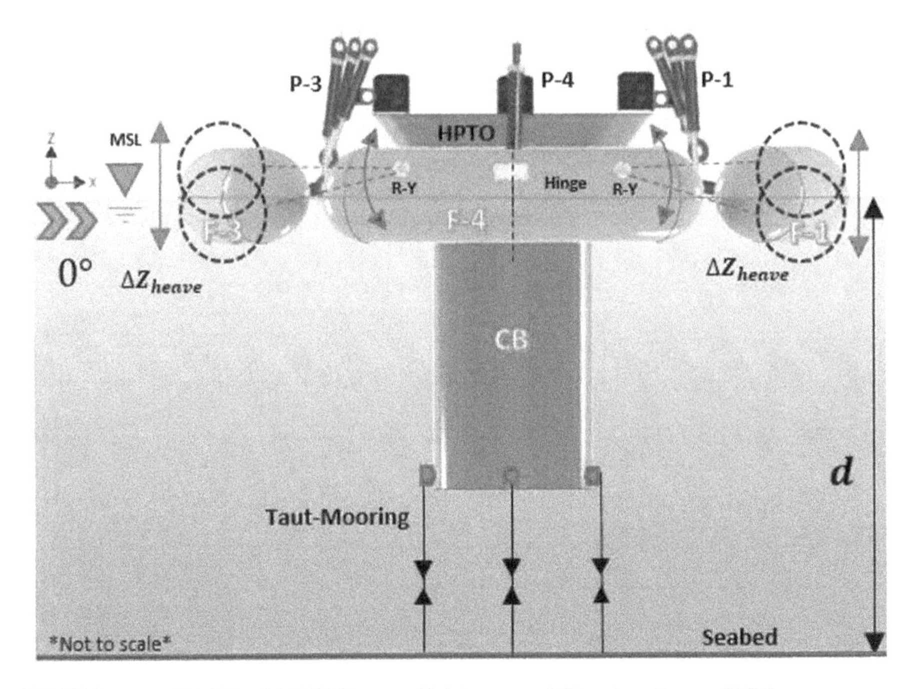

FIGURE 3.30 FWEC with HPTO setup (Sricharan and Chandrasekaran, 2021)

approached in two steps: the first step is to select the parameters in regular waves for obtaining a first-hand estimate of the power characteristics by maximizing the absorbed power; the second step is to fabricate the hydraulic PTO based on the nominal values obtained from the earlier step.

There are three conversion stages: hydrodynamic to hydraulic, hydraulic to mechanical, and mechanical to electrical. The dynamic response of the hydraulic PTO influences the performance of the device. The inclusion of the gas accumulators makes the FWEC response non-linear due to the adiabatic expansion process of the inert gas; it can be conveniently analyzed using the time-domain modeling. As discussed earlier, the radiation memory effects and the nonlinear effects of the HPTO are handled using the convolution method. The numerical simulations include the time-varying parameters of the gas accumulators, such as volume, pressure, and compressibility. The forcing function that acts on the PTO will now become a function of nonlinear displacement (or velocity) of the components of the PTO; vital components are, namely, check valves, low-pressure and high-pressure accumulators, piston, hydraulic motor, and an electric generator.

A few parameters that control the performance of a check valve are discharge coefficient (C_d), maximum valve area (a_{max}), minimum valve area (a_{min}), the maximum pressure difference across the valve (p_{max}) and the minimum pressure difference across the valve ($p_{min} \approx 0 or 0.75 p_{max}$), and the oil viscosity (ρ_o). A low-pressure accumulator's governing parameters are pre-charge pressure (P_{IL_pre}) and initial volume (V_{ILo}). These are calculated using the following ideal gas equations assuming the adiabatic process (γ= 1.4) throughout the operation. The following relationships give the upper limit and lower limits of pressure:

$$P_{IL\_upper} = \frac{4}{3} P_{rL} \tag{3.75}$$

$$P_{IL\_lower} = 0.5 P_{IL} \tag{3.76}$$

The pre-charge pressure is given as:

$$P_{IL\_pre} = 0.9 P_{IL\_lower} \tag{3.77}$$

The following relationships give the maximum and the minimum initial volume:

$$V_{IL\_max} = V_{ILo} \left(1 - \left(\frac{P_{IL\_pre}}{P_{IL\_upper}} \right)^{1/1.4} \right) \tag{3.78}$$

$$V_{IL\_min} = V_{ILo} \left(1 - \left(\frac{P_{IL\_pre}}{P_{IL\_lower}} \right)^{1/1.4} \right) \tag{3.79}$$

Similarly, a high-pressure accumulator also has a pre-charge pressure (P_{IH_pre}) and initial volume (V_{IHo}), which are calculated using the ideal gas equations. The upper limit and the lower limit of pressure are given as:

$$P_{IH\_upper} = \frac{4}{3} P_{rH} \tag{3.80}$$

$$P_{IH\_lower} = 0.5 P_{IH} \tag{3.81}$$

$$P_{IH\_pre} = 0.9 P_{IH\_lower} \tag{3.82}$$

$$V_{IH\_max} = V_{IHo} \left(1 - \left(\frac{P_{IH\_pre}}{P_{IU\_upper}} \right)^{1/1.4} \right) \tag{3.83}$$

$$V_{IH\_min} = V_{IHo} \left(1 - \left(\frac{P_{IH\_pre}}{P_{IH\_lower}} \right)^{1/1.4} \right) \tag{3.84}$$

where P_{rL} and P_{rH} are the rated pressures of low- and high-pressure accumulators, respectively. The governing equation based on the torque-balance between the hydraulic motor and electric generator is given as follows:

$$J_m \dot{\omega}_m = D_m \left(P_h - P_l \right) - C_g \omega_m - f \omega_m \tag{3.85}$$

where, J_m is the combined moment of inertia of the motor and the generator, D_m is the fixed motor volume, (P_h, P_l) are the pressure in the high and low-pressure accumulators, respectively, (ω_m, ω_d) are the speed of the motor speed and the desired speed, respectively, and f is the friction coefficient of the electric motor.

The mechanical design of a generator (or choice of a suitable generator) is carried out using a look-up table constituting a few vital parameters: generator efficiency (η_g) against the rated torque (T_r) or rated angular speed (ω_r), torque (T_m) and the motor speed (ω_m). For simplicity, η_g is assumed to be a constant (= 90%). The generator damping, C_g is given as follows:

$$C_g = \frac{D_m \left(P_h - P_l\right)}{\omega_d} \tag{3.86}$$

All the pistons in the circuit are modulated to have the same sizing and properties. For a compressible, double-acting, equal-area cylinder A_{ep}, the piston volume is taken as $V_p = 15 * A_{ep}$. Throughout the analysis, A_{ep} is kept constant. Pressures in the upper and lower chambers of the cylinder are considered equivalent to that of the upper and lower limit accumulators' pressures. Effective bulk modulus (β) is incorporated as a measure of compressibility. The HPTO force on any given float, F_{PTO} is given by the pressure difference between both the ends of a double-acting cylinder:

$$F_{PTO} = (P_t - P_b)A_{ep} \tag{3.87}$$

where, P_t and P_b are top and bottom end pressure of A_{ep}. The damping of HPTO, referred to as effective damping, C_{EPTO}, which influences the energy absorption of floats is given by:

$$C_{EPTO} = \left(\frac{A_{ep}}{D_m}\right)^2 C_g \tag{3.88}$$

The efficiency of the HPTO is given by:

$$\eta_H = \frac{P_{E_{BFWEC}}}{P_{TA_{BFWEC}}} * 100 \tag{3.89}$$

The average absorbed power at each piston-cylinder arrangement (P_A), mechanical power (P_M), and electric power from the generator $(P_{E_{BFWEC}})$ are given by the following relationships:

$$P_A = \frac{1}{T}\int_0^T F_{PTO}\, v_p \, dt \tag{3.90}$$

$$P_{M_{BFWEC}} = D_m \left(P_h - P_l\right)\omega_m \tag{3.91}$$

$$P_{E_{BFWEC}} = D_m \left(P_h - P_l\right) \omega_m \eta_g \tag{3.92}$$

where, v_p is the piston velocity, and η_g is the generator efficiency. Total mechanical power, $P_{TA_{BFWEC}}$ combining all the floats in the four-float configuration is given by:

$$P_{TA_{BFWEC}} = \sum_{F=1}^{4} P_{A_{FWEC}} \tag{3.93}$$

The overall efficiency, η_{WEC} is given by:

$$\eta_{WEC} = P_{E_{WEC}} * CWR * 100 \tag{3.94}$$

Effective damping is a function of the piston area, motor volume, and generator damping. In a practical design, it is difficult to vary C_g. Hence, varying either A_{ep} or D_m would be wiser to assess the influence on the PTO's performance. However, for every value of either A_{ep} or D_m, C_g is varied automatically due to its dependence on the motor volume and its pressure difference. Therefore, to fix this problem, the viscous damping coefficient is chosen to vary the hydraulic PTO's damping criteria. It is important to note that variations in the C_{lpto} vary the pressure indirectly across the cylinders of all the floats. This, in turn, influences the pressure difference across the motor. It affects the oil flow due to the change in the resistance offered by the motor. To choose a set of practical governing parameters, one can vary a wide range of values for (A_{ep} and D_m) while the parameters of the accumulators, check valve, and the generator values are taken from the standard design catalogs.

3.12 PRACTICAL GUIDE TO DESIGN OF HYDRAULIC POWER TAKE-OFF SYSTEM

Rated power and speed of 1 kW and 150 RPM are chosen as initial values of the design. The absorbed power is obtained using a PTO as a viscous damper; however, the HPTO behaves more like a Coulomb damper. Hydraulic components of the HPTO are sized to deliver the rated power and rotational speed ($\omega_r = 2\pi N / 60$).

$$T_r = \frac{P_r}{\omega_r} \tag{3.95}$$

The generator damping coefficient is given by:

$$C_{ng} = \frac{T_r}{\omega_r} \tag{3.96}$$

The nominal piston area is given by:

$$A_{np} = \frac{D_m \omega_r}{\dot{\eta}} \tag{3.97}$$

For the motor displacement (D_m = 5 cm^3), axial piston pump is found to satisfy the torque and speed constraints. For a given rated power and the torque of an electric generator, the following parameters are estimated:

$$A_{np} = 0.000356\,\text{m}^2 \tag{3.98}$$

$$C_{ng} = 8.\text{NS}/\text{m} \tag{3.99}$$

$$\omega_r = 10\,\text{rad}/\text{s} \tag{3.100}$$

The accumulator's sizes are chosen based on the Bosch Rexroth (HAB/ RE 5017/ 02.12) catalog.

3.13 NUMERICAL STUDIES ON HYDRAULIC POWER TAKE-OFF SYSTEMS

Numerical studies on the FWEC with HPTO are carried out under regular waves. Simulations are carried out for a total time of 500 s with a step interval of 0.1 s. Wave parameters considered for the HPTO calculations are given in Table 3.6. Numerical-based governing parameters of the HPTO are evaluated over a wide range of the piston area and the motor displacement. Further, optimal damping coefficients are estimated.

Fig. 3.31 shows the average absorbed power against the variation of the piston area, while Fig. 3.32 illustrates the variation in the hydraulic motor displacement. In both cases, there is an increase in the absorbed power with the increase in the parametric values, which subsequently reduces once it reaches its nominal value. It is also observed that the nominal values obtained by this numerical process are close enough with the theoretical values (28% and 20%) of the piston area and motor displacement, respectively. The rest of the parameters are directly taken from the catalog depending on the nominal values. All the estimated parameters are mentioned in Table 3.7.

The pressure acting on the piston increases with the increase in the piston area. Hence, there must be a trade-off between the maximum permissible pressure and the power-generation capability. Also, note that there will be no appreciable increase in the accumulator pressure for piston-areas beyond optimal values, reducing overall power. In the case of the motor displacement, the flow rate increases with its value, leading to a reduction in the flow into the accumulators. It further leads to the power fluctuations for oversized motor values. Ideally, the accumulator's volume can be

TABLE 3.6
Wave Parameters for HPTO Simulations

S. No	Wave Height (*H*)	Wave Period (*T*)
1	0.3	2, 2.4, 2.8, 3.2
2	0.5	2, 2.4, 2.8, 3.2
3	0.7	2, 2.4, 2.8, 3.2

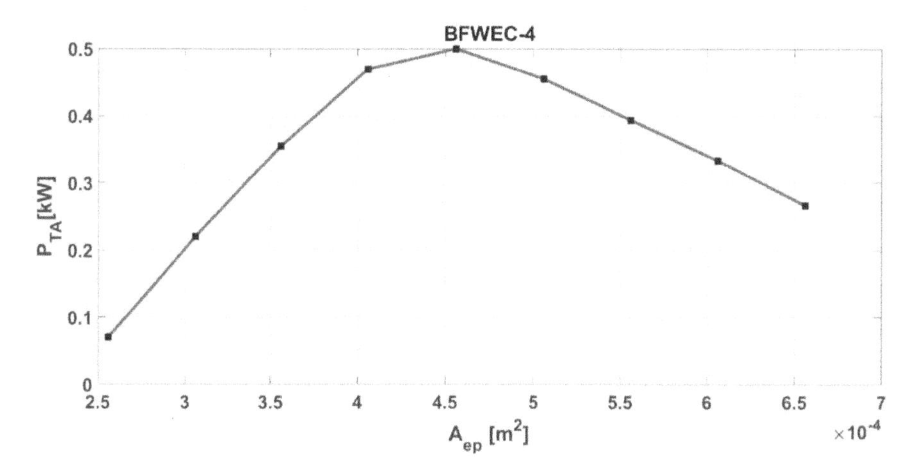

FIGURE 3.31 Average absorbed power against piston area

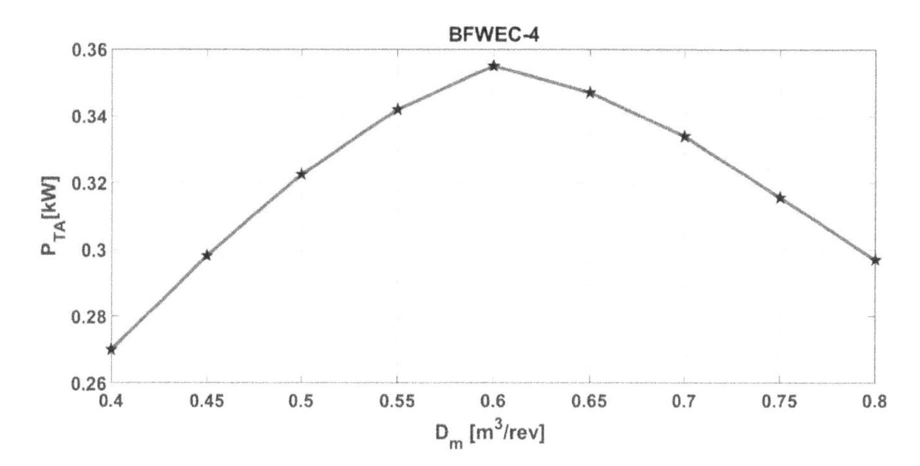

FIGURE 3.32 Average absorbed power against motor displacement

chosen as large as possible for smoothening, but selecting the least size with maximum smoothening is economical. As the accumulator volume reduces, the power-storing capacity reduces, resulting in a higher power with the improved motor flow.

A wide range of damping coefficients is considered to take 1000 Ns/m as a base value, which is optimal for a viscous damper. The absorbed power is a direct function of the force acting on the piston and its velocity. For a given optimal piston area, increasing the damping value increases the resistance in the circuit. It is advantageous till the absorbed power reaches its peak value. Further increase in the resistance is not desired as it may reduce the absorbed power since the float and the piston velocity will be almost zero. Fig. 3.33 shows the total average absorbed power of the FWEC with a four-float configuration for the variation in the wave periods. It is useful to assess its

TABLE 3.7
Hydraulic PTO Parameters

Parameters and Units	Value
a) Double acting hydraulic piston	
Area	0.000456 m^2
Limited Stroke length	±0.3 m
b) Rotation to the linear conversion mechanism	
Crank radius	0.3 m
Rod initial length	0.8 m
Offset	0.05 m
c) Low-pressure accumulator	
Initial volume	$3*10^{-3}$ m^3
Rated working pressure	1.6 MPa
d) High-pressure accumulator	
Initial volume	$4.25*10^{-3}$ m^3
Rated working pressure	3.0 MPa
e) Hydraulic motor Inertia	5 kg-m^2
f) Electric generator	
Desired speed	2.5 rad/s
Rated torque	14.2 N-m
Generator efficiency	95%

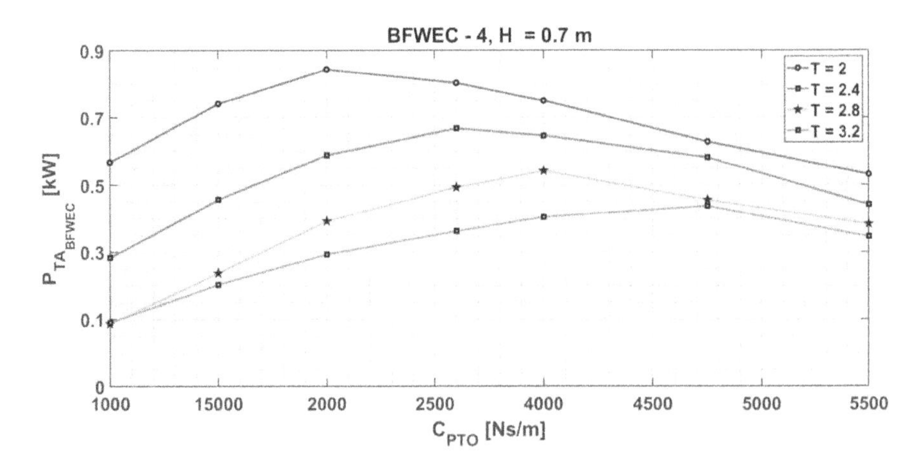

FIGURE 3.33 Average total absorbed power variation with PTO damping

sensitivity to the damping values. It is seen that as the wave period increases for the constant wave height, the absorbed power reduces with a shift in the optimal value of damping to the right. An increase in the wave period requires a higher damping value to absorb power. The optimal damping values are, in general, independent of variation in the wave height and therefore not considered in the analysis.

Fig. 3.34 illustrates the piston force-time history for various damping coefficients for fore-float. The observed pattern is close to a square wave with a different slope or plateaus at its crest and trough. As the wave force of the floats is transferred, the piston moves either up or down depending upon the float motion. After reaching a certain distance, wave forces driving the piston can no longer lift the float. As a result, the slope of force variation decreases, showing a reduction in the wave force. It is desired that this phase of slope change must be minimum to draw a continuous power. Therefore, choosing optimal damping and piston size is very crucial. It can be seen from the figure that the plateau band is small (less than a second), verifying the fact that the chosen values in the design are almost optimum. The estimated optimal damping, the variation of the piston force of all floats are illustrated in Fig. 3.35. It is observed that a similar trend of slope at crests and troughs is observed for all the

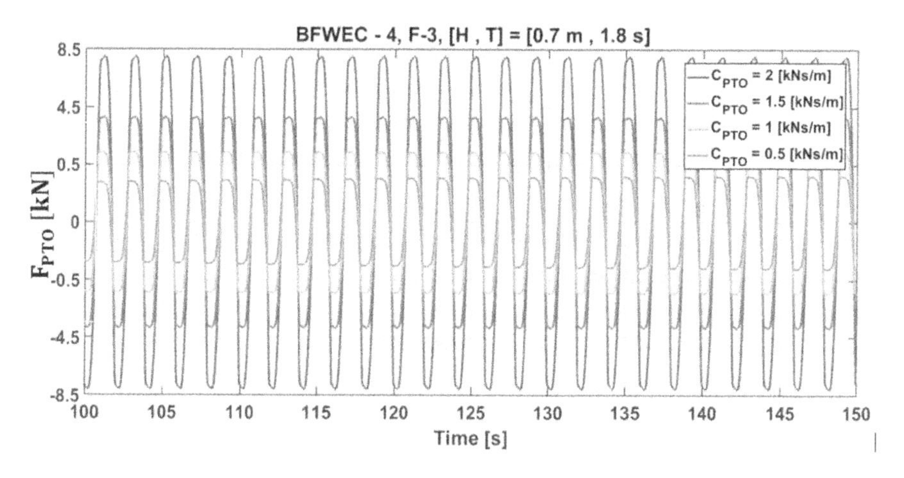

FIGURE 3.34 Piston force-time history with the influence of PTO damping coefficients

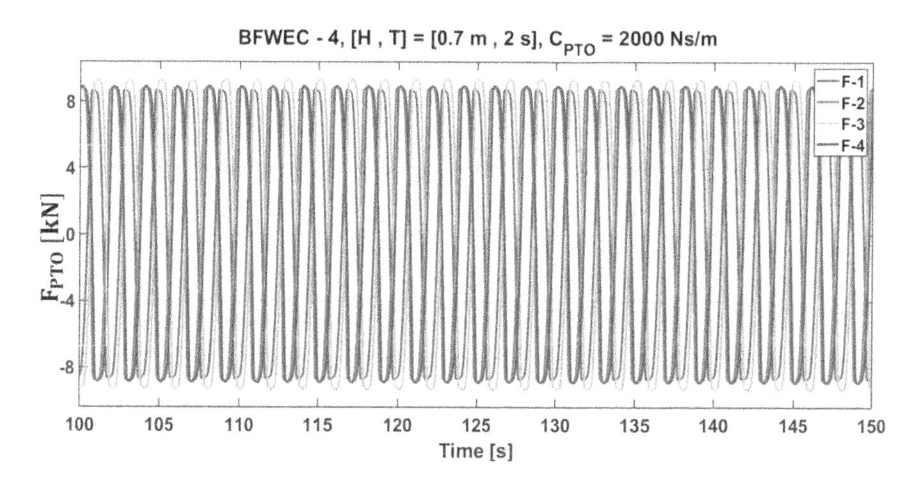

FIGURE 3.35 Piston force-time history of all floats at optimum PTO damping

floats. There is no major difference in the force observed for different floats. It is because, for a given wave period of 2 seconds (wavelength = 1.56 T^2 = 6.24 m), the maximum end-to-end width of the FWEC is only about 4.2 m; It is lesser than the incident wavelength and hence more than half of the device is either under the crest or trough under the incident wave. It has amounted to a small amount of phase shift between the floats. The chosen design is a typical example of a point absorber with lesser device dimensions than the incident wavelengths.

Irrespective of the HPTO design, there is an optimum piston force for each wave period, producing maximum power. It is different for monochromatic and irregular waves, and hence there is a steady-state response in regular waves, where the stall periods are very low. Low flows with large motor volume are undesirable as it affects the PTO efficiency adversely with low motor displacements. Hence, a smaller motor that can take most of the frequency bandwidth and amplitudes is recommended with suitable control systems. It has led to investigations using digital displacement motors and pumps to cover up the part-load efficiency discrepancies.

Further, a detailed analysis is carried out under a regular wave (0.7 m, 2 s) to observe the time histories of the piston position, velocity, and power of the floats. Figs. 3.36 and 3.37 show the piston velocity variation and average power absorption for the optimum PTO damping. It can be seen from the figures that the fore-float (F-3) is showing a higher piston force, velocity, and amplitude due to higher power output. However, the side floats (F-2 and F-4) also show similar performance, confirming that the device is directionally insensitive.

Fig. 3.37 shows the power absorption in various stages of conversion. The plot shows the total absorbed, mechanical, and electric power (P_A, P_M, P_E). As the power is converted from one form to another, its magnitude reduces due to compressibility effects and the generator conversion efficiency. Including other losses such as friction, hose bends shall further reduce the absorbed power.

For different sea states, the average values of absorbed and electrical power at the optimum damping coefficient are given in Table 3.8. The capture width and CWR

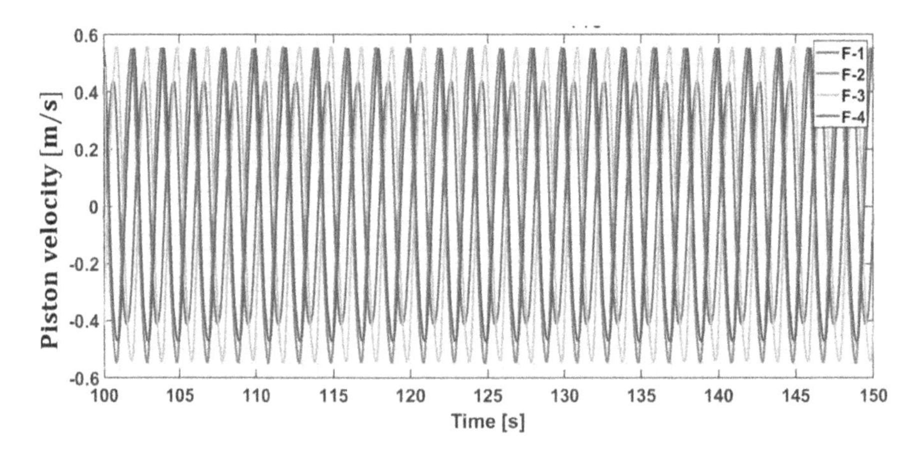

FIGURE 3.36 Piston velocity variation of all floats at optimum PTO damping (2000 Ns/m)

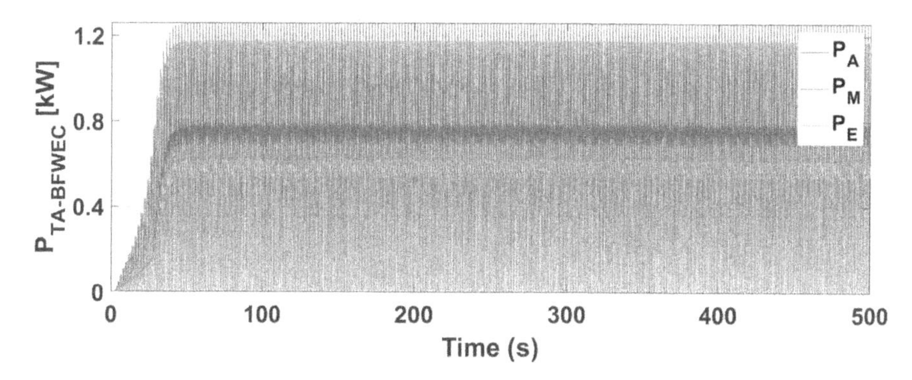

FIGURE 3.37 Total average power at optimum PTO damping (2000 Ns/m)

TABLE 3.8
Absorbed and Electric Power of FWEC with Four-Floats and HPTO

Wave Period, *T* (s)	Wave Height, *H* = 0.3 m		Wave Height, *H* = 0.5 m		Wave Height, *H* = 0.7 m	
	Absorbed Power (kW)	Electric Power (kW)	Absorbed Power (kW)	Electric Power (kW)	Absorbed Power (kW)	Electric Power (kW)
2.0	0.55	0.43	0.76	0.57	0.86	0.63
2.4	0.48	0.37	0.64	0.49	0.83	0.61
2.8	0.41	0.32	0.56	0.44	0.63	0.48
3.2	0.35	0.28	0.49	0.39	0.55	0.42

TABLE 3.9
Capture Width and CWR of FWEC with Four-Floats and HPTO

Wave Period, *T* (s)	Wave Height, *H* = 0.3 m		Wave Height, *H* = 0.5 m		Wave Height, *H* = 0.7 m	
	Capture Width (m)	CWR (%)	Capture Width (m)	CWR (%)	Capture Width (m)	CWR (%)
2.0	1.76	101.53	1.08	62.21	0.69	39.59
2.4	0.95	59.94	0.50	33.06	0.41	23.79
2.8	0.52	30.24	0.32	18.38	0.20	11.68
3.2	0.32	18.25	0.20	11.43	0.12	7.20

(in %) are summarized in Table 3.9, while Table 3.10 highlights the efficiencies of the HPTO and the overall device. It is seen that the variation of these parameters shows a similar trend to the change in the wave parameters considered for the study. As seen from the tables, the variation in the HPTO efficiency is not significant over the range of cases considered for the study.

TABLE 3.10
HPTO Efficiency and Overall Efficiency of FWEC with Four-Floats

Wave Period, T (s)	Wave Height, H = 0.3 m		Wave Height, H = 0.5 m		Wave Height, H = 0.7 m	
	η_{HPTO}	$\eta_{overall}$	η_{HPTO}	$\eta_{overall}$	η_{HPTO}	$\eta_{overall}$
2.0	76.92	78.1	74.63	46.42	73.02	28.9
2.4	77.52	42.59	76.34	25.24	74.05	17.62
2.8	78.74	23.82	78.25	14.38	75.53	8.83
3.2	80.64	14.72	78.93	9.03	77.18	5.56

3.14 FLOATING WAVE ENERGY CONVERTER RESPONSE WITHOUT POWER TAKE-OFF SYSTEM

A set of experiments are carried out in the wave basin of the Department of Ocean Engineering, Indian Institute of Technology Madras, India. The authors acknowledge the administrative support extended by the Dept. and the Institute to carry out the experimental studies. Results of the studies are shared in this chapter with the intention of capacity building in the ocean energy sector. As very few detailed studies are available in the literature illustrating the design, fabrication, and analysis of FWEC devices, the set of discussions presented in this section will be very useful. Fig. 3.38 shows the experimental setup of the FWEC device commissioned in the testbed. Four wave gauges are used, of which two are placed in the front of F-3, one at an offset of 2 m towards the float F-4 and one at the back of the float F-1, as shown in the figure. For measuring the response, uni-axial and bi-axial inclinometers are placed as a pair on all the floats, including the central buoy. All sensors are connected to the data acquisition system on the carriage, whose data logger is integrated with the in-house MATLAB® interface program. The 1:10 scaled model of the FWEC is commissioned at the center of the wave basin using a custom-made mooring-anchor arrangement. Clamp weights are placed on the base frame to counter the motion of the FWEC, ensuring a minimal load on the mooring lines. The floats are hollow, weighing 100 kg in dry conditions with a default draft of 0.05 m. The reference draft of the floats is half of its diameter, which is 0.3 m. The dead weights of 200 kg are mounted on each float by equally distributing 100 kg on each float's end. A closer view of the ball joint used to connect the floats with the central buoy is shown in Fig. 3.39; it is fabricated using stainless steel. In the absence of a suitable PTO, the response behavior of the FWEC under regular waves at 0° is assessed.

Based on the detailed numerical studies conducted, the specifications for the fabrication are determined. It consists of different stages: fabrication, assembling, towing, installation, instrumentation, logging, and decommissioning. Each float has two ball joints, as a single ball joint will be vulnerable to failure under concentrated loads. The bean-shaped floats are fabricated by cutting and bending the tapered, semi-circular contours and welding them together. The hemispherical cups are fabricated using cold forming and subsequently welded on both sides of the tube to form a bean-shaped

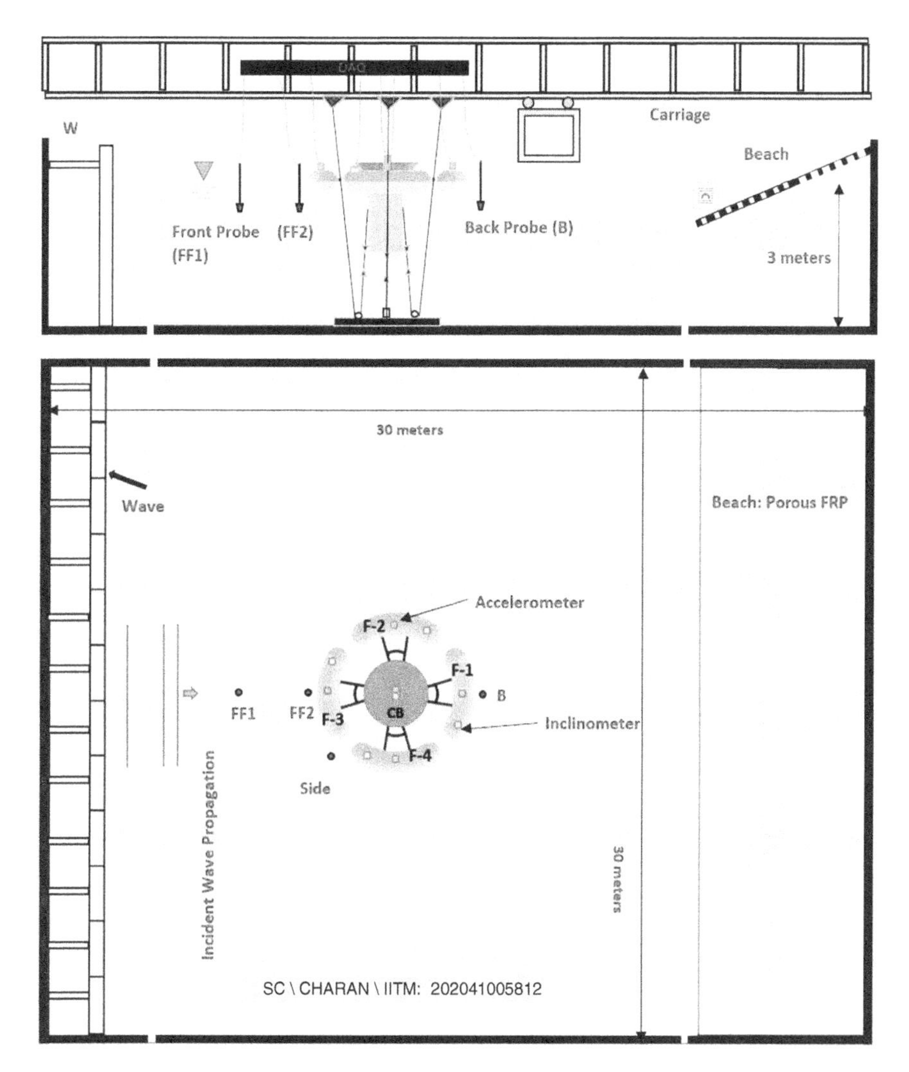

FIGURE 3.38 Experimental setup of the FWEC model (four-float configuration)

float. The central buoy is towed to the target coordinates and ballasted (see, Fig. 3.40 and 3.41). With the help of a skilled workforce, the floats are connected to four sides of the central buoy. As the central buoy is positively buoyant, even without mooring pre-tension, they remain afloat. Fig. 3.42 shows the working visuals of the device.

Free-decay tests are conducted to estimate the damped natural period and the associated properties of the FWEC; plots are shown in Fig. 3.43. The estimated heave natural period has a good agreement with the non-linear time-domain model, which is determined to be 1.3 s (4.83 rad/s). Further, the natural pitch period about which the floats oscillate gives rise to the net heave and is found to be 1.3 s from the experiments; it is also equal to that obtained from the numerical studies.

FIGURE 3.39 Enlarged view of ball joint

The heave-RAO is calculated from the measured wave amplitude and heave motion of the floats. As the PTO offers no damping for lower periods in the F-2 and F-4, the numerical simulation underestimates the performance compared to the experimental studies; the numerical results are overestimated for higher periods. Hence, the viscous effects are important to estimate, at resonance, particularly where the variations are greater due to large oscillations. It cannot be captured under the assumption of linear wave theory during numerical simulations. Fig. 3.44 shows the absorbed wave power for different wave periods and amplitudes under regular waves. Fig. 3.45 shows the comparison of RAO in the heave motion as obtained from the numerical and experimental studies. It is seen that the heaving amplitude increases with an increase in the wave amplitude at resonance; a maximum RAO of 1.6 is observed in F-3 at resonance in the presence of external damping. However, the maximum value would have been reduced suppose a PTO system was present during the experiments. Even in the real scenario with viscous losses, friction, and wave nonlinearities, the heave response is 1.5 times higher than the incident wave amplitude. It may be attributed to the multi-body interactions causing a positive interference. Further, due to the deliberate positioning of float's center of gravity (0.318 m from mean sea level) above its center of buoyancy (0.15 m below mean sea level) with a meta-centric height of (-0.24 m) and a righting moment of - 11.98 kN-m/°. As the negative righting moment is generated, the moment created due to the wave excitation force (disturbed waves) will further rotate the floats resulting in improved oscillations as truly desired by any FWEC.

The floats heave time history is shown in Fig. 3.46 for three-wave conditions with increasing incident wave period and wave height. For a short period (1.3 s), the response of F-2 and F-4 is suppressed as their length is lengthier than wavelengths, nullifying the wave effects. F-3 has the maximum oscillations with amplitudes more than that of wave amplitude. Response of the side floats increases with the increases in the wave period. It is seen from the above plots that the crests and troughs of the response are quite sharp

FIGURE 3.40 Central buoy: fabrications, assembly, and towing

FIGURE 3.41 Towing and installation of floats

FIGURE 3.42 Working visuals of the FWEC (four-float configuration)

even under monochromatic waves with equal crest and trough magnitudes. The heave amplitudes fluctuate with the influence of active periods and the associated frequencies generated from the interactions of the floats within the vicinity of the natural period. The presence of multi-body has a net positive interference effect due to floats and their relative

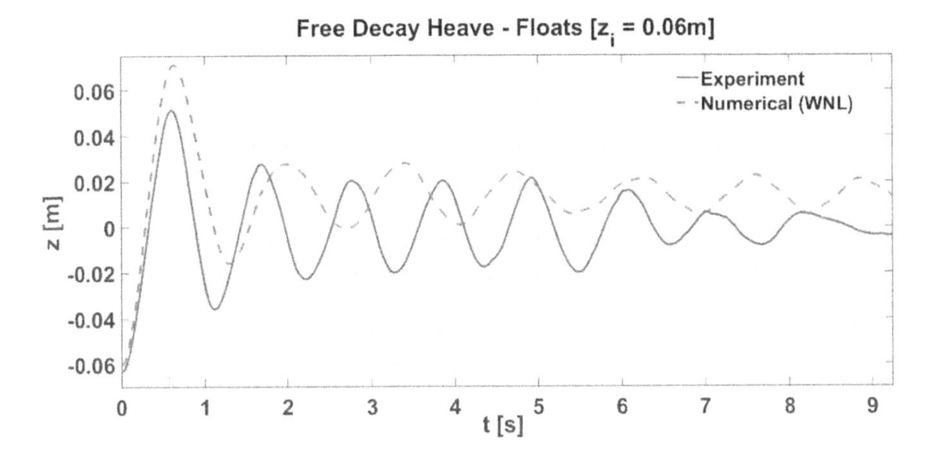

FIGURE 3.43 Free decay test in heave

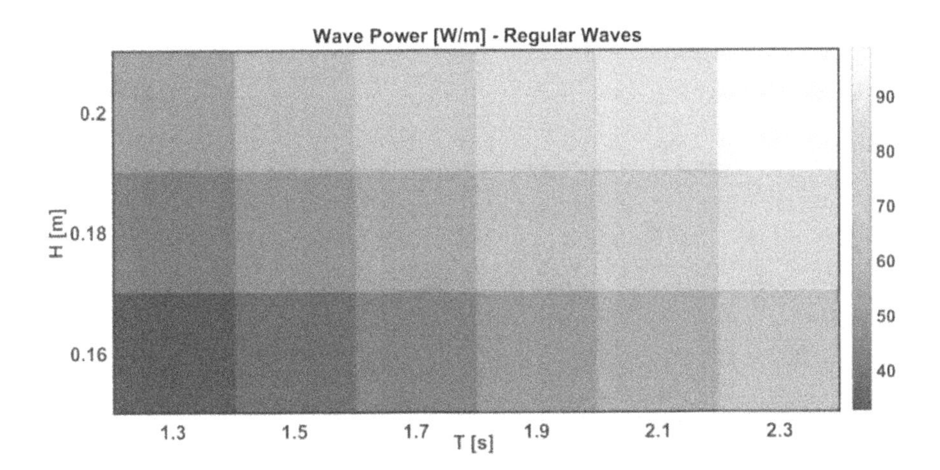

FIGURE 3.44 Wave power matrix of FWEC under regular waves

motion with that of the central buoy. As the floats oscillate upwards, their wetted areas decrease to the minimum at the wave crest.

On the other hand, when the floats move downwards, they intrude into the waves with a small cross-section, causing the least resistance. Thus, the floats quickly recede into the water, as indicated by the sharp peaks; it verifies that the float's shape significantly affects the oscillation. Hence, the proposed bean shape is useful in justifying the above discussion. Instantaneous average absorbed power from the experiments using the relative rotational velocity in the pitch degree of the floats measured during experiments. Fig. 3.47 shows the total absorbed power measured from the experimental studies. Fig. 3.48 shows the total CWR measured from experimental studies.

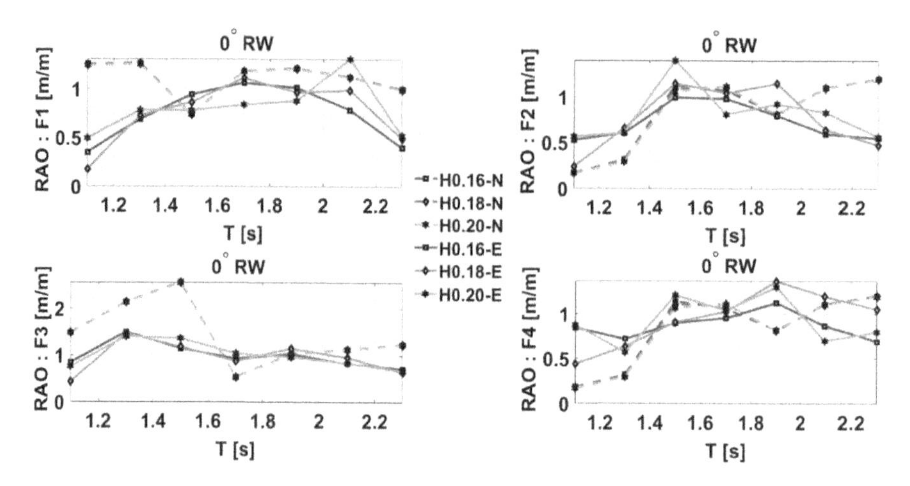

FIGURE 3.45 Have RAO comparison of numerical simulation and experiments

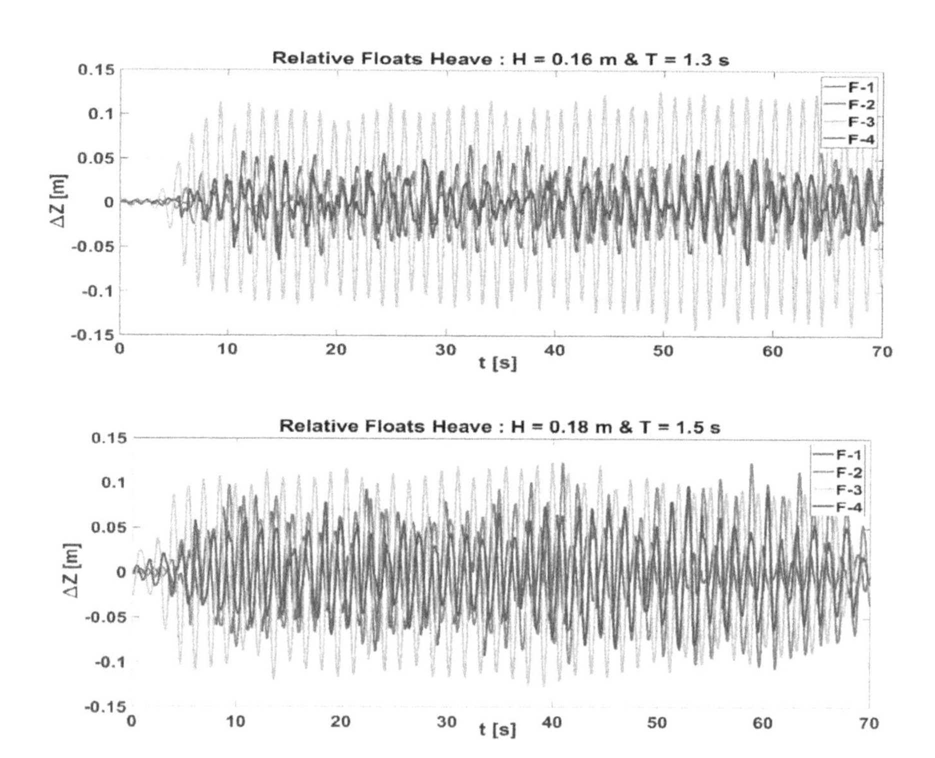

FIGURE 3.46 Relative heave motion of the floats

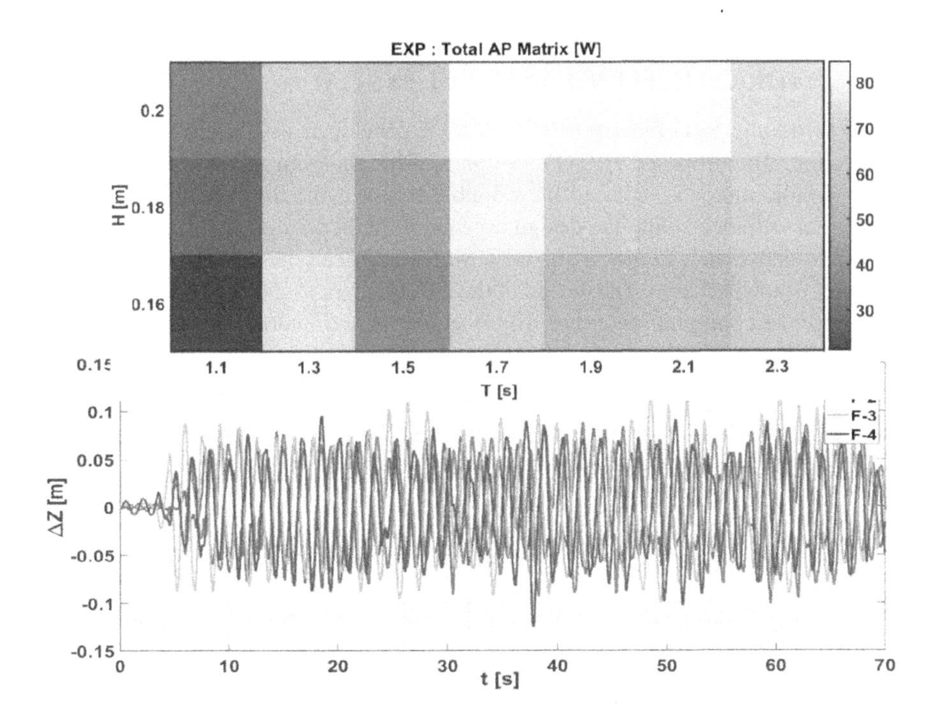

FIGURE 3.47 Total average absorbed power of FWEC from experiments

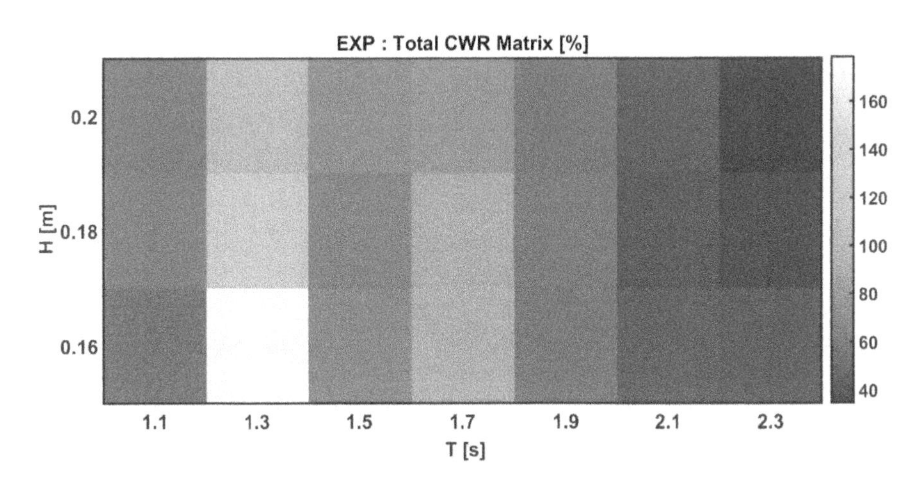

FIGURE 3.48 Total CWR of FWEC from experiments

The total absorbed power and CWR are the maximum at the resonance period. They increase with the wave height, except at 1.3 s, where there is a reduction in the values. Despite the exception at 1.3 s, the algebraic mean of total average power over the wave periods is estimated to be 65.24 W, 73.46 W, and 87.77 W, respectively for H = 0.16 m, 0.18 m, and 0.20 m. The algebraic mean of the CWR is 102.81%, 89.8%, and 85.03%, indicating a reduction with the increase in the wave height.

3.15 FLOATING WAVE ENERGY CONVERTER WITH A NOVEL HYDRAULIC POWER TAKE-OFF SYSTEM

A new hydraulic PTO circuit[Copyright:SC/Charan/IITM], developed in-house, is presented in this section. The proposed HPTO's design is different from the conventional ones. The hydraulic motor is replaced by a double-acting hydraulic cylinder (referred to as a master cylinder). Only the design aspects of the hydraulic circuit are discussed; associated conversion to the rotational power using gearbox and electricity using a generator is not discussed. The design of the HPTO refers to choosing the components of appropriate configuration either from the commercial market or custom-made to suit the proposed HPTO circuit. A CAD modeling in Autodesk Inventor is prepared, and then HPTO is fabricated in-house. Fig. 3.49 shows the CAD model of novel

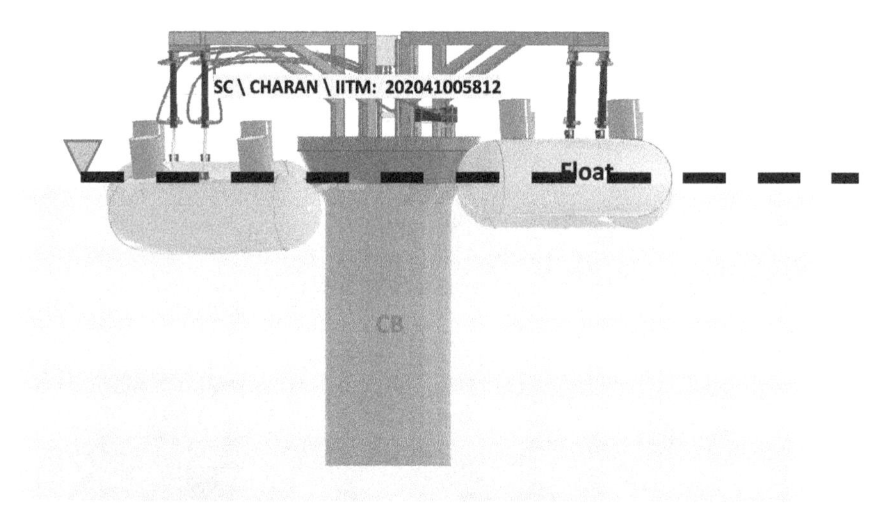

FIGURE 3.49 CAD model and experimental setup of a novel HPTO[copyright]

HPTO, which is commission integrated with the four-float configuration FWEC. The circuit is divided into two parts, namely the float-side and the central buoy-side. While the hydraulic components directly associated with the floats are included in the floats-side circuit, others are included in the central buoy circuit. As seen in the figure, each float is mounted with two double-acting hydraulic cylinders. A master cylinder is fixed on the top side of the central buoy. The float-side cylinders and the master cylinder communicate with the hydraulic hoses through which oil is circulated. There are eight float cylinders, one master cylinder, one oil reservoir, eighteen non-return valves (check valves), thirty-four hoses, eight analog pressure gauges, and two digital pressure gauges. Figs. 3.50a and 3.50b show a close view of the typical float side novel HPTO connections as a CAD representation and the installed setup.

Fig. 3.50 shows the CAD model and the installed HPTO with the float-side details. Fig. 3.51 shows the top-view of the installed HPTO. The circuit diagram of the novel HPTO is shown in Fig. 3.52. With reference to the figure, one can observe that each float-side comprises double-acting cylinders with four ports: two at the rod end and the others at the bore end. The master cylinder is customized with nine ports at the piston-end and eight ports at the bore-end. Eight bore-side ports and rod-end ports of

a)

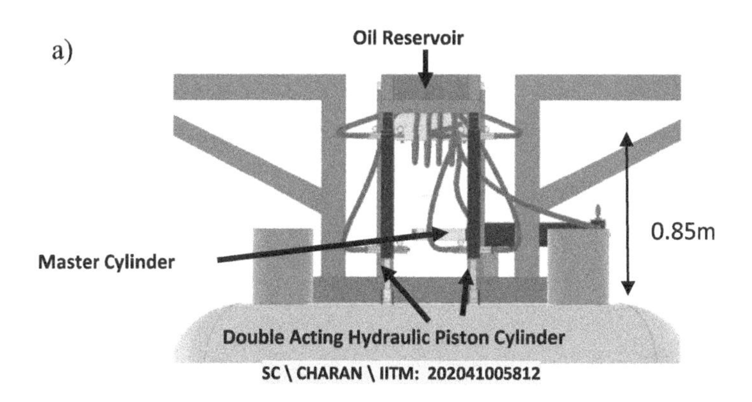

b)

FIGURE 3.50 A typical connection of float-side HPTO[copyright]

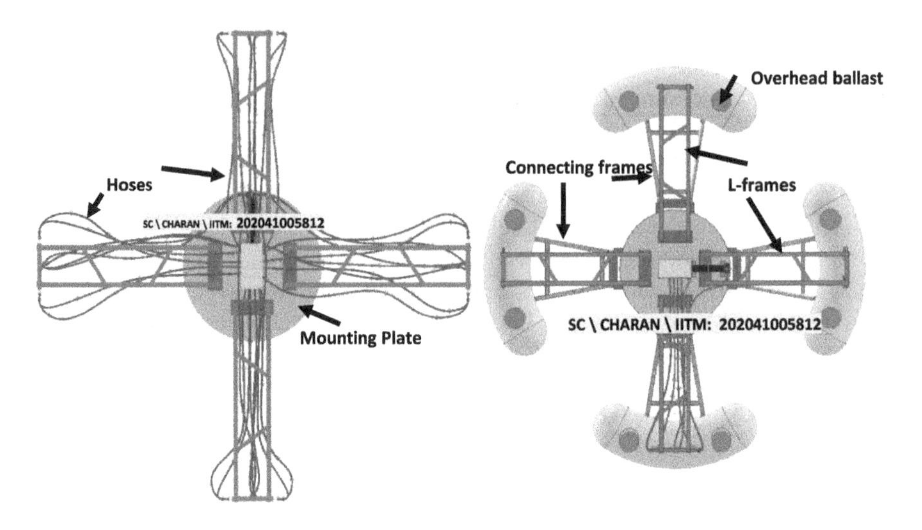

FIGURE 3.51 Top view of HPTO[copyright]

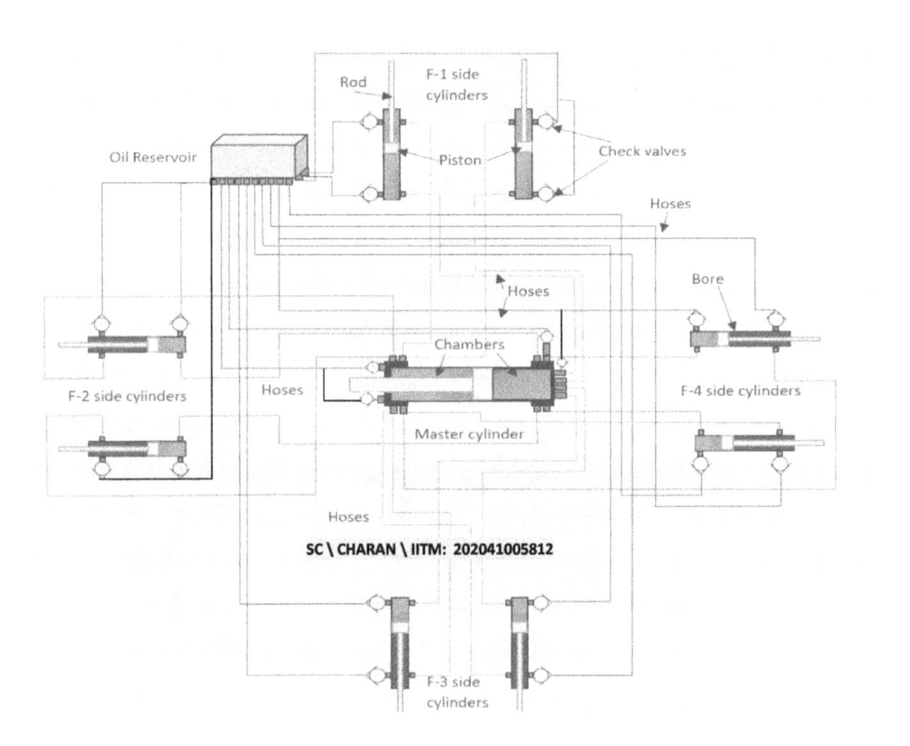

FIGURE 3.52 Circuit diagram of HPTO[copyright]

the float cylinders (inner-side ports) are connected to the corresponding eight bore-end and rod-end ports of the master cylinder using hoses. The hoses are connected to the float-side and the master cylinder through male couplings and appropriate leak-proof sealing.

The external ports of all float-side cylinders (both bore-end and rod-end) are connected to the bottom of the oil reservoir. There are twenty ports: the oil reservoir has sixteen to accommodate hoses connecting outside ports of all float-side cylinders and the others to account for both ends of the master cylinder. All hoses are connected through non-return valves to the reservoir at the piston-end to avoid reverse flow during suction. When the floats oscillate, oil is pressurized in the float-side cylinders; oil reaches the master cylinder through the inside ports. When the floats move upwards, the piston of the float-end cylinders pressurizes oil in the upper chamber; check valves do not allow the chamber oil from the float-side cylinders to the reservoir through the outside ports. As a result, the pressured oil gets discharged through the bore-end ports (inner ports) upward and through the rod-end ports downward. Oil flows into the corresponding bore-end ports of the master cylinder from the bore-side ports of the float-side cylinders in the upward stroke. Thus, the piston of the master cylinder expands. When the floats are placed in the trough of the incoming waves, the float-side pistons expand, pressurizing oil in the bottom chamber of the cylinder. As the check valves on the external ports of these cylinders will not allow the flow, the pressurized oil flows out through the rod-end inner ports; it is further directed to the corresponding rod-end ports of the master cylinder, creating contraction of the piston in the master cylinder.

As the float oscillations are dynamic under the action of the incident waves, the extent of their reciprocation is controlled by the motion of the float. A smooth control is achieved by directly coupling the floats to the hydraulic cylinders. To ensure smooth working of the master cylinder, two dedicated ports, one at each end, are connected to the oil reservoir through check valves. The check valves are positioned so that the spring inside the check valve compresses when the pressure inside the master cylinder falls below the oil reservoir pressure. This allows oil to flow into the master cylinder and smoothens the output. A check valve is shown in Fig. 3.53, indicating the direction of the oil flow.

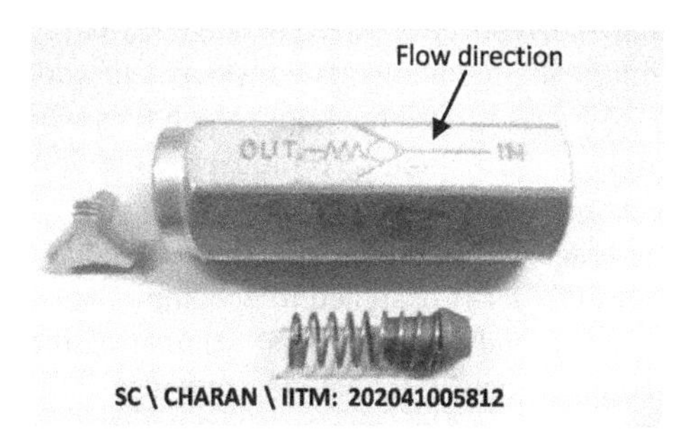

FIGURE 3.53 Un-assembled view of a check valve (non-return valve)

By design, the head of the reservoir is about 0.85 m (= 72 bar). When the pressure in the master cylinder is increased excessively under the extreme wave forces, oil is directed to the reservoir through the other two dedicated ports, whose direction of the check valve is different. It regulates the pressure in the hoses and the circuit within the desired level. Under the fluctuations of the absorbed energy caused by a maximum phase difference of 2 s (as observed from the experiments), side-pistons compress (or expand) with the time lag of 2 s, avoiding the opposite movement of one cylinder to the other. Thus, the reservoir acts as a pressure stabilizer. Irrespective of the flow into the master cylinder, the pressure from the master cylinder is equally distributed to the float-side cylinders. This ensures a balanced reciprocation between the low (back floats) and high-energized floats (front floats). One typical cycle is completed when oil from one end enters the corresponding end of the master cylinder, and the flow from the opposite end of the master cylinder enters the other end of the float-side cylinders.

The external ports of the float-side cylinders are provided to suck the oil from the reservoir only when the flow from the master cylinder is insufficient to reciprocate the float-side cylinder, which is a rare case. So, the inner side ports of the float-side cylinders are the active ports, and the master cylinder's ports leaving behind the connections to the reservoir as semi-active. Due to manufacturing constraints, the rod-end of the master cylinder has only eight ports. A T-coupling is used to provide necessary additional slots to connect the hose to the reservoir. In addition to this, two more T-couplings are fastened, one at each end of the master cylinder to provide an additional option to connect the digital pressure gauge. Similarly, the float-side cylinders' inner sides have analog-type pressure gauges fastened to the T-couplings. The oil reservoir has two ports on the top for filling oil in the event of deep-drying of oil.

Each component of the circuit is customized based on the preliminary experiments and the standard HPTO simulations. Based on the wave-power matrix presented earlier, the float-side piston's stroke is chosen as 0.4 m. At the start of the experiments, all the pistons are adjusted to ensure equal stroke in and outside the cylinder's bore, which is about 0.2 m; enough clearance is ensured to avoid knocking. The stroke length is chosen as about 1.4 times the maximum anticipated wave amplitude. The technical specifications of the components of HPTO are summarized in Table 3.11 to 3.13; details of a float-side cylinder, master cylinder, and other miscellaneous components are summarized, respectively. The average wave force acting on the floats, obtained from the numerical simulations, is considered the operating pressure. The average relative float's velocity is used to choose the NRVs whose spring stiffness is parametrized accordingly. Figs. 3.54 and 3.55 illustrate the float and master cylinder dimensions, respectively.

Based on the total average stoke volume of eight float-side cylinders and the operating pressure, a master cylinder is designed to accommodate all the hoses from the cylinders. Working pressure and the chambers' volume of the master cylinder are chosen as about 8–10 times that of a float-side cylinder. The hoses are chosen to sustain a maximum operating pressure of about 180 bar. Fig. 3.56 shows the master cylinder assembly details. Experimental studies are carried out on the FWEC with the new HPTO. The response of the floats and the central buoy in pitch motion under 0° incident waves is measured and plotted in Fig. 3.57.

TABLE 3.11
Specifications of Float-Side Cylinders

Action	Double acting with the single rod
Fluid	Hydraulic oil ISO 22 (862 kg/m^3)
Proof pressure	5 MPa (50 bar)
Maximum allowable pressure	3.5 MPa (35 bar)
Operating pressure	2.5 MPa (25 bar)
Theoretical force output at bore-end	3104 N
Theoretical force output at the rod-end	2505 N
Bore size	40 mm
Bore area	1256 mm^2
Stoke length	400 mm
Rod size	18 mm
Rod area	1002 mm^2
Available volume at the bore-end	250 ml (half stroke)
Available volume at the rod-end	52 ml (half stroke)
Rod- and cylinder-end	Clevis type
Stroke speed	8 to 300 mm/s
Cylinder tube, Rod cover, Headcover	Aluminum alloy (hard black anodized)
Piston rod and bore	Medium carbon steel, casehardened, hard chrome-plated, and polished
Piston seal, gaskets, and rod seal	Nitrile rubber (NBR)

TABLE 3.12
Specifications of the Master Cylinder

Action	Double acting with the single rod
Fluid	Hydraulic oil ISO 22 (862 kg/m^3)
Total ports	17
Proof pressure	75 MPa (750 bar)
Maximum allowable pressure	40 MPa (400 bar)
Operating pressure range	10 MPa-30 MPa (100–300 bar)
Theoretical force output at bore-end	52450 N
Theoretical force output at the rod-end	34050 N
Bore size	102 mm
Bore area	8108 mm^2
Stoke length	500 mm
Rod size	80 mm
Rod area	5027 mm^2
Available volume at bore-end	2000 ml (half stroke)
Available volume at the rod-end	600 ml (half stroke)
Rod- and cylinder-end	Clevis type
Stroke speed	8 to 300 mm/s
Cylinder tube, rod cover, head cover	Aluminum alloy (hard black anodized)
Piston rod and bore	Medium carbon steel, casehardened, hard chrome-plated, and polished
Piston seal, gaskets, and rod seal	Nitrile rubber (NBR)

TABLE 3.13
Specifications of Miscellaneous Components

Non-Return Valve (Check Valve)	
Code	NRV10
Cut-off pressure	2 MPa (20 bar)
Discharge	600 LPM (0.01 m^3/s)
NRV length	0.05 m
NRV diameter	2.54 m
Spring stiffness	100 N/m
Spring diameter	0.003 m
Spring coil thickness	0.00025 m
Material and spring	Stainless steel
Hydraulic flexible hose	
Code	Polyhose ISO 1436 RI AT-06 (3/8")
Weight	0.23 kg/m
Nominal diameter	0.0095 m (inner), 0.0174 m (outer)
Maximum working pressure	18 MPa (180 bar)
Minimum burst pressure	72 MPa (720 bar)
Bend radius	0.13 mm
Tube	Oil resistant synthetic rubber
Reinforcement	High tensile steel wire braid
Cover	Abrasion and weather-resistant synthetic rubber
Oil reservoir	
Capacity	15 L
Mean head	80 bar

It can be observed from the figures that the pitch response consists of multiple frequencies due to the presence of multi-body interactions. The maximum response is obtained at resonance, which is about 1.5 s. It is observed that the side floats perform well under longer wave periods ensuring directional insensitivity. Figs. 3.58 and 3.59 show the pressure variations in the master cylinder at the bore-side and rod-side, respectively.

As seen in the above figures, the chamber pressure of the master cylinder varies in the square waveform. The hydraulic power is estimated from the responses until 30 sec; however, plots are extended beyond to illustrate the response decay. As observed in the plots, different plateaus confirm that input pressure to the master cylinder is not sufficient to reciprocate; the wider the plateau, the higher the deficit in the input pressure. The pressure variation at 1.1 s is minimal, and it increases with the increase in the wave period. Three distinct plateaus of 0.5 s, 1 s, and 2.1 s are observed for 1.3 s and are periodic throughout the duration. In certain cases, at the bore-side, the plateau is about 5 s, indicating there is not enough oil from the float-side cylinders. It is also verified that there is no compensation from the reservoir to smoothen it out. Similarly, the rod-side pressure variations also have plateaus, indicating the regions where the

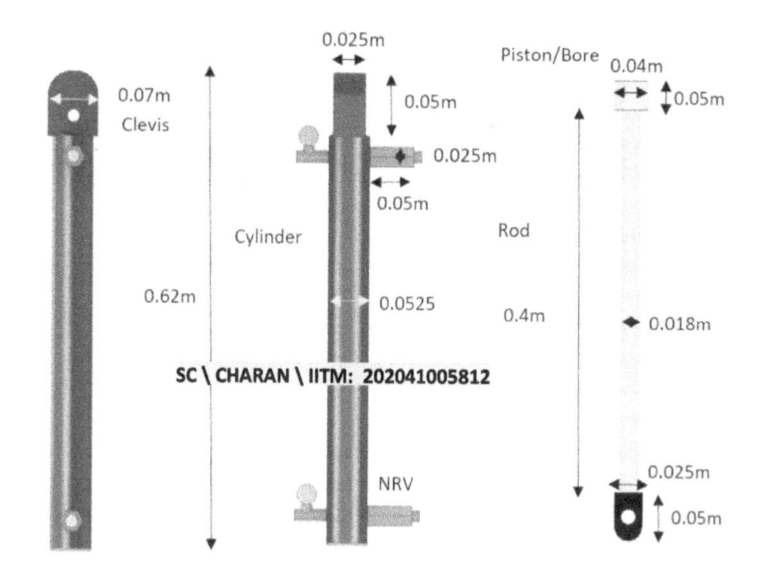

FIGURE 3.54 Design summary of float-side cylinder

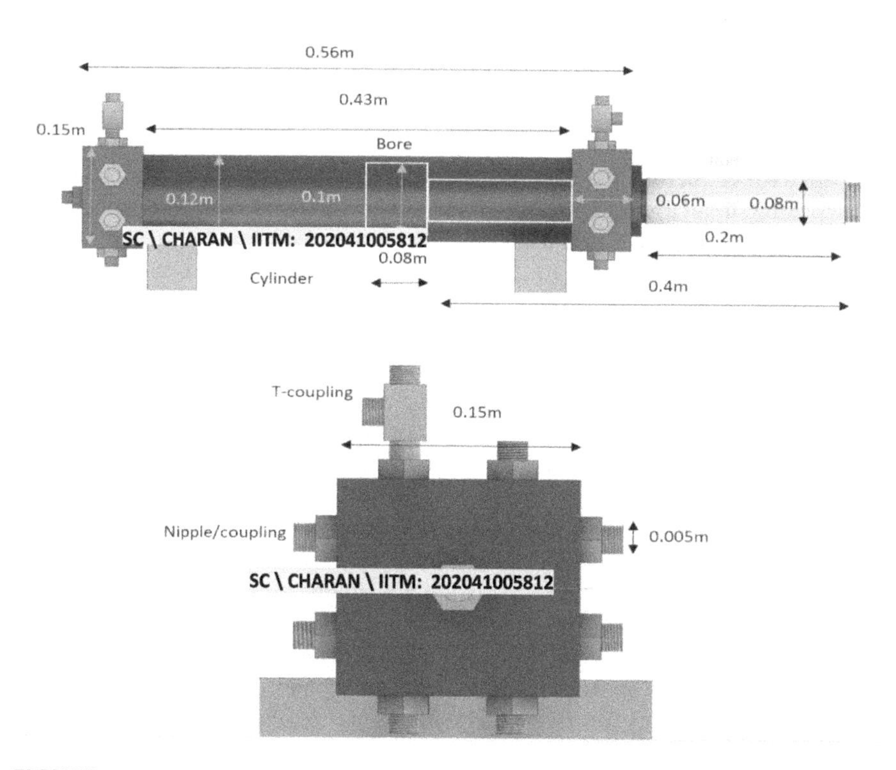

FIGURE 3.55 Design summary of master cylinder

FIGURE 3.56 Master cylinder assembly on FWEC with HPTO

piston stagnates and subsequently extends or retracts, depending on the inlet pressure. Due to the inconsistency of the pressure variations across the duration, it is difficult to understand the behavior of the master cylinder qualitatively. The salient pressure values at the bore-side across all wave periods and for 0.16 m wave height are 13.7, 23, 25.6, 26.58, and 15.39 kg/cm^2. The corresponding values at the rod-side are 10.55, 13.38, 23.92, 20.29, and 12.96 kg/cm^2.

The average force matrix is shown in Fig. 3.60, which illustrates the force exerted under each wave condition. The force in each pixel varies across the wave periods and wave heights, showing its sensitivity to the change in the wave parameters. A maximum average force of 650 N is obtained under a regular wave (0.2 m, 1.5 s) with a secondary peak at 1.7 s. It is seen that the force increases with the increase in the wave amplitude across the wave periods; this trend continues until resonance, after which there is a reduction. The range of force matrix is about (55–650 N), amounting to an average force of about 270 N. The reciprocation displacement and velocity are also measured using an accelerometer mounted on the piston tip of the master cylinder. Figs. 3.61 and 3.62 show the displacement and velocity-time histories of the master piston under regular waves of 0.16 m wave height and a wide range of periods. As seen from the plots, these responses are not consistent due to the lack of smoothening effect from the reservoir. The maximum response is observed neat to the resonance period (= 1.5 s) throughout the duration. Overall maximum, average, and minimum velocities of the master piston are estimated to be 0.1313 m/s at 1.5 s, 0.056 m/s (average of all velocities combined), 0.005 m/s at 2.1 s, respectively.

The average hydraulic power matrix is shown in Fig. 3.63, illustrating the hydraulic power observed by the HPTO. Fig. 3.64 shows the average efficiency of the novel HPTO. The intensities of the pixel variation are similar to the force matrix with a maximum at the resonance. The experiments' maximum average hydraulic power and efficiency are about 73 W and 67%, respectively, under regular waves (0.2 m, 1.5 s). The lower and upper limits of hydraulic power are 2 and 73 W, respectively; PTO efficiency is 3% and 67%, respectively. Longer waves beyond 1.9 s generate a poor response. The FWEC with the novel HPTO showed an average hydraulic power and efficiency of 20 W and 23%. The overall average efficiency and hydraulic power can be enhanced significantly by providing gas-charged accumulators.

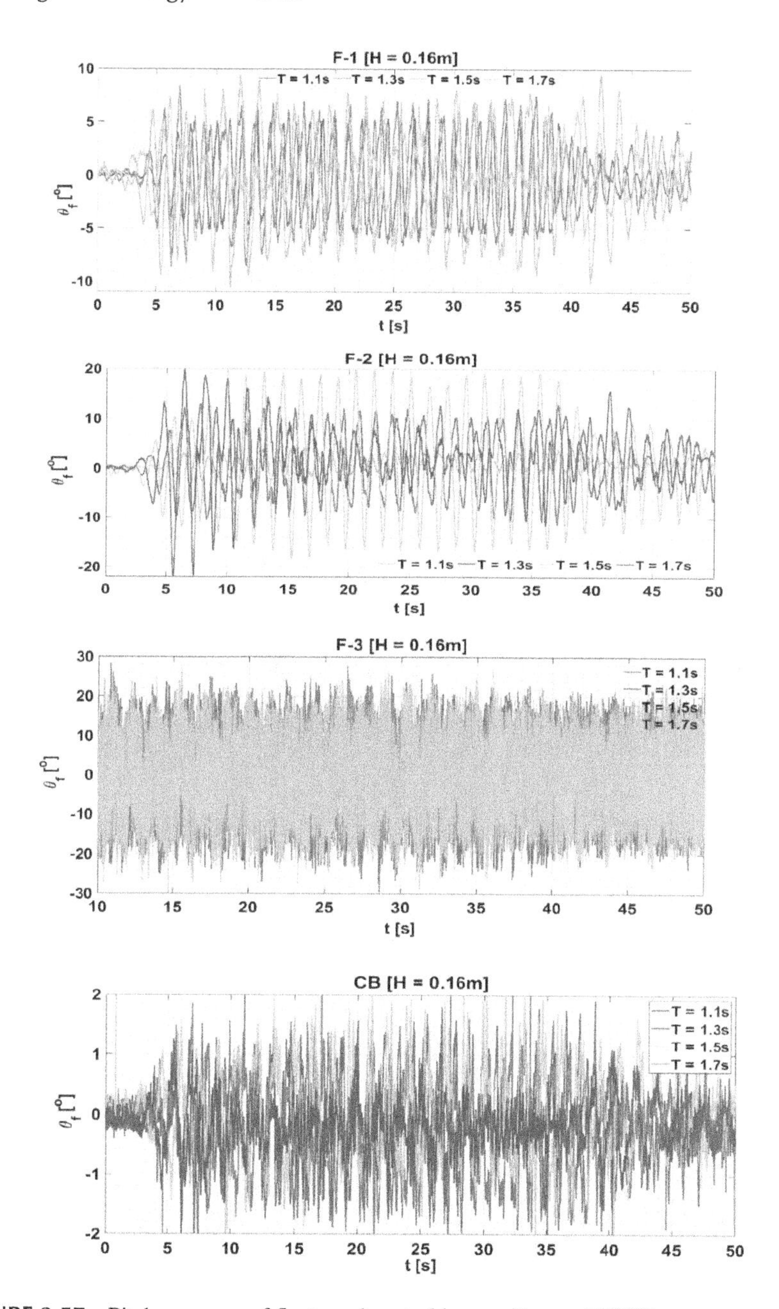

FIGURE 3.57 Pitch response of floats and central buoy with novel HPTO

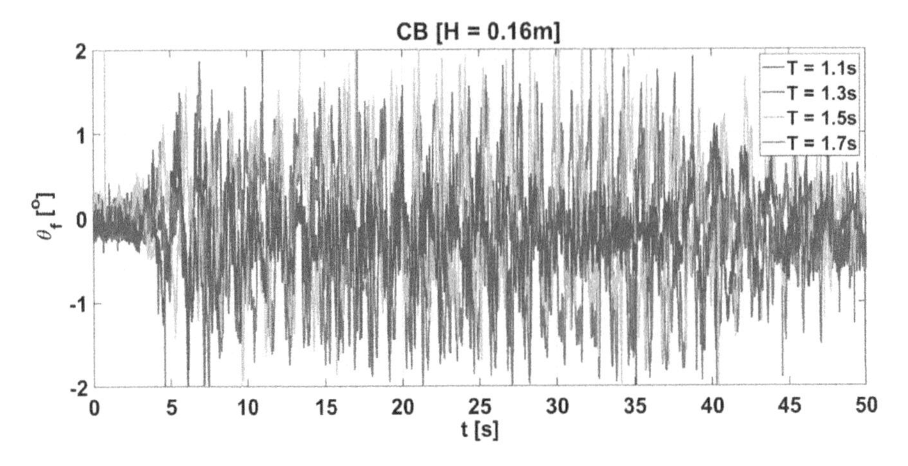

FIGURE 3.58 Pressure variations in the master cylinder at bore-side

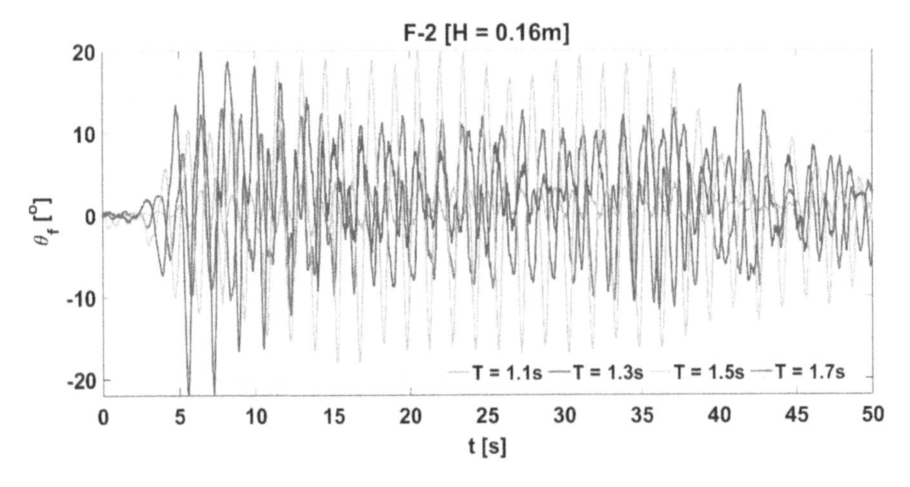

FIGURE 3.59 Pressure variations in master cylinder at rod-side

3.16 FAILURE ASSESSMENT OF FLOATING WAVE ENERGY CONVERTER

Every component of the novel HPTO and the FWEC is examined by failure mode and effect analysis (FMEA). This is a probabilistic tool used to identify the possible failure modes and then prioritize the risk involved in the conceptual design; a specific risk priority number (RPN) is associated with each device's mechanical component. RPN is a product of three variables: occurrence, severity, and detectability of the mechanical and electrical components represented numerically (1–10). The variables used to compute the RPN are assessed using the components' specifications and the manufacturers' manual. The frequency of failure of each of these components, the occurrence of failure, and remedial measures to avoid those failures are obtained

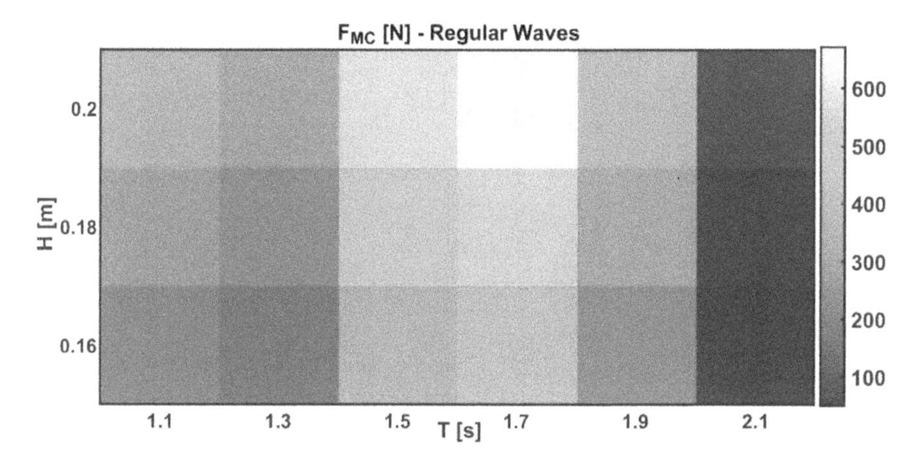

FIGURE 3.60 Average force matrix of master cylinder

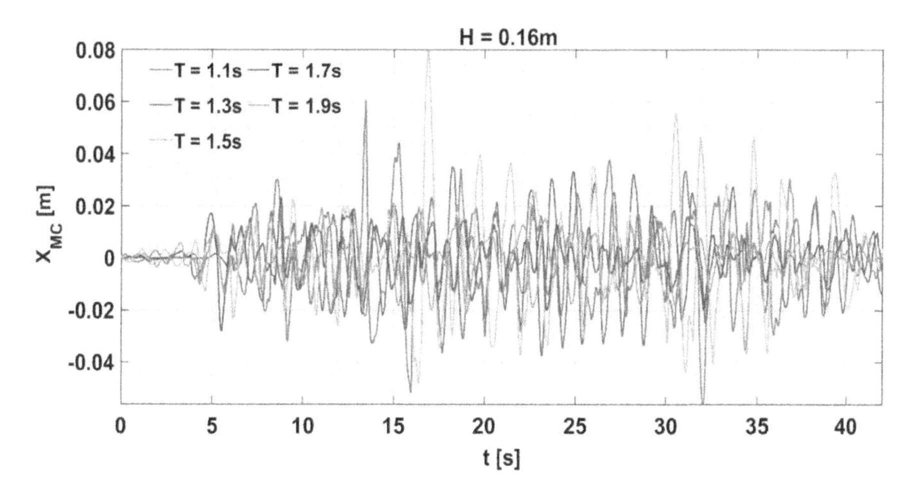

FIGURE 3.61 Displacement of master cylinder under regular waves

from these manuals. One of the analytical methods that follow the DNV-RP-A203 code provisions is to assess the FWEC according to the prevailing industry standards. A MATLAB® program is developed for accommodating the components, their data, and the values of occurrence (OCC), severity (SEV), and detectability (DET). RPN is obtained for each component and then summarized for comparison. For convenience, a graphical format is shown in Table 3.14.

The FMEA is applied at the design stage itself, following the recommendations suggested in the literature (Chandrasekaran, 2015; 2016a). In addition, the manufacturer's manuals and DNV-RP-A203 are used to carry out FMEA and obtain RPN values (Reliability-Analysis-Center, 1968; Anleitner, 2011; Chandrasekaran and Harender, 2012, 2015; Ambühl et al., 2015; Chandrasekaran, 2016a, 2019b;

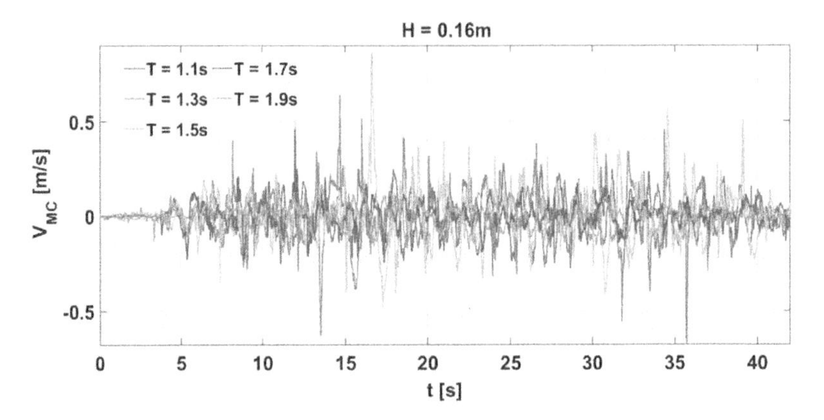

FIGURE 3.62 Velocity of master cylinder under regular waves

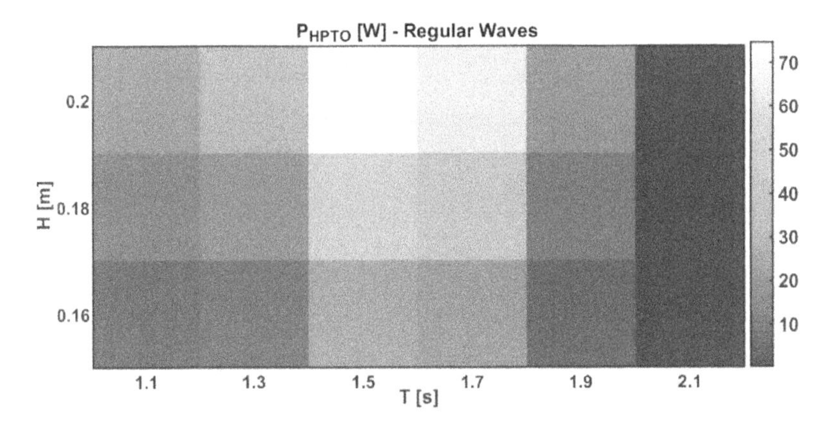

FIGURE 3.63 Average hydraulic power matrix of novel HPTO

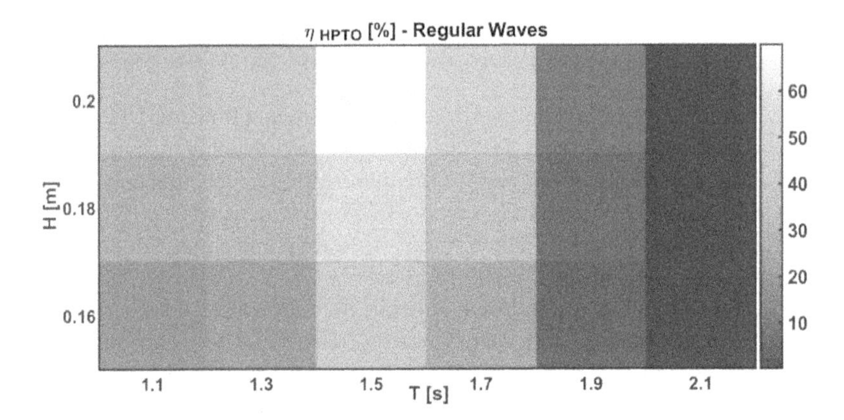

FIGURE 3.64 Average efficiency matrix of novel HPTO

TABLE 3.14
Risk Assessment of FWEC and Its Components

Process Step	Failure Mode	Effect on the User	SEV	Potential Causes	OCC	Process Control	DET	RPN
Check valve (NRV)	Fail to open (no flow)	Loss of hydraulic conversion efficiency: Over pressure on the piston head and the cylinder valves + No increase in the pressure in the hose and the system	10	High cut-off pressure (high spring stiffness) or Clogging of impurities or spring failure: Mechanical failure	4	Measuring pressure difference across the check valve	10	400
Check valve (NRV)	Back flow	Loss of system pressure	8	Jamming of the spring due to mechanical failure	4	Measuring pressure difference across the check valve	10	320
Check valve (NRV)	Leakage	Wear	8	Low spring stiffness (wrong selection): Mechanical failure	4	Measuring pressure difference across the check valve	10	320
Ball and socket joint	Wear and tear	Squeaky, knocking, clunky noise	10	Insufficient lubrication or the wrong lubrication leading to friction (less viscosity)	4	Double check the joints during dry tests and commissioning	7	280
Ball-and-socket joint	Deteriorate	Squeaky, knocking, clunky noise	10	Wrong lubrication or presence of impurities (dirt or water) or leakage of grease	4	Leak proof sealing	7	280

(*Continued*)

TABLE 3.14 (Continued)
Risk Assessment of FWEC and Its Components

Process Step	Failure Mode	Effect on the User	SEV	Potential Causes	OCC	Process Control	DET	RPN
Hydraulic cylinder	Over pressure (piston hitting the cylinder ends)	Tapping sound (exceedance of design stroke)	10	Excess pressure due to extreme wave interactions or Incorrect piston selection/design specifications	6	Pressure transduces and relief valve	3	180
Ball-and-socket joint	Unfastening/slip stud at either on CB or the arms	Ejection of the float(s) from the CB	10	Installation error	3	Proper training of the technician	5	150
Hydraulic cylinder	Leakage	Oil leakage: Sudden or gradual depending on the severity	8	Faulty (worn) piston sealing, rod creep, and over pressure exceeding the design value	3	Replace the sealing's accounting maximum possible excess pressures during design	6	144
Mooring	Partial slackening of tethers	Significant YAW motion is observed	7	Insufficient dead weight on the base frame & Uneven mooring load distribution	4	Extra dead weight on the base frame (total dead weight = 35% of buoyancy)	5	140
Hydraulic cylinder	Chocking	Jamming of piston	8	Air lock (negative pressure): Any leakage through piston seals or hoses or fastenings in the circuit or ports blockage	5	Remove airlock in the dry tests and ensuring filtered hydraulic oil (no impurities)	3	120
Hose	Hose burst	Catastrophic loss of operation	10	Over pressure	4	Install pressure sensors and relief valve	3	120

(Continued)

Hydraulic cylinder	Rod deformation	Jamming of piston	10	Faulty mounting: misalignment	2	Trained technician: Fitting and Inspection	5	100
Mooring	Rupture of mooring lines	Loss of recentering ability (However the system stays afloat but cant be functional)	10	Excess dynamic tension & Inadequate mooring line properties (size, stands etc.) & Material damage (manufacturing defect)	1	Extra dead weight on the base frame (Total dead weight= 3.5% of buoyancy)	8	80
Ball-and-socket joint	Deformation or Fracture	Ejection of float(s) from the CB	10	Excessive rotation (leading to non-uniform load distribution) or accidental impact loads during commissioning and decommissioning	1	Choosing and conducting appropriate design method (ULD, fatigue etc.) and analysis	8	80
Ball-and-socket joint	Jamming	Reduced floats motion thereby overall efficiency of WEC & Temperature rise due to churning of grease	10	Wrong lubricant selection (excess viscosity)	1	Double check the rotation joints during dry tests and commissioning	7	70
Mooring	Pull out of mooring lines (complete loss in pre-tension)	Loss of re-centering ability/station-keeping (however, the system stays afloat but can't be functional)	10	Excess dynamic tension	1	Sufficient ballast and extra mooring clamps at each mooring line	5	50
L-Frame + bracings set	Twisting/warping/ permanent deflection	Misalignment of the hydraulic cylinder, reduced stroke lengths, loss of efficiency	10	Structural error	3	Additional braces/ stiffeners/redesign	2	60

(Continued)

TABLE 3.14 (Continued)
Risk Assessment of FWEC and Its Components

Process Step	Failure Mode	Effect on the User	SEV	Potential Causes	OCC	Process Control	DET	RPN
Hose	Leakage	Pinching	4	Oil drip	7	Rerouting of hoses avoiding pinch points	2	56
Oil tank/ reservoir	High level	Overflow/spill/excess pressure	8	Human error	3	Following filling procedure	2	48
Oil tank / reservoir	Leakage	Loss of head and circuit pressure & Contamination & Air suction	7	Faulty fixing of hose nobes and faulty vent hole	3	Double check the fastenings and wounding sufficient Teflon tapes	2	42
Central buoy	Instability	Undesired oscillations (usually increased motion): Failure of static stability	10	Human error (faulty design)	2	Recheck the design calculations	2	40

Kenny et al., 2017; Coe et al., 2018). RPN values help understand the overall risk of every component by arranging in highest to lowest risk rankings. As observed in Table 3.14, novel HPTO has sub-system assemblies, such as the float-end and the central buoy, tether-anchoring system, hydraulic system, and joints. Each of these sub-systems consists of several components. A total of 10 sub-systems are identified with one or more potential causes of failure. The components with RPN greater than 200 are considered the most critical and are highlighted in Table 3.14.

RPN components in the range (100–200) are categorized as moderately-critical, and those below 100 as least-critical. Further, the recommendations or the monitoring methods are described in detail for all the components. Based on the FMEA, ball-and-socket joints and the NRVs have higher RPN values, requiring close and periodic monitoring. In addition, the categorizations helps to choose components with better quality and functionality, improving the reliability of the FWEC. Failure of the critical components will result in a catastrophic functional failure of the FWEC. It also provides comprehensive information by diagnosing the root causes and it suggests the necessary monitoring actions to enhance the reliability of the FWEC.

EXERCISES

1. List a few wave-energy devices used for harnessing wave energy. Discuss their working principles, in brief.
2. How do you define the capture width ratio in the WEC concept? What is its significance in harnessing wave power?
3. Why are point absorbers more popular in harnessing wave energy?
4. Is it possible to harness wave energy at the resonating condition of the wave energy device? If so, explain the working principle of such devices.
5. Which degree of freedom is harnessed in capturing wave power?
6. Explain the potential of wave power in terms of green energy and compare it with other renewable energy sources/methods.
7. Draw a schematic diagram to explain the working principle of FWEC, as discussed by the author.
8. What are the factors that govern the mechanical design of a power take-off system?
9. What is a hydraulic power take-off system? Explain in detail while comparing with a mechanical PTO.
10. What is the role of WEC-Sim in the design of PTO, as presented by the author?
11. How are bean-shaped floats advantageous?
12. Explain how wave energy is harnessed in both heave and pitch motion using FWEC, as discussed by the author.
13. What are the structural characteristics of the FWEC presented by the author?
14. How will you compute the efficiency of a wave energy device? Explain the input power used to calculate the efficiency.
15. Plot a PM spectrum and state its limitations of use in terms of sea states.
16. What is FMEA? Explain briefly.

17. How is RPN calculated? What are the parameters used to compute RPN? Explain them in detail.
18. In reference to FMEA, how is the risk associated with the mechanical components of a device computed?
19. Name a few other methods useful to compute the risk of mechanical systems. Compare them with FMEA and highlight the advantages of using FMEA.
20. Discuss a summary of different wave energy devices used to harness wave power.

4 Double-Rack Mechanical Wave Energy Converter

4.1 INTRODUCTION

Ocean waves are a power-dense, predictable source of clean and sustainable energy that have not yet been harnessed to a significant extent elsewhere in the world (Dunnet, 2009). The rising price of oil and the growing awareness of the depletion of fossil fuels made the renewable energy resources of nature attractive. Several resources such as solar radiation, wind, surface waves, tide, ocean thermal gradient, ocean currents, and bio-decomposition are examined as possible economic alternatives for a good power supply (Panicker, 1976; Antonio, 2010). In the evolution of renewable energy, wave energy has an emerging and promising prospect. Although wave-energy harvesting technology is relatively new and currently not economically competitive with others such as wind energy, the interest from governments and industries is steadily increasing. The most important feature of sea waves is their high energy density, the highest among comparable renewable energy sources (Alain et al., 2002; Falnes, 2004).

Wave energy is a concentrated form of wind energy produced by the progressive transfer of energy from the wind as it blows over the sea surface (John et al., 1976; Michel, 1982; Panicker, 1976). It has economic, environmental, and technical advantages that set it apart from other renewable energy resources. Several researchers have explored many existing technologies such as Pelamis wave power, Aqua Buoy, and Wave Dragon and found these to be not very successful from a commercial perspective, except for Pelamis, which seems to be promising in the pre-commercial stage. In the case of power requirements of offshore platforms, wave-energy devices attached to the platforms can be useful for deriving auxiliary power. A mechanical wave energy converter (MWEC) is useful to extract energy from waves through the heave motion of a float (buoy). It has many advantages over existing technologies: it is simple and easy to fabricate, economical to produce and maintain, and has a lower failure rate. Mobility and compatibility with offshore platforms add more attraction to the preference of MWEC. Although wave surging gives maximum energy, complexities in wave characteristics in actual sea states impose severe constraints to these devices under operation. Alternatively, other parallel technologies involving turbine application require a uniform high-pressure fluid flow to maintain the conversion at maximum efficiency. In addition, hydraulic and pneumatic power take-off systems require precision machining and have a high maintenance cost. Such devices also suffer from variable tide and other wave conditions. Derived from the advantages of existing technologies, MWEC is primarily designed using a point absorber as a wave-energy capturing device.

DOI: 10.1201/9781003281429-4

To protect the environment for future generations, we must move rapidly to a more sustainable lifestyle, reducing greenhouse gas emissions and consuming limited resources. Offshore wave energy has the potential to be one of the most environmentally benign forms of electricity generation with a minimal visual impact from the shore. The power density of wave energy is higher than wind and solar energies (Chandrasekaran and Harender, 2014; 2015; 2012). Wave energy could play a major part in the world's efforts to combat climate change, potentially displacing one to two billion tons of CO_2 per annum from conventional fossil fuel generating sources (Cruz, 2008). The following are the benefits of wave energy:

- The currents and tides are about 800 times denser than the air that drives wind turbines.
- Wave energy harvesting can shorten the time required for research and development by re-engineering the existing and well-established technologies deployed in the offshore oil and gas, wind power, and shipbuilding industries.
- Wave energy is the most concentrated form of all renewable energies. There are many ideal locations in Europe, North and South America, Africa, the South Pacific Ocean, and Asia, where high power densities exist close to highly populated areas.
- Wave power is predictable and dependable, with the ability to forecast the wave-power spectrum within an acceptable level of accuracy.
- Wave power is environmental-friendly and non-polluting: there is no fuel, no exhaust gases, and no noise. It causes no visual impact.
- Wave power is scalable up to large power stations and it is feasible to connect to power grids.

4.2 MECHANICAL WAVE ENERGY CONVERTER

A preliminary geometric design of a novel MWEC concept is shown in Fig. 4.1 (Chandrasekaran and Harender, 2014). It consists of a floating buoy with rack and pinion arrangement employed for converting reciprocating (vertical) motion into oscillatory (rotary) motion. The floating buoy is connected to the gear rack through the shaft. It is assembled with the vertical gear rack, encircled by various guideposts to protect the floating buoy. Ball bearings support driving sprockets mounted on shafts. The secured pinion gears are subsequently connected to the free-wheel sprocket. The whole assembly is transversely mounted on a shaft supported by ball bearings securing an RPM multiplier connected to the shaft of an electric generator. The overall assembly of MWEC is fixed on a floating vessel or a platform deck that is anchored to the seabed. Under the waves, when the floating buoy moves upwards, the toothed gear rack attached to the buoy rotates the pinion gear (5) clockwise while the other pinion gear (6) rotates anti-clockwise; free-wheel sprocket prevents the interference with rotation of the pinion gear (5). When the floating moves downwards, the toothed gear rack attached to the buoy rotates the pinion gear (5) anti-clockwise while the pinion gear (6) rotates clockwise due to the free-wheel sprocket (17). Power transmission is through pinion gear (5) in the upward motion while it is through the

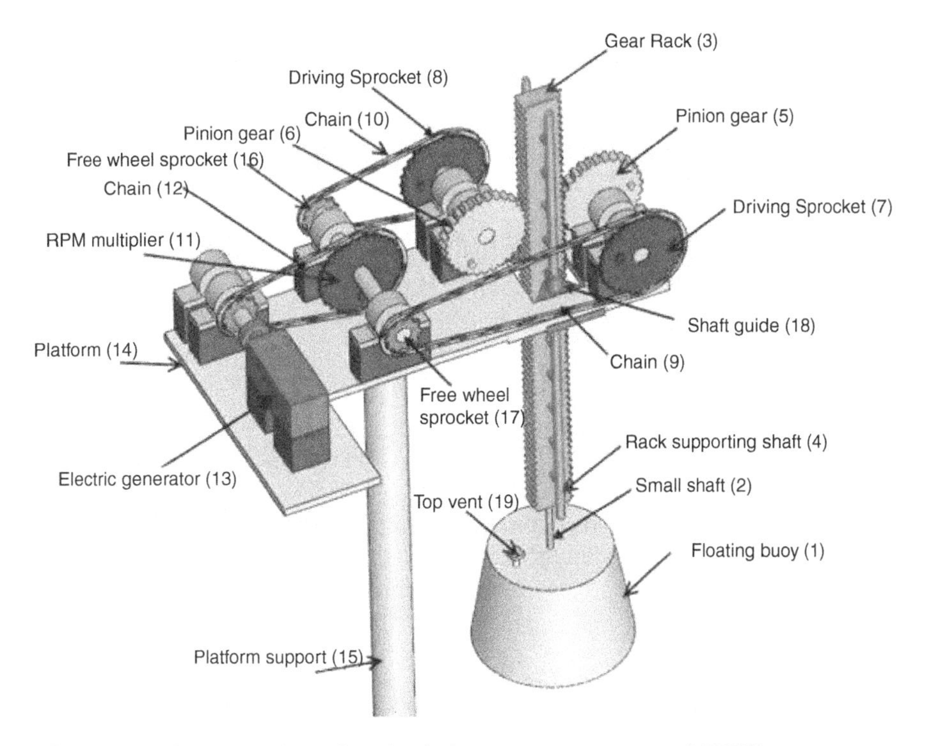

FIGURE 4.1 Conceptual idea of mechanical wave energy converter (MWEC)

pinion gear (6) in the downward motion; to enhance the RPM of the output shaft, an RPM multiplier (11) is also deployed.

4.2.1 Equation of Motion

The floating buoy is restricted to heave motion. For the buoy position defined by a vertical coordinate (x) ($x = 0$ in calm water), the equation of motion is given by the following relationship:

$$[M]\{\ddot{x}\} = \{F_f\} - \{F_{PTO}\} \tag{4.1}$$

where FPTO is the force exerted by the power take-off system, F_f is the force present in the incident wave field, M is the mass of the floating buoy $\ddot{x}$ is heave acceleration. Force in the incident wavefield comprises the excitation force (F_{ext}), which is also the diffraction force, radiation potential force (F_r), and hydrostatic restoring force (F_s). These vectors are given by:

$$\{F_{ext}\} = Re\{X\ e^{i\omega t}\} \tag{4.2}$$

$$\{F_r\} = -[A(\omega)\ddot{x} + B(\omega)\dot{x}] \tag{4.3}$$

$$\{F_s\} = C\{x\} \tag{4.4}$$

Where, $A(\omega)$ is added mass coefficient, $B(\omega)$ is the damping coefficient, $C\,(=\rho gs)$ is the hydrostatic coefficient, and (s) is the waterplane area at the designed draft in still water. After substitution, Eq. (4.1) reduces to the following form:

$$[M + A]\{\ddot{x}\} = \{F_{ext}\} - [B]\{\dot{x}\} - [C]\{x\} + \{F_{PTO}\} \tag{4.5}$$

The mean absorbed power per wave cycle is given by:

$$P_{m=}\overline{F_f(t)}\,\overline{\dot{x}(t)} \tag{4.6}$$

where $\overline{F_f(t)}$ is the time average of the force acting on the buoy and $\overline{\dot{x}(t)}\ \overline{\dot{x}(t)}$ is the time average of the velocity of the buoy. Averaging shall produce non-zero contributions only from those force components in phase with the velocity; contributions from the added mass and hydrostatic forces will be insignificant. The following relationship gives mean hydrodynamic power generated by the body in heave motion:

$$P_m = \frac{1}{8B(\omega)}F_{ext}^{\ 2} - \frac{B(\omega)}{2}\left[U - \frac{1}{2B(\omega)}(F_{ext})\right]^2 \tag{4.7}$$

For a given incident wave field, diffraction exerting force (F_{ext}) and the damping coefficient ($B(\omega)$) shall be determined using WAMIT (Lee and Newmann, 1997). In the absence of an external power take-off mechanism, the oscillation amplitude of the buoy and the complex velocity amplitude are given by the following relationships:

$$\xi = (RAO)\ A_m \tag{4.9}$$

$$U = \xi\omega \tag{4.10}$$

where RAO is the response amplitude operator and A_m is the incident wave amplitude. If the buoy is connected to a power take-off mechanism, its complex velocity amplitude decreases. This decrease is assumed as a percentage of the initial velocity amplitude when the buoy is free from PTO. As a part of the preliminary design, the mean power absorbed by the buoy is calculated for various reduction percentages of the velocity amplitude. For a complex velocity amplitude of 0.5 m/s, which is about 71% of that of the buoy in the absence of PTO, the net power absorbed by the buoy (P_m) is 28 kW for a wave height and period (2.5 m, 7.4 s). The diffraction exerting force (F_{ext}) and damping coefficient ($B(\omega)$) are estimated as 111.41 kN and 2038.27 Ns/m, respectively. The following relationship gives wave power per meter of wavefront:

$$P = \frac{1}{2}\frac{\gamma H^2}{8}C_0\text{kg/s} \tag{4.11}$$

For the incident regular waves considered in the preliminary study, power is computed as 28 kW/m wavefronts at a water depth of 30 m for a wave height and period as (2 m, 9 s), respectively. For a circular buoy of 4.4 m diameter, the total power available (P_t) is about 123.2 kW, which estimates the efficiency of the buoy as 22.7%.

4.2.2 Power Take-Off System Design

Various components of the proposed PTO system are discussed in detail. The gear design (AGMA, 1984; Earle, 1988) is carried out for a pitch line velocity of 0.5 m/s and an overload factor (K_0) of 1.2. Hence, the design power is estimated as 33.6 kW (= 1.2 x 28). The following relationships give the diameter of the pinion and the speed:

$$D_P = N_P m = 20\, x 10 = 200\ mm \tag{4.12a}$$

$$n_P = \frac{60\ u}{2\pi r} = \frac{0.5\ x 60}{2\ \pi 0.1} = 48\ rpm \tag{4.12b}$$

where m is the module (which is taken as 10 mm), N_p is the number of teeth in the input pinion gear (taken as 20), u is the pitch line velocity, and r is the radius of the pinion. Transmitted load is then computed as below:

$$W_t = \frac{1000\ P}{u} = \frac{1000\ x 33.6}{0.5} = 67200\ N \tag{4.13}$$

Recommended face width, (F) is 120 mm (= 12 x m = 12 x 10). Based on the AGMA standards, for the preliminary design, the following design parameters are considered, namely: i) size factor (k_s) as 1.20; ii) load distribution factor (k_m) as 1.26; iii) rim thickness factor (k_b) as unity; iv) dynamic factor for bending strength (k_v) as 1; and v) the geometric factor for pinion (J_p) as 0.33. The following relationship gives the tensile stress number (S_{tp}):

$$S_{tp} = \frac{W_t\ K_0\ K_s\ K_b\ K_m\ \ K_v}{F\ m\ J_p} = \frac{67200\ x 1\ x 1.2\ x 1\ x 1.26\ x 1}{120\ x 10\ x 0.33} = 256.58\ MPa \tag{4.14}$$

Stress value computed from the above equation is found to be well within the reasonable stress level for a carburized steel whose properties are namely: i) AISI number 4320; ii) condition SOQT300; iii) Su = 1500 MPa; iv) percentage elongation is 13; and v) hardness is 429. The elastic coefficient (C_p) and the geometric factor (J_p) are obtained as 191 and 0.33, respectively (Mott, 2004). The contact stress number (S_{cu}) is computed as given below:

$$S_{cu} = C_P \sqrt{\frac{W_t\ K_0\ K_s\ K_m\ \ K_v}{F\ D_p\ J_p}} = 191 \sqrt{\frac{67200\ x 1 x 1.2\ x 1.26\ x 1}{120\ x 200\ x 0.33}} = 683.87\ MPa \tag{4.15}$$

The above value is well within the reasonable limits of the Grade 3 steel. Fig. 4.2 illustrates various terminologies used in the rack and pinion gear drive analysis.

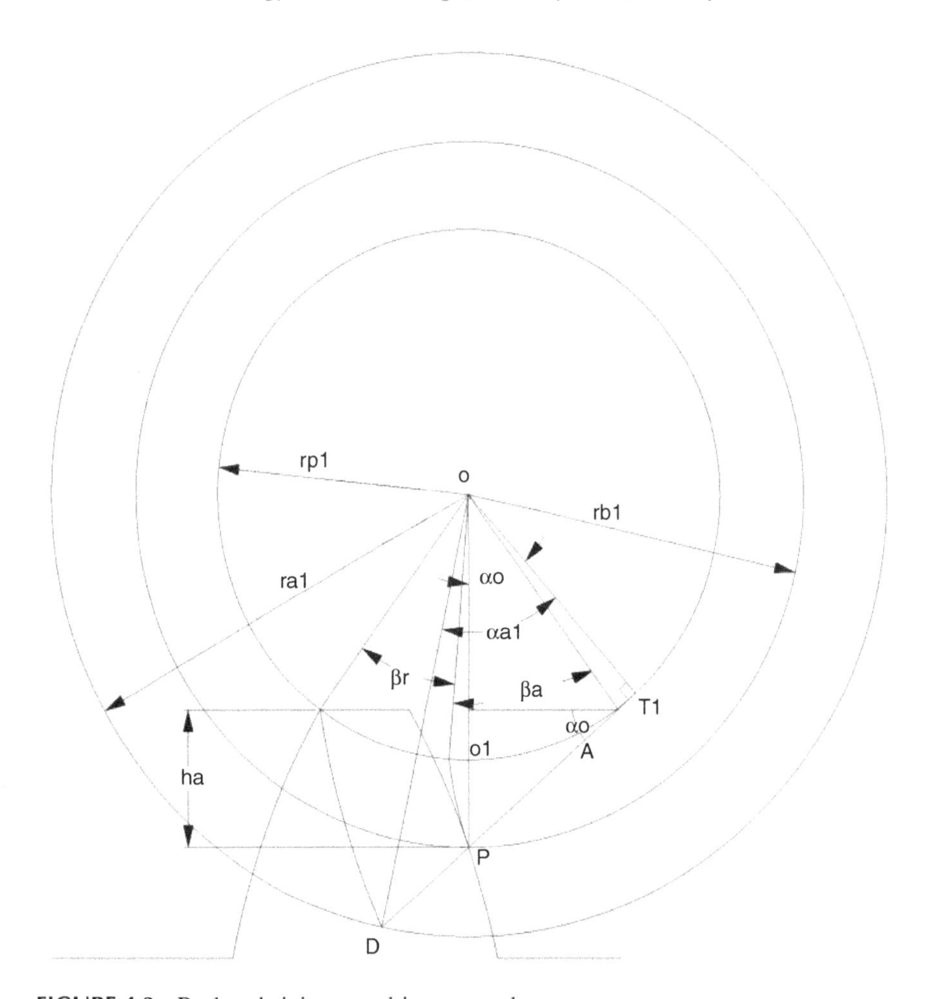

FIGURE 4.2 Rack and pinion gear drive nomenclature

In the above figure, the following terminology is important: (h_a) is the addendum; (r_{b1}) is the base circle radius; (r_{p1}) is the pitch circle radius; (r_{o1}) is the outside radius; (β_a) is the angle of approach; (β_r) is the angle of recess; (α_o) is the pressure angle; (α_{a1}) is the pressure angle at the point of disengagement. The following relationship gives the efficiency for the gear power transmission:

$$\eta = 1 - \left[\frac{\left(1 + \frac{1}{m}\right)}{\left(\beta_a + \beta_r\right)} \right] \frac{f'}{2}\left(\beta_a^2 + \beta_r^2\right) \tag{4.16}$$

From the above figure, the following relationships can be inferred:

$$\sin\left(\alpha_0\right) = \frac{O_1 P}{PA} \tag{4.17a}$$

$$PA = \frac{h_a}{\sin(\alpha_0)} \quad (4.17b)$$

$$\beta_a = \frac{PA}{r_{p1}} = \frac{h_a}{r_{p1}\sin(\alpha_0)} \quad (4.17c)$$

$$\beta_r = \frac{DP}{rp1} \quad (4.17d)$$

$$DP = T_{1D} - T_1P = r_{a1}\sin(\alpha_{a1}) - r_{b1}\sin(\alpha_0) \quad (4.17e)$$

where h_a is an addendum, r_{bl} is base circle radius, r_{pl} is pitch circle radius, r_{al} is the outside radius, m is gear ratio/velocity ratio, and f' is coefficient of friction. Design parameters for the gear considered in the study are, namely: i) h_a = 10 mm; ii) r_{pl} = 87.5 mm; iii) r_{bl} = 100 mm; and iv) r_{al} = 110 mm.

Further, the following relationships also hold:

$$r_{b1}\cos(\alpha_0) = r_{a1}\cos(\alpha_{a1}) \quad (4.18a)$$

$$\cos(\alpha_{a1}) = \frac{r_{b1}}{r_{a1}}\cos(\alpha_0) = \frac{100}{110}\cos(20) \quad (4.18b)$$

$$\alpha_{a1} = 31^0 21' 2'' \quad (4.18c)$$

Other design parameters are derived as follows:

$$\beta_a = \frac{PA}{r_{p1}} = \frac{h_a}{r_{p1}\sin(\alpha_0)} = \frac{10}{87.5\sin(20)} = 0.33 \quad (4.19a)$$

$$DP = ra1\sin(\alpha_{a1}) - rb1\sin(\alpha_0) = 110\sin(31^0 21' 2'') - 100\sin(20) \quad (4.19b)$$

23.03 mm

$$\beta_r = \frac{DP}{rp1} = \frac{23.03}{87.5} = 0.26 \quad (4.19c)$$

Considering the coefficient of friction (f') as 0.02 for pitch line velocity of 0.5 m/s and the gear ratio of rack and pinion as ∞, efficiency of the rack and pinion gear mechanism of the PTO is given by the following relationship:

$$\eta_g = 1 - \left\{\left[\frac{1}{\beta_a + \beta_r}\right]\frac{f'}{2}\left(\beta_a^2 + \beta_r^2\right)\right\} \quad (4.20a)$$

$$\eta_g = 1 - \left\{ \left[\frac{1}{0.33 + 0.26} \right] \frac{0.02}{2} \left(0.33^2 + 0.26^2 \right) \right\} = 99.7\% \quad (4.20b)$$

The chain drive design is based on the power output of the pinion gear (AGMA, 1984; Gustan, 1978). The following factors influence the design of the chain drive: i) power transmitted for the design purpose, which is considered as 33.26 kW (= 33.6 x 0.99); ii) RPM of the driver sprocket (large), N_1 as 48; iii) RPM of the driven sprocket (small), N_2 as 200; iv) transmission ratio as 4.16 (= N_2/N_1). Let the number of teeth on the smaller sprocket (Z_1) and the chain's pitch be 17 and 38.1 mm, respectively. Then, the pitch line velocity (u) is given by the following relationship:

$$u = \frac{38.1 \text{ x} 17 \text{ x} 200}{60 \text{ x} 1000} = 2.16 \text{ m/s} \quad (4.21)$$

The number of teeth on the driver sprocket (Z_2) is computed as 71 (= 17 x 4.16). The design shall further be modified based on the service factor (k_s), which depends on various parameters, namely: i) a factor for the variable load with a mild shock, k_1 (= 1.25); ii) a factor for the distance regulation with the adjustable supports, k_2 (= 1); iii) a factor for the center distance of sprockets, k_3 (= 1); iv) a factor for the position of sprockets with an inclination up to 60°, k_4 (= 1); v) lubrication factor for periodic maintenance, k_5 (= 1.5); and vi) the rating factor, considering a single shift of 8 hours operation, k_6 (= 1). Based on the above parameters, service factor and breaking load (Q) are computed as below:

$$k_s = k_1 k_2 k_3 k_4 k_5 k_6 = 1.25 \, x 1 \, x 1 \, x 1 \, x 1.5 \, x 1 = 1.875 \quad (4.22)$$

$$Q = \frac{102 \, n \, k_s \, P}{u} = \frac{102 \, x 7.8 \, x 1.875 \, x 33.26}{2.16} = 22970.19 \text{ kgf} \quad (4.23)$$

In the above equation, n is the factor of safety, which is taken as 7.8 as per the standard design practices. The chain of configuration ISO/DIN-24B-3TR3825 with pitch (p) of 38.10 mm is assessed to be safe for the above power transmission. For a transmission ratio of 4.16, as in the current design, a minimum center of $1.3a'$ is required, which is given by the following expression:

$$a' = \frac{d_{a1} + d_{a2}}{2} \quad (4.24)$$

where d_{a1} and d_{a2} are the tip diameter of the small and large sprockets, respectively. Based on the power transmission of 33.36 kW, they are 226 and 883 mm, respectively. It amounts to a minimum center distance of 722 mm (= 1.3 x 555). Optimum and maximum center distances are estimated as 40P and 80P, respectively. The efficiency of the chain drive is estimated as 97.3% for the chosen geometric configuration. Power available at the driving sprocket of the unidirectional chain assembly is 27.92 kW (= 28 x 0.997). Power available at the driving sprocket of RPM multiplier is 27.17 kW (= 27.92 x 0.973). Using the above chain drive as RPM multiplier, the available RPM at the generator shaft is 843 RPM. Power available at the input of the generator shaft is 26.43 kW (= 27.17 x 0.973). For efficiency of 95%, the output power of the generator is 25.11 kW. The following relationship gives the overall efficiency of the device:

$$\eta_o = \eta_b \, \eta_g \, \eta_c \, \eta_r \, \eta_{gen} = 0.227 \text{ x} 0.997 \text{ x} 0.973 \text{ x} 0.973 \text{ x} 0.95 = 20.35\% \quad (4.25)$$

It can be compared with the overall efficiency of the actual power transmission, which is computed as below:

$$\eta_o = \frac{\textit{power output from generator}}{\textit{wave power input to the buoy}} = \frac{25.11}{123.2} = 20.38\% \quad (4.26)$$

Efficiencies, computed from the mechanical system design and the overall performance efficiency, as computed from Eq. (4.26), are in close agreement.

The design parameters of various components of the PTO are summarized below:

a) Design parameters of the buoy:
The shape of the buoy is taken as a truncated cone with the following dimensions:
Top circular plate diameter = 4 m
Bottom circular plate diameter = 4.8 m
Total height = 2.2 m
Thickness = 50 mm
Mass of the mild steel buoy = 230 kN
Additional mass due to rack on buoy = 50 kN
The calculated draft is 1.747 m

For wave height and period of (9 s, 3 m) and at a water depth of 30 m, the reaction force, damping force, and velocity are estimated from the numerical model using WAMIT; the corresponding values are 111.41 kN, 2038.27 Ns/m, and 0.5 m/s, respectively. Hence, the net power absorbed by the buoy is estimated as 28 kW. The velocity amplitude of the buoy is assumed under the condition that the buoy is not connected to a PTO.

b) Gear design parameters:
Safe module: 10 mm
Number of teeth in pinion gear: 20 mm
Number of teeth in rack = 300
Thickness of rack = 12 × module = 12 × 10 = 120 mm
Recommended face width: 120 mm
Material: Casehardened, carburized steel, AISI Number – 4320 with condition SOQT300, Grade 3

c) Design of chain drive
The chosen chain ISO/DIN-24B-3TR3825 with pitch 38.10 mm for the power transmission.
Pitch = 38.10 mm
Roller diameter = 25.40 mm
Width between inner plates = 25.50 mm
Pin body diameter =14 .63 mm
Bearing area = 16.64 cm^2
Weight per meter length of chain = 20 kgf
Diameter of the smaller sprocket = 226.68 mm
Diameter of the larger sprocket = 883.35 mm
No of teeth on smaller sprocket = 17
No of teeth on larger sprocket = 71
RPM available at generator shaft = 834 RPM

4.2.3 Experimental Studies

A scaled model of the preliminary device of MWEC is experimentally investigated to assess its performance characteristics under regular waves. A model of 1:8.8 is investigated. Fig. 4.3 shows the scaled model used in the present study.

FIGURE 4.3 Scaled model of mechanical wave energy converter (Chandrasekaran and Harender, 2011)

A floating buoy with a 500 mm diameter and height of 500 mm is housed in the platform's frame. The gear rack is connected to the unidirectional chain assembly, which is connected to the output shaft of the generator. An RPM multiplier is used to enhance the RPM of the device, as shown in the figure. Figs. 4.4 to 4.6 show the response parameters of the device under regular waves; power variation, the influence of wave height, and wave period on the power absorbed by the float are presented, respectively. It is seen from the figure that power absorbed by floating buoy is maximum at 2.5 s wave period for a 30 cm wave height. It corresponds to a wave height and period of (2.64 m, 7.4 s) of the prototype of the device. Power absorbed by the float for different wave heights and wave periods shows that the float absorbs the maximum power for a wave height of 30 cm and 2.5 s wave period; this corresponds

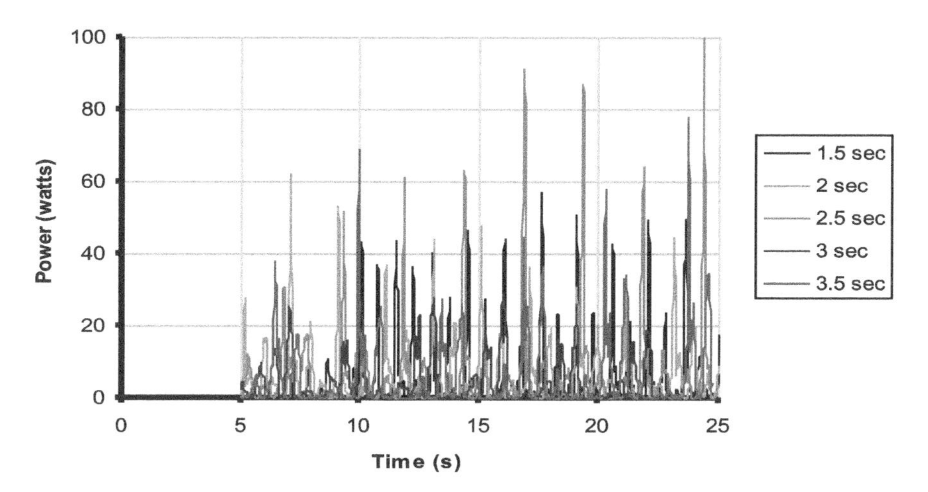

FIGURE 4.4 Power measured for different wave periods in regular waves

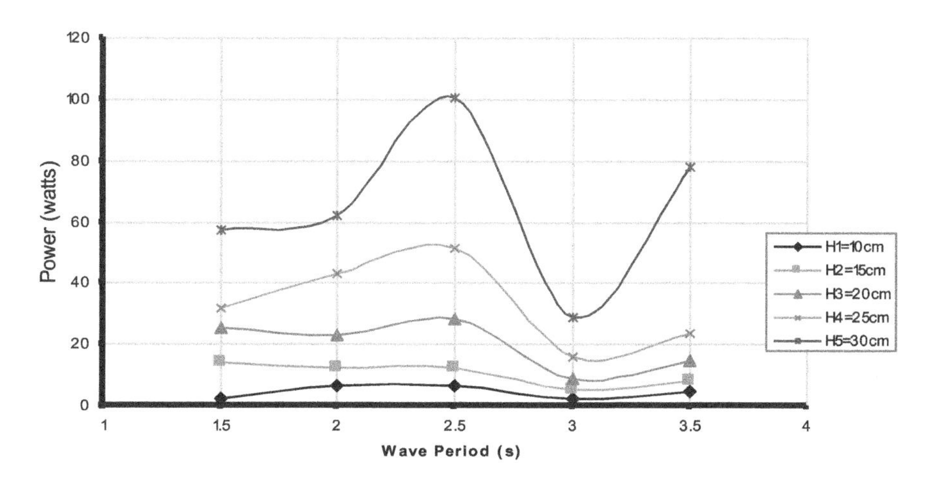

FIGURE 4.5 Power measured for different wave periods in regular waves

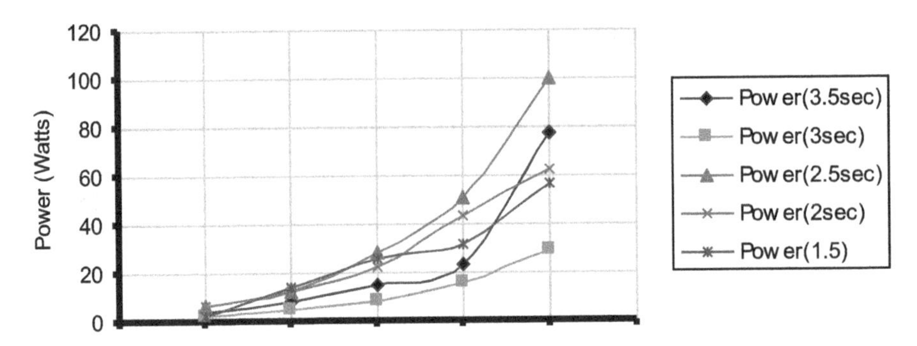

FIGURE 4.6 Influence of wave height on the power absorbed by float

TABLE 4.1
Average Mechanical Power (Watts) Produced by MWEC

Wave Height (cm)/ Wave Period (s)	2.3 s	2.5 s	2.7 s
18	3	9	11
21	13	14	18
24	15	19	23

TABLE 4.2
Mechanical Efficiency of the Device Based on Average Power Produced

Wave Height (cm)	Input Power (W)	2.3 s		2.5 s		2.7 s	
		Output Power (W)	η (%)	Output Power (W)	η (%)	Output Power (W)	η (%)
18	77	3	3.90	9	11.69	11	14.29
21	105	13	12.38	14	13.33	18	17.14
24	137	15	10.95	19	13.87	23	16.79

to 2.64 m and 7.4 s in the prototype, respectively. In particular, the figures show a nonlinear variation in the power absorbed by the floats under the influence of wave period and height; it increases with an increase in the wave height and period. Tables 4.1 and 4.2 show the average mechanical power and mechanical efficiency of the scaled model of the MWEC.

As the overall mechanical efficiency of the MWEC is only about 15%, the preliminary design needs modifications to address the factors that influence the device's performance. A double rack mechanical wave-energy converter (DRMWEC) is subsequently designed and developed in-house, whose details are discussed in the following sections. A few problems associated with the MWEC are addressed as a part of revising the design. A unidirectional chain assembly is replaced by a double-rack mechanism with pinion gears, as the friction losses are expected to be higher in the former case. The RPM multiplier is replaced with 1:3 step-up pairs of gear assembly to overcome the additional friction losses that arise from the RPM multiplier. Vertical guide support replaces the rack supports; rollers attached to the floating buoy enable a smooth reciprocating motion of the buoy along with the vertical guide. A flywheel is also attached to the output shaft to enable continuous power output. The buoy is also stiffened using vertical stiffeners to improve its resistance against lateral loads.

4.3 DOUBLE RACK MECHANICAL WAVE ENERGY CONVERTER

The modified design incorporating major modifications is shown conceptually in Fig. 4.7. The proposed device is modified to a portable design, which can be mounted on a fixed offshore platform to support auxiliary power to the platform. Kindly note that the originally developed device was a stand-alone model mounted on a supporting frame. The proposed device consists of a floating buoy connected with two gear racks through a mini-shaft and a mild steel plate. The floating buoy is supported by guides and a roller mechanism to protect from surge and other lateral forces. The gear racks (1,2) are messed with pinion gears on either side to enable smooth transmissibility. The output shaft of the rack and pinion mechanism is connected to an electrical generator through a step-up gearbox (1:3) and a flywheel. The proposed device is new and

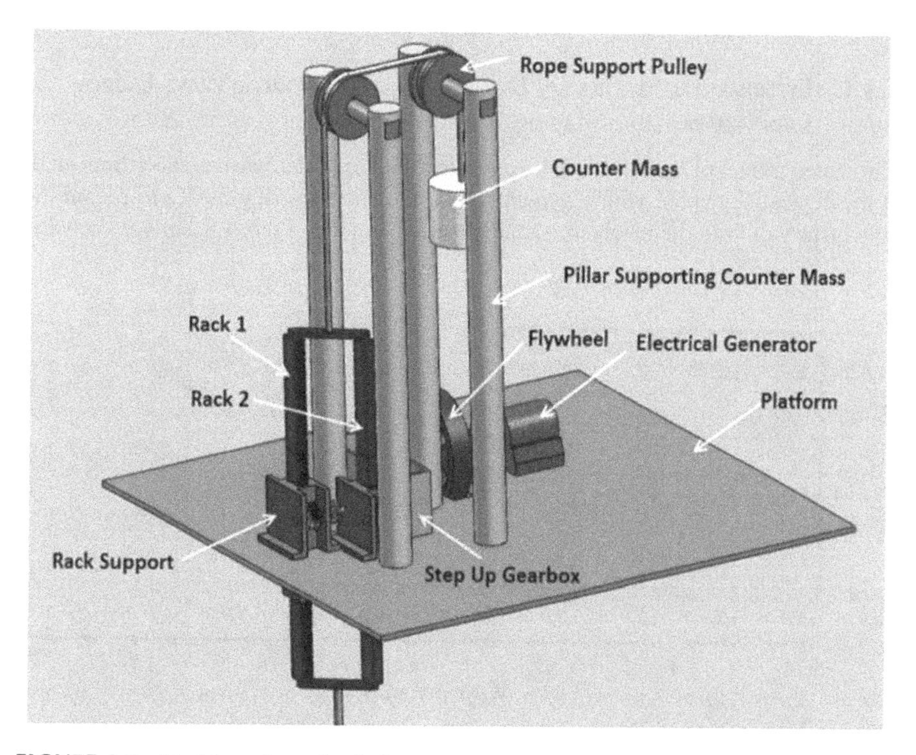

FIGURE 4.7 Double-rack mechanical wave energy converter: concept

innovative that can be housed in any existing fixed platform with little modifications. The wave-energy conversion takes place in four different stages: The first stage is the heave motion of the floating buoy in the vertical plane under the incident waves; the second stage converts the reciprocating motion of floating buoy to alternative rotation by a rack and pinion arrangement; in the third stage, alternative rotation is converted into a continuous unidirectional rotation using one way bearing inside pinion gears; the last stage of conversion takes care of converting the unidirectional rotation into any usable form of energy like mechanical or electrical energy.

When the approaching wave moves the floating buoy upward, the toothed gear rack (1) attached to the buoy rotates the pinion gear (1) in the clockwise direction. The gear rack (2) also rotates the pinion gear (2) in the same direction under the presence of a free-wheel, which is inbuilt in the bearing. It, therefore, does not counteract or interfere with the rotation of pinion gear (1). On the other hand, when the wave lowers the floating buoy, the toothed gear rack (2) attached to the buoy rotates the pinion gear (2) in the anti-clockwise direction, while the gear rack (1) remains idle. In simple terms, in the upward motion, power transmission is achieved through the pinion gear (1), while in the downward motion, it is through pinion gear (2). Thus, the power transfer is effective in both the upward and downward motion of the buoy. The main advantages of the revised design are simplicity in the design and expected improvement in mechanical efficiency. Salient advantages of the revised design are: simple to fabricate, install and maintain, failure probabilities are lesser as it deploys proven and time-tested technologies, high mobility, and compatibility.

4.3.1 Experimental Studies on Double Rack Mechanical Wave-Energy Converters

The scaled model of the device is investigated under regular waves. The output shaft of the device is coupled with a dynamo and a torque measuring device to record the time history of both the mechanical and electrical outputs. Figs. 4.8 and 4.9 show the

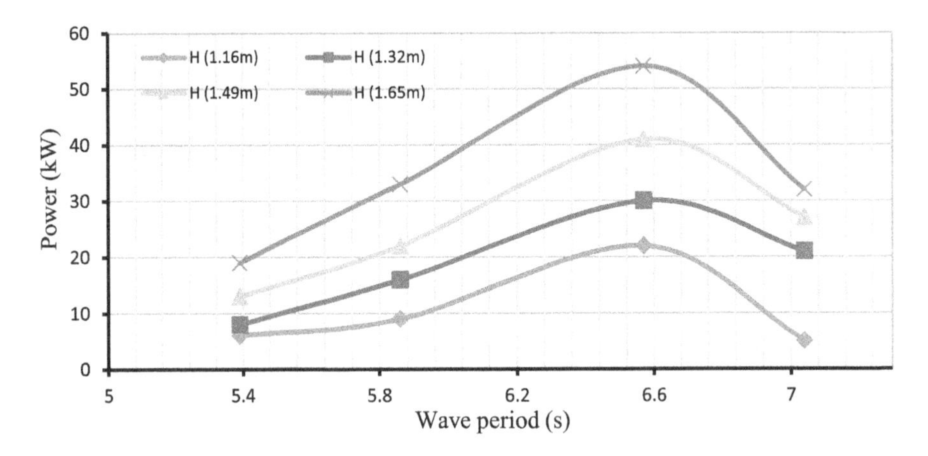

FIGURE 4.8 Influence of wave period on power generated by DRMWEC

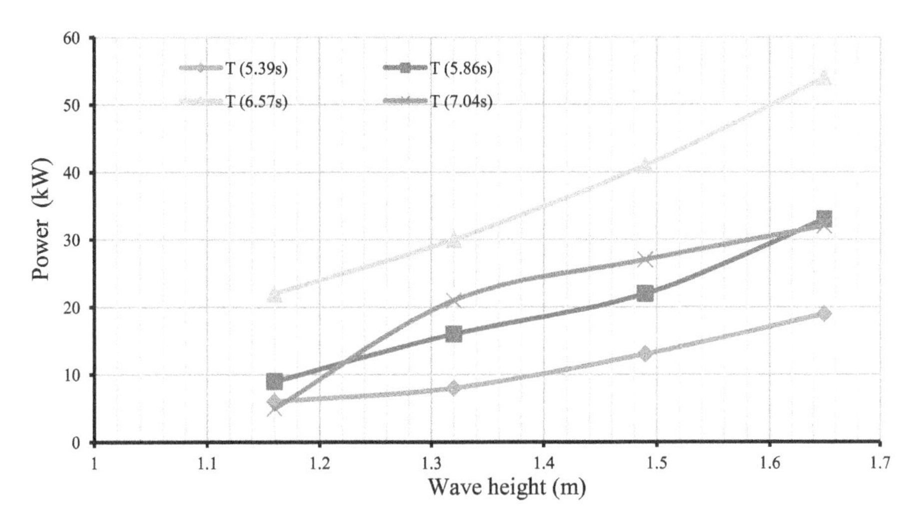

FIGURE 4.9 Influence of wave height on power generated by DRMWEC

influence of wave period and height on power generation. It is seen from the figures that the power generated is maximum when the wave height is 1.65 m and the wave period 6.57 s for the prototype. A significant increase in the power generated with respect to the wave height is also observed. This characteristic behavior, as seen in MWEC, is maintained in the revised design. At the same time, improvement in the efficiency and the operational band of sea states are made wider through modifications in the design (use of flywheel is one of them).

The average mechanical power produced by DRMWEC for an 800 mm float is summarized in Table 4.3. The overall mechanical efficiency of the device, computed based on average mechanical power produced by the device, is shown in Table 4.4. In contrast, Table 4.5 shows the device's efficiency at the resonance period based on the average power produced. It is seen from the tables that the mechanical efficiency increases with the increase in the wave height, but beyond a threshold limit on the wave height, there is a marginal decrease in the efficiency. However, it can be easily seen that the revised design (DRMWEC) showed a significant improvement in mechanical efficiency (MWEC showed a maximum efficiency of only 17%). The desired improvement and smooth output are achieved by deploying a flywheel to the output shaft of the device. Due to the phase lag encountered by the float and non-instantaneous reaction of the buoy to the encountered waves, reserve energy stored in the flywheels is utilized to ensure a continuous mechanical and electric output at the device's shaft.

Further increase in the generation of output power is seen at 30 cm wave height when the device operates at resonance band with that of the encountered wave period (2.8 s), resulting in the overall mechanical efficiency of about 25%; continuous operation of the device at this band of wave period is not desirable, and therefore the device shall be ceased to operate at such sea states which correspond to 6.57 s for a prototype. The summary of the results can be seen in Table 4.5.

TABLE 4.3
Average Mechanical Power Produced (in Watts)

H(cm)\T(s)	2	2.3	2.5	2.8	3
21	30	27	22	34	27
24	34	28	33	39	34
27	41	35	38	46	39
30	48	39	45	58	42

TABLE 4.4
Mechanical Efficiency Based on Average Power Produced

Wave Height (cm)	Period: 2.5 s			Period: 2.7 s		
	Output Power (W)	Input Power (W)	η (%)	Output Power (W)	Input Power (W)	η (%)
21	22	135	16.30	34	135	25.19
24	33	177	18.64	39	177	22.03
27	38	224	16.96	46	224	20.54
30	45	257	17.51	58	257	22.57

TABLE 4.5
Mechanical Efficiency at Resonance Period Based on Average Power Produced

Wave Height (cm)	Output Power (W)	Input Power (W)	η (%)
21	34	135	25.19
24	39	177	22.03
27	46	224	20.54
30	58	257	22.57

Power generated by the device is computed for the appropriate prototype using Froude's scaling laws. A comparison of the mechanical power produced by the DRMWEC is shown in Table 4.6. It is seen from the tables that the operational sea state (1.65 m, 6.57 s) are safe sea states at which the device shall perform to the desired efficiency. The average power produced by an individual unit of the wave energy device shall be about 23 kW, which is quite reasonable.

4.4 FAILURE ASSESSMENT

There is always an increasing demand to assess the newly developed mechanical devices for their reliability, performance, and maintainability. Fault analysis can

TABLE 4.6
Comparison of Average Mechanical Power

Model (Watts)					Prototype (kW)				
H(cm)*T*(s)	2.3	2.5	2.8	3	*H*(cm)*T*(s)	5.39	5.86	6.57	7.04
21	27	22	34	27	1.16	11	9	13	11
24	28	33	39	34	1.32	12	13	15	13
27	35	38	46	39	1.49	14	15	18	15
30	39	45	58	42	1.65	15	18	23	16

provide sufficient information about the occurrence of a fault, its quick and accurate detection. A fault occurs in a mechanical system before it leads to failure. 'Fault' refers to an undesirable deviation from the design intent, which suggests that the performance characteristics of at least one of the mechanical system components should be unacceptable. 'Cause' refers to the source of failure, which initiates the failure and is responsible for the system's performance degradation. 'Symptom' refers to the actual deviation that occurred in the system concerning the intended performance of the system. It is important to note the mechanical system may continue to perform its intended function under the presence of any symptom but with a deviation from its intended function. 'Failure' refers to permanent damage caused to the system, resulting in an inability to perform the intended function of the system (Chandrasekaran, 2016a). Replacing the failure component shall ensure the smooth functioning of the mechanical system. Please note that the fault analysis is a component-level analysis useful for identifying or detecting the probable component-level failure in advance. It shall aid preventive maintenance and is helpful in avoiding a shutdown of the system.

A successful design of any system is testified after rigorous use, but this is a post-validation of the product. Many mechanical systems are recalled in the current era due to poorly designed products (see, for example, recall invitations of automobile products in the passenger-car manufacturing sector). These failures are referred to as incapability of the manufacturers and designers; in turn, this affects future investment in the technology transfer of the product to large-scale manufacturing. Failure mode and effects analysis (FMEA), is a methodology that allows the research organizations to anticipate failure during the design stage itself. It helps identify possible major failures in a design or manufacturing process. Developed in the 1950s, FMEA is one of the earliest structured reliability-improvement methods, which is still a highly effective method of reducing the possibility of failure. FMEA is a structured approach to identify the potential failures within the design of a product or a process. Failure modes are how a product can fail; 'failure' refers to non-functioning the components in an intended manner and does not refer to a complete failure. Effects refer to the consequences of these failures, which may cause serious defects in the product's intended functioning or affect its efficiency. FMEA helps identify, prioritize and limit such failure modes; it enhances the knowledge of the FEED team to review the design by assessing its risk.

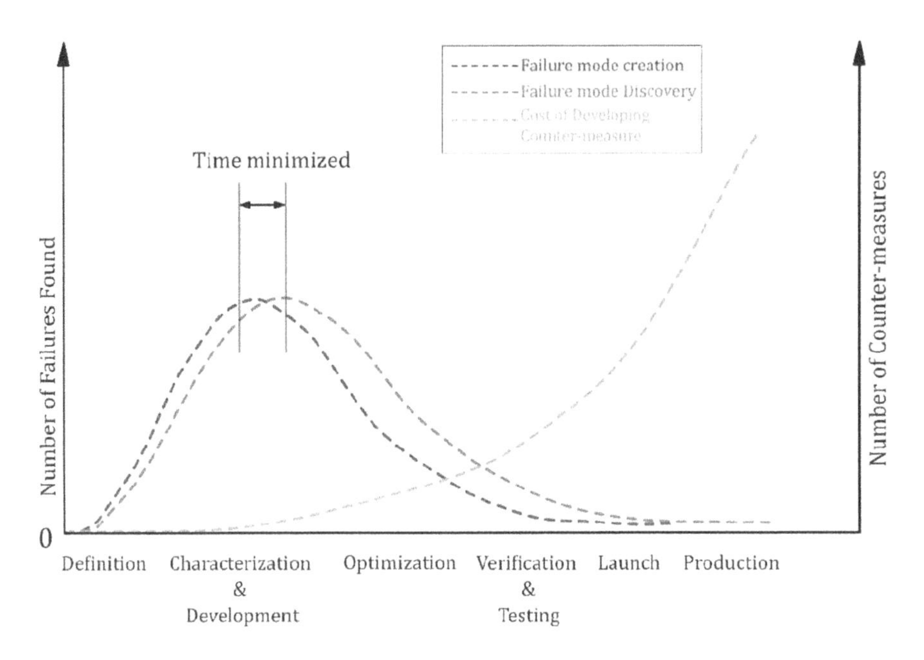

FIGURE 4.10 Earlier failure mode

There are two categories of FMEA, namely design FMEA and process FMEA. Design FMEA addresses the issues related to the products, namely malfunctioning, reduced service life, and unintended outcome. They are focused on addressing concerns arising from material characteristics, geometric design and tolerances, scantling details, and interaction of components. Process FMEA is focused on exploring failure related to the quality of outcome, defective process, and possible environmental hazards that arise from the defective process. They pay more attention to human factors, processing methods and techniques, machines used, and other environmental factors connected with the respective process. Figs. 4.10 and 4.11 show the comparison of detecting the failure modes in the earlier design stage with that of the later stage. A greater delay in identifying the failure modes shall increase the number of failures and the cost of the end-product development.

FMEA can be performed on several occasions; namely, i) during the development of a new product; ii) while planning for a deviation in the process; iii) to improve the quality of an existing product; and iv) to improve the failures in an existing process. In addition, FMEA is also carried out regularly to improve the performance of a product based on its performance under the actual service conditions. For example, revised versions and variants in automobile models are classic examples of design FMEA.

4.4.1 Conducting Design Failure Mode and Effect Analysis

FMEA is performed in seven steps, as discussed below. Each step has a specific objective, and hence team members shall also be chosen to fulfill the respective objectives.

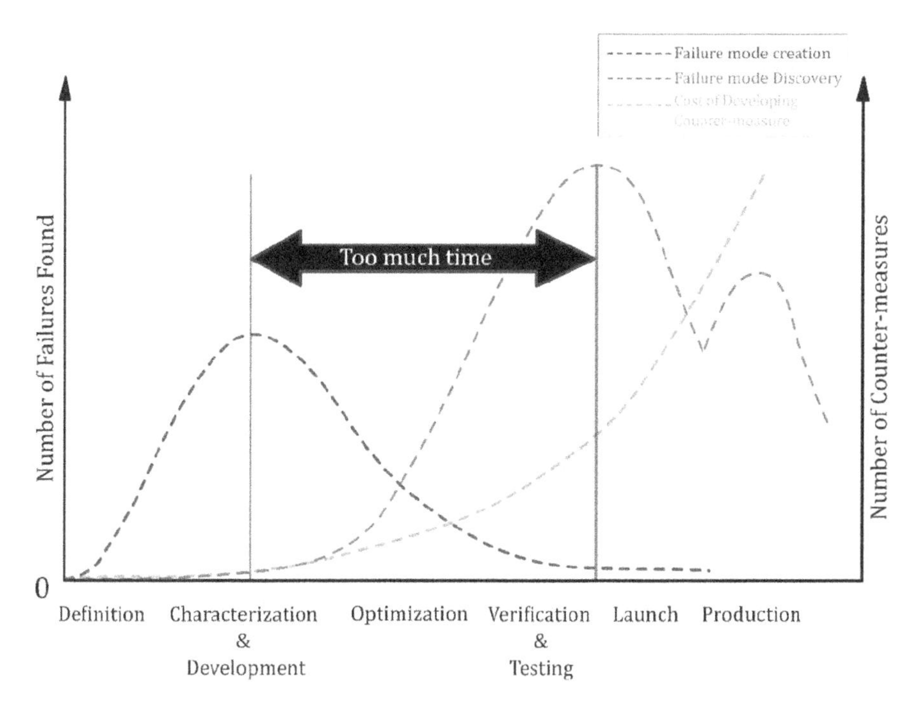

FIGURE 4.11 Late failure mode

- FMEA pre-work and choosing the FMEA Team
- Development through severity ranking
- Development through occurrence ranking and subsequently identifying the potential causes and prevention controls.
- Development through detection ranking and subsequently establishing the test methods and detection controls.
- Action priority and assignment
- Design review
- Re-ranking the risk priority number (RPN)

The pre-work team collects (or prepares, as the case may be) all the key documents about the product. It can be based on the investigations reports of the past failures of the same or similar products. These documents include the following, namely: i) block diagram (explaining the product manufacturing layout); ii) parameter diagram (explaining various control parameters in the production/manufacturing perspective); and iii) critical failure modes and their avoidance. One of the effective ways of performing this task is through a pre-work checklist. This should include design assumptions, bill of materials, potential causes from component interfaces, design alternatives, and other environmental factors controlling production and cost.

Development through severity ranking consists of identifying the failure functions, modes of failure, and their order of severity. They will be governed by a few factors, namely material specifications and the specific characteristics that control the design/

manufacturing, government regulations that govern the design and product development (for example, use of non-biodegradable materials should be avoided), total and partial function failure, unintended function failures and their consequences on the severity ranking, actions required to control those items with high severity ranking (greater than nine on a scale of 10). Development through occurrence ranking subsequently identifies the potential causes and prevention controls. Causes for a potential failure are identified based on the experience of a similar design/product development; it can also be performed based on the post-diagnosis of a failure report. It is equally important not to oversight the prevention controls already exist in the design layout. Effective use of the existing control mechanisms shall lead to a workable report. Actions are needed to be identified to address the high-risk severity and occurrence combinations. It can be done by preparing a criticality matrix.

Development through detection ranking helps establish test methods and detection controls. The major contribution is towards developing additional control methods while the effective use of the existing controls is reviewed in the earlier step. It is focused on identifying the detection controls and their ranking. If the identified risks in the earlier steps are significant, then specific actions are recommended to address them; more focus is placed on the prevalence of testing methods and quality assurance techniques. The outcome of this step is the design verification plan (DVP), which is one of the key factors in a successful design FMEA. Actions identified in the earlier steps are revisited and assigned with a risk priority number (RPN). The RPN is the product of severity, occurrence, and detection ranking, whose maximum value is 1000. The RPN is calculated for each potential failure, cause, and control combination. It is important to note that there are no standard guidelines to assign relevant numbers to inputs, namely severity, occurrence, and detection ranking; the maximum number for each category is 10, however the numeric value should be assigned based on experience. Hence, after conducting a thorough inspection and studying the design method, a well-trained team should suggest these values as they affect the overall picture of FEMA. Under this step, the team should also recommend RPN for additional follow up, if any. Actions should be assigned to due review dates, closed upon the proper action taken against each recommendation. As the main objective of design FMEA is to identify and mitigate risk, FMEAs that do not find risk are considered weak and do not add value.

Design review is an important step, which closely monitors the actions recommended under various steps and their incorporation into the modified design. A successful design review will repeat all the earlier steps, as discussed. After successful confirmation of risk mitigation actions, the core team shall re-rank the appropriate ranking value (severity, occurrence, or detection) to obtain an RPN for each mode of failure. The original RPN is compared to the revised versions, and their relative importance to the design revision is confirmed in this last step of design FMEA.

4.5 FAILURE MODE AND EFFECT ANALYSIS OF MECHANICAL WAVE-ENERGY CONVERTER

Design FMEA is most effective when applied before the product design is released; the focus is on failure prevention and not detection. This procedure examines different components' functions and sub-systems and identifies incorrect material choices, inappropriate specifications, etc. FMEA worksheet helps to characterize the components that are at the experimental stage critically. Rated on a scale of 10, 'severity' is a rating of the effect of potential failure mode; 'occurrence' corresponds to the rate at which a first level cause and its resultant failure mode occurs over the design life of the device. Control measures employed to detect the failure mode even before the product is released for design are assessed as detection on a 10-point scale. The risk priority number (RPN) identifies the greatest areas of concern. There are no absolute values for a high RPN; FMEA is often viewed relatively (the highest RPN is addressed first). Detailed failure mode and effect analysis (FMEA) on the proposed mechanical wave energy converter is seen in Table 4.7. Failure analysis indicates the advantageous features of the proposed device. Design FMEA, carried out on the device, shows that the floating buoy is the most vulnerable with the highest RPN. Recommended action mostly focuses on rigorous testing because the product is in the development stage.

4.6 DEEP-OCEAN WAVE ENERGY CONVERTER: CONCEPTUAL DESIGN

Deep water offshore structures have access to very powerful ocean waves under their location and site conditions. Should the energy possessed by these waves be harnessed, it can be one of the popular green energy systems. Suppose suitable devices are developed to capture offshore wave energy in deep-water locations. In that case, it can supply power to partially or fully meet the energy demand of these offshore structures and thereby reduce operational costs. An attempt is made to design and develop a novel device fitted on an offshore semi-submersible platform and produce electricity to meet their operational energy demands (Chandrasekaran et al., 2014c; Chandrasekaran and Raghavi, 2015). The concept is to harness wave energy through the heave motion of the buoy, which is subsequently converted to mechanical work by deploying a hydraulic cylinder and motor.

4.6.1 Geometric Design of the Device

Experimental investigations are conducted to study a cylindrical wave float's motion response and force characteristics at different drafts. The influence of the change of float shape is also studied by attaching a fin to the cylinder. For coupling the wave-energy device to an offshore platform, the desired qualities of the device are, namely: i) the device should generate enough power to meet air-conditioning and lighting requirements of the platform; ii) floor space on any offshore platform is valuable and hence the device must be compact in shape and size; iii) motions transferred to the platform must be minimal, and within permissible limits; iv) the system should be reliable, demanding less maintenance and inspection schedules as offshore repairs and labor are expensive; and v) it should be cheap enough to be feasible.

TABLE 4.7
FMEA for the Proposed Device

Part or process name: Mechanical wave energy converter | **Supplies and plant affected: Experimental stage**

Design responsibility: XXX company | **Model date: MM/DD/YYY**

Other areas involved: Power generation from ocean wave energy | **Engineering change level: Not applicable**

Component	Function or Process	Failure Mode	Effects	Sev(S)	Oce(O)	Cause of Failure: Potential Reasons	Occ	Controls	Dt	(SxO)	RPN	Recommended Action
Buoy	Heave motion	Does not give desired displacement & force	Power at low voltage or No power generation	4	4	Faulty design or less wave energy	3	Check the buoy design & wave energy properly	7	16	112	Rigorous testing in lab as well as in wave basin required
Pinion Gears	Convert linear to rotary	Broken teeth	No power generated	5	3	Manufacturing faults or not messed properly with rack	3	Check the design	6	15	90	Rigorous testing in lab
Anti-friction bearings	Help in smooth working	Damaged ball	Efficiency reduced	3	5	Manufacturing faults	3	Check design properly	5	15	75	Rigorous testing in lab
Sprockets	Transfer power	Damaged teeth	Efficiency reduced by 50%	4	3	Poor material selection	3	Check the quality of material properly	6	12	72	Select material acc to power to be transmitted

Chains	Power transmission from pinion to generator shaft	Broken link	Efficiency reduced by 50%	2	2	Manufacturing faults or lubrication not proper	2	Apply lubrication properly	5	4	20	Check up lubrication pump
Free whell sprocket	Transfers power	Broken link	Efficiency reduced by 50%	3	2	Manufacturing faults	2	Check design properly	6	6	36	Proper inspection required
Electrical generator	Power generation	Faulty wiring	Less power or no power	2	2	Manufacturing faults	2	Check wirings properly	6	4	24	Proper inspection required

Sev: Severity; Occ: Occurrence; Dt: Detection; RPN: Risk Priority Number; (Severity x occurrence x detection)

4.7 WORKING PRINCIPLE OF DEEP OCEAN WAVE ENERGY CONVERTERS

The proposed wave-energy converter is a point absorber type, as shown in Fig. 4.12. Every unit consists of a buoy fitted to a link, moving in a vertical arc about a pivot. A hydraulic cylinder is attached to the other end of the link. Under the action of the waves, the buoy's motion activate the link, which drives the hydraulic cylinder. Hydraulic PTO is used to convert buoy motions to rotary motions. The rotary output is coupled to a generator to generate electricity, subsequently used to supply auxiliary power to run the air-conditioning system and lights. The excess energy is stored in batteries for use during periods of low wave energy. A total of 12 devices will be fitted on three sides of the platform, as shown in Fig. 4.13. The fourth side is left free for crawler deployment and offloading operations. Each buoy has a double-acting hydraulic cylinder attached to it. The compressed hydraulic fluid will be stored in a single central hydraulic accumulator, driving a single hydraulic motor-generator set. The main components of the proposed device are the buoy, mechanical linkages, and the power take-off system.

The hydraulic power take-off system consists of a piston, a hydraulic pump, and a hydraulic motor, as shown in Fig. 4.14. The linear wave motion moves the piston up and down, which pumps pressurized hydraulic fluid through the hydraulic pump. The pump then feeds the hydraulic motor, which initiates a rotary motion to drive an electric generator. By coupling the hydraulic motor to a generator, the conversion process is complete.

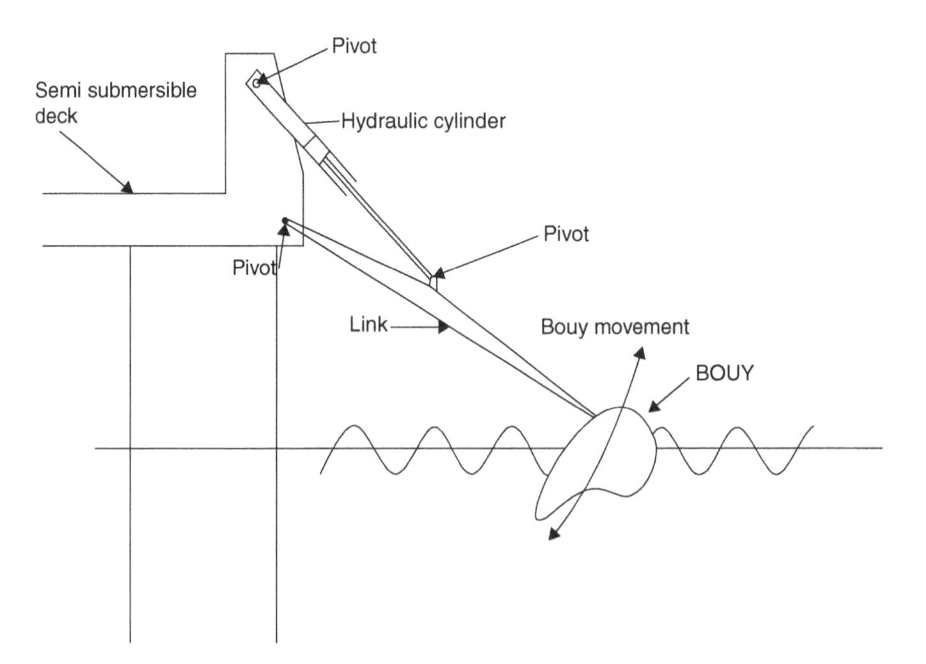

FIGURE 4.12 Conceptual design of deep ocean wave energy converter (Chandrasekaran et al., 2014c)

FIGURE 4.13 Device mounted to an offshore platform: concept (Chandrasekaran et al., 2014c)

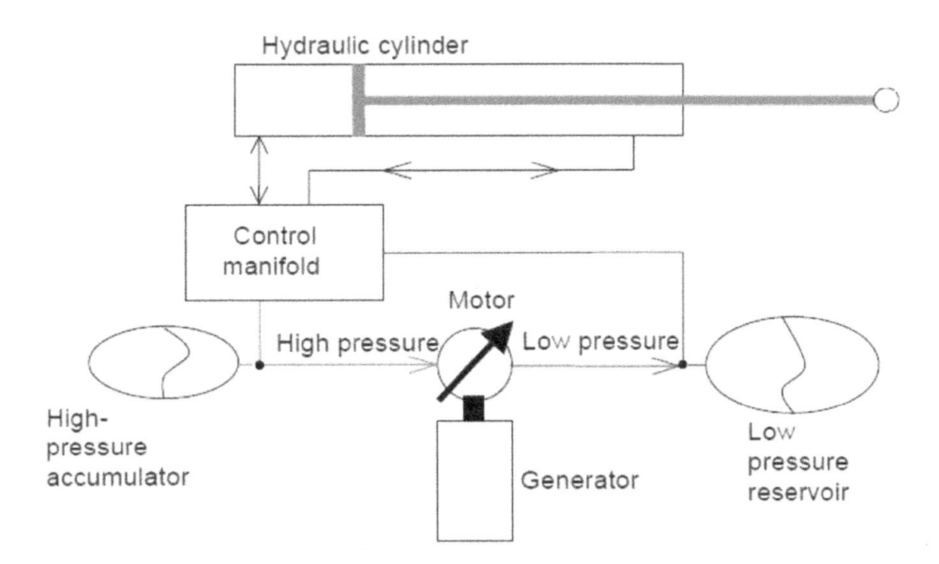

FIGURE 4.14 Hydraulic power take-off system

Waves apply large forces slowly, which is compatible with the hydraulic systems ability to absorb energy. Hydraulic systems are well known for their versatility and use as opposed to direct drives. Further, hydraulic systems are less expensive to design and build than direct drives. Hydraulic systems possess many favorable

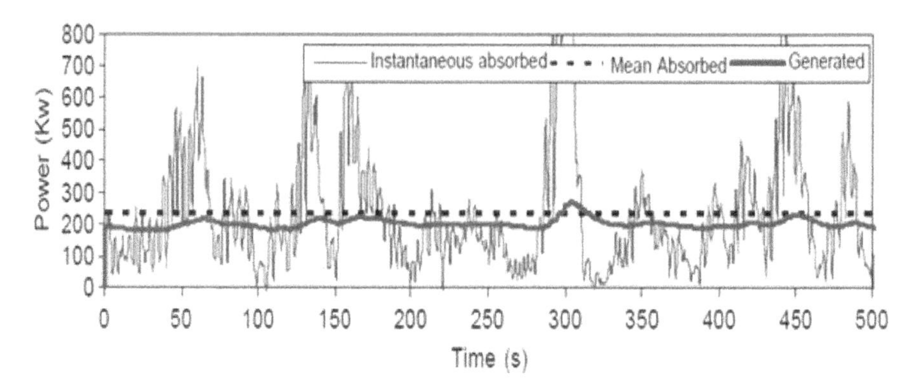

FIGURE 4.15 Typical output of the hydraulic system

characteristics which make them a favorite choice for wave-energy converters. It is important to note that the hydraulic power take-off method is mechanically inefficient as the conversion process is indirect. Losses that occur during pumping and driving the hydraulic motor invoke losses in addition to those that arise from the generator and inverter; more moving parts of a hydraulic system are also a cause of concern. Issues related to maintenance of moving components, breaking a seal or a valve, causing leaking are some of the main concerns.

Further, the hydraulic systems are designed to work at a speeds lower than those experienced by a wave-energy converter; typically, high-pressure gas accumulators can be deployed to achieve short-term energy storage and smooth electricity production. As shown in Fig. 4.15, a plot sample emphasizes the necessity to achieve smooth electricity production using high-pressure gas accumulators.

4.8 EXPERIMENTAL INVESTIGATIONS ON DEEP OCEAN WAVE ENERGY CONVERTERS

Experimental studies are carried out on a scaled model of the deep ocean wave energy converter. Since hydraulic components are not readily available to cater to the power output of the scaled model, no power take-off system is installed. Instead, experiments are conducted to determine the force acting on the buoy under the fixed position and free to move about the pivot. The model, shown in Fig. 4.16, consists of a platform mounted on the rails of the wave flume, a movable frame, and two buoys. A shaft on the movable frame ran through the pillow bearings fitted on the platform. The shaft is free to rotate inside the bearing. The experimental matrix focuses on testing the buoys of two geometric shapes, namely a cylinder, without and with a fin, as shown in the figure; buoys are 1.5 m long and 0.2 m diameter, and the fin length is 0.2 m.

Pivoted at its center, the buoy is free to move about an arc, whose displacement time history is measured using an inclinometer fitted on the movable frame. The displacement of the buoy is calculated from the angular displacement and the arc radius. Fig 4.17 shows a typical time history of the cylindrical buoy. Strain gauges are mounted on the movable frame in a half-bridge configuration such that bending

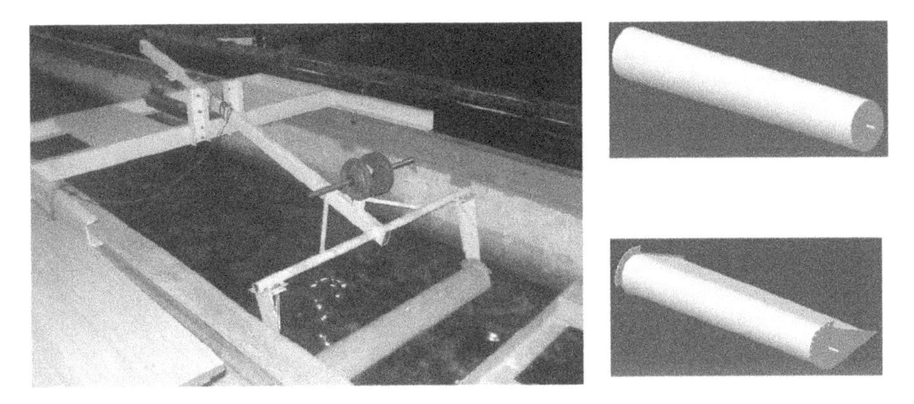

FIGURE 4.16 Scaled model of the deep ocean wave energy converter

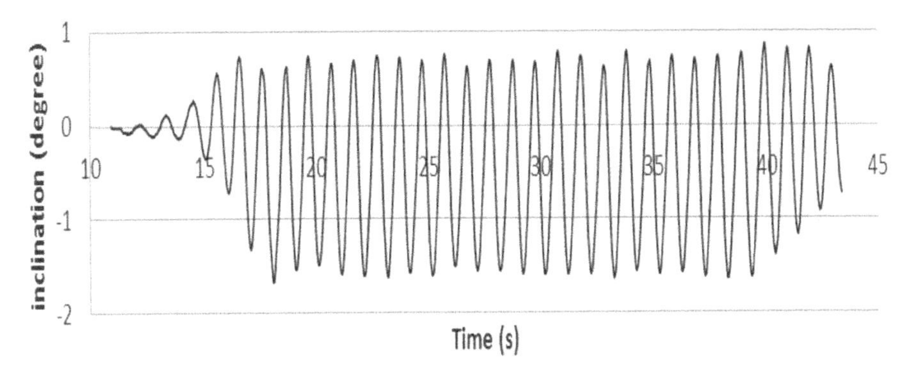

FIGURE 4.17 Inclination of the cylindrical buoy (0.1 m, 1 s)

forces are measured while eliminating the axial forces; bending forces are of interest as only the force perpendicular to the frame will be transmitted to the PTO. Fig. 4.18 shows the test setup used to measure the forces, while Fig. 4.19 shows the strain time history.

The following relationship gives the power produced by the device:

$$P = F\,v \tag{4.27}$$

Where P is power, F is force, and v is velocity. When the buoy is free to move, its displacement is the highest, but the force acting on it is enough to overcome the inertia and the bearing friction; hence usable power is almost negligible. However, when the buoy is fixed, the force acting on it becomes the maximum, but power is negligible as there is no motion. Hence, the maximum power is captured when PTO offers the maximum resistance such that the product of force and displacement becomes a maximum. The value of force and displacement at the maximum power condition cannot be determined correctly with the current experimental setup. However, the force and displacement at the condition of maximum efficiency will be proportional to the

FIGURE 4.18 Test setup to measure force in the rigid link

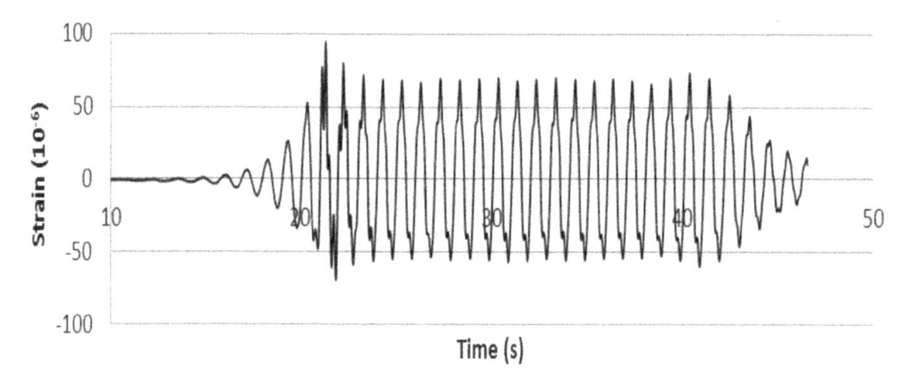

FIGURE 4.19 Strain time history of the fixed arm to measure bending force

maximum force and displacement, respectively. Applying this logic, the following equations are derived, which hold good:

$$P_{max} = \frac{8}{\pi}\frac{8}{\pi}\frac{F_\eta S_\eta}{T} \tag{4.28a}$$

$$F_\eta = a\ F(0 < \mathrm{a} < \mathrm{t}) \tag{4.28b}$$

$$S_\eta = b\ F(0 < b < 1) \tag{4.28c}$$

Substituting, we get the following relationship:

$$P_{max} = \frac{8}{\pi}\frac{8ab}{\pi}\frac{F\ S}{T} \tag{4.28d}$$

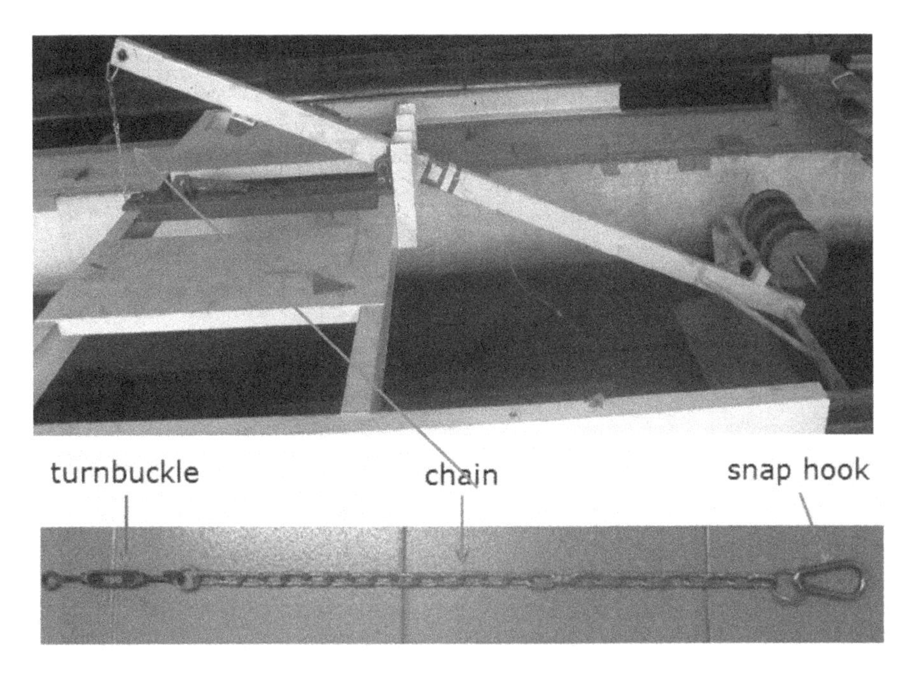

FIGURE 4.20 Experimental setup to study fully submerged buoy

Rewriting, we get:

$$P_{max} \propto \frac{F\,S}{T} \tag{4.28e}$$

where F_η is the force amplitude at a condition of maximum power capture, F is the force when the buoy is fixed, S_η is the displacement amplitude at the condition of maximum power capture, (a, b) are constants, and T is the wave period. Under the fully submerged condition of the buoy, slapping and slamming forces shall remain negligible; under such conditions, buoys can be designed for lesser loads as they shall encounter a lower risk of damage and shall possess a higher life span. Fig. 4.20 shows the test setup under the fully submerged condition of the buoy by adjusting the chain length to suitable links using a turnbuckle. The chain allows only the upwards motion of the buoy while the downward movement has to be constrained to prevent it from sinking to the sea floor. Experiments were conducted to determine the forces acting on the buoy and its displacement when the buoy is fixed and free to move, respectively; buoys were positioned at different drafts and angles of the approach of the input waves. Table 4.8 summarizes the resisting force measured on a cylindrical buoy for the different wave heights, period, and angle of attack for a constant draft of 0.05 m.

TABLE 4.8
Resisting Force on a Cylindrical Buoy

Wave Height (m)	Wave Period (s)	Angular Displacement (deg)	Force (N)
0.1	1	1.15	54.07
	1.5	2.55	56.52
	2	2.15	73.73
	2.5	1.95	75.20
	3	1.875	98.30
0.15	1	1.25	83.56
	1.5	3.425	90.93
	2	2.7	105.67
	2.5	2.625	95.84
	3	2.625	122.88
0.2	1.5	4.25	125.33
	2	3.625	137.62
	2.5	3.5	130.25
	3	3.5	149.91
0.25	1.5	3.65	124.35
	2	3.05	152.86
	2.5	3	135.65
	3	2.8	158.26

4.8.1 Cylindrical Float

Fig. 4.21 shows the displacement response amplitude operator (RAO) and force RAO for a cylindrical float with a draft of 0.05 m. The influence of the wave period and height on the absorbed power is shown in Fig. 4.22. It is seen that force and displacement RAO decrease with wave height, and the variation is nonlinear. It is seen that the power output increase with the increase in wave height. The maximum power output occurs at 1.5 s for all wave heights conducted for the study. For all wave heights conducted during the study, the maximum displacement is observed at 1.5 s; force values increase with the period. Force values increase with wave height until a wave period of 2 s and then decrease, indicating a closer maximum output range.

Fig. 4.23 shows the displacement RAO and the force RAO for a cylindrical float with a draft of 0.15 m. The influence of the wave period and height on the absorbed power are shown in Fig. 4.24. As seen in the earlier case, power output increases with the increase in wave height. It is seen that force and displacement RAO decrease with wave height, and the variation is nonlinear. The maximum power output occurs at 1.5 s for all wave heights considered in the study. When the buoy is fully submerged, no peak is observed for the power output. Power increases with the wave height for all periods, but the variation is nonlinear. Fig. 4.25 shows the displacement RAO and the force RAO for a fully submerged cylindrical float. The influence of the wave period and height on the absorbed power are shown in Fig. 4.26.

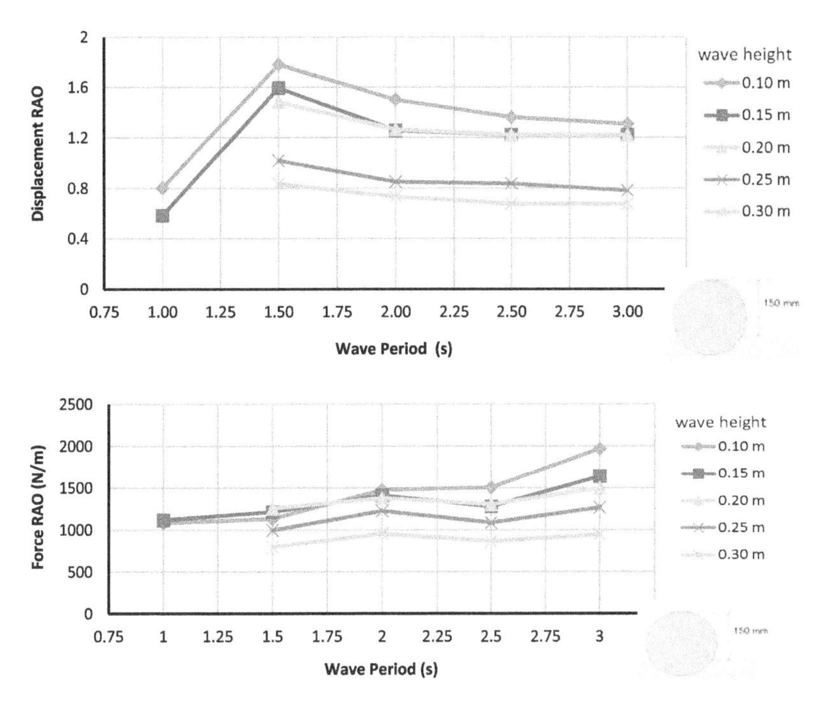

FIGURE 4.21 Performance of partially submerged cylindrical buoy (draft, 0.05 m)

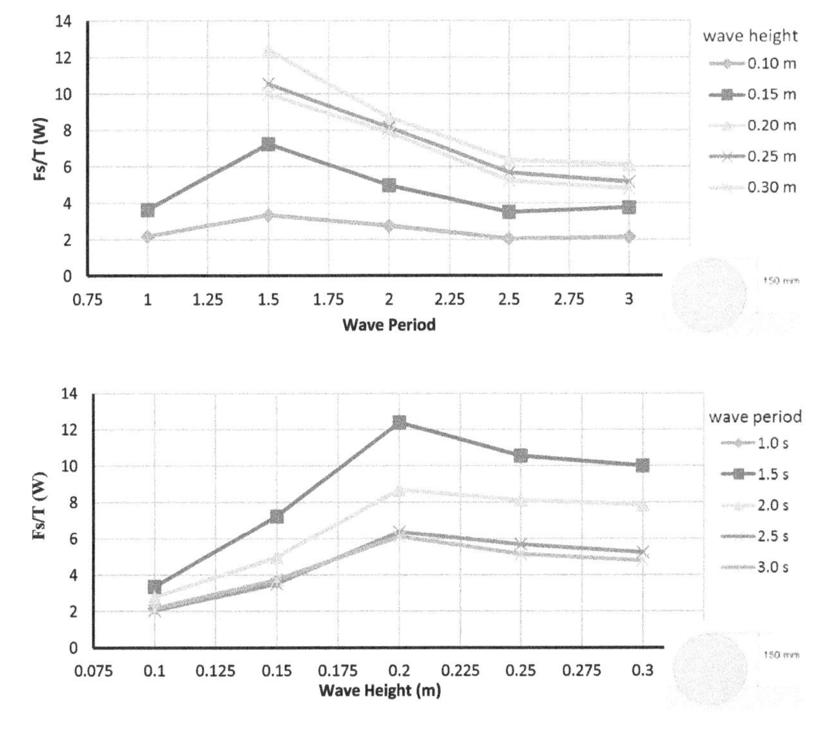

FIGURE 4.22 Power absorbed by partially submerged cylindrical buoy (draft, 0.05 m)

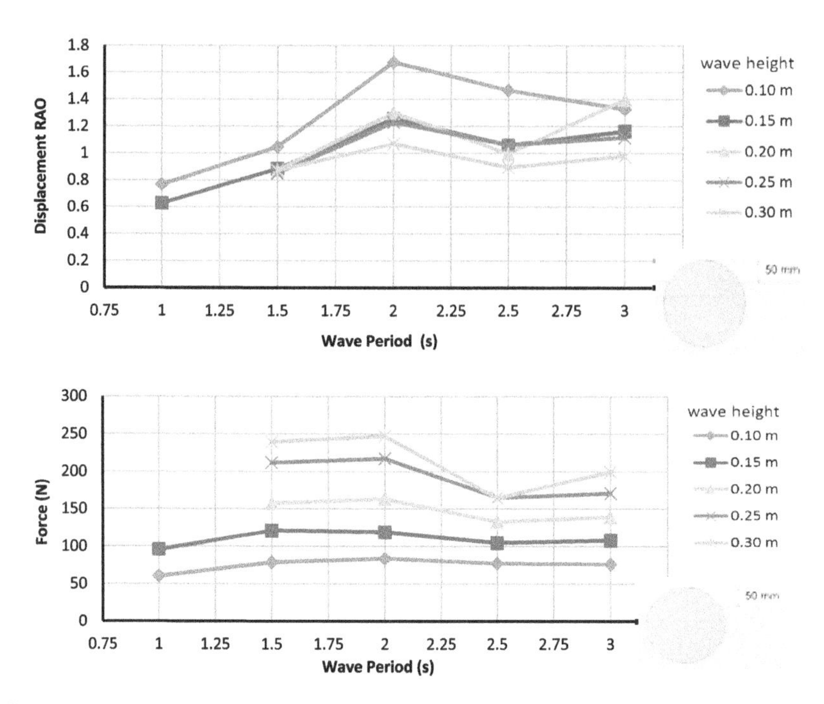

FIGURE 4.23 Performance of partially submerged cylindrical buoy (draft, 0.15 m)

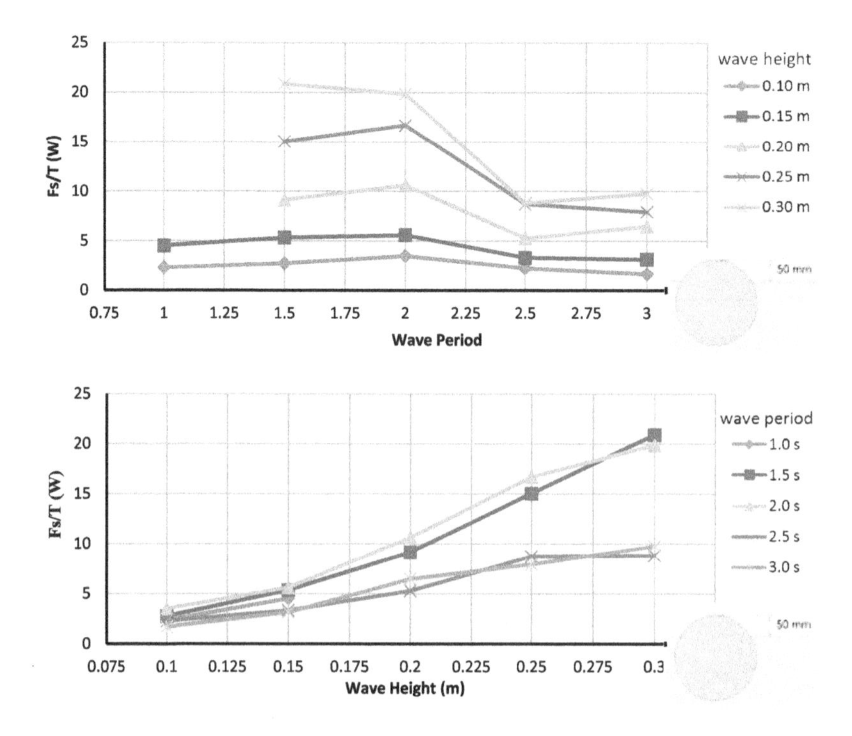

FIGURE 4.24 Power absorbed by partially submerged cylindrical buoy (draft, 0.15 m)

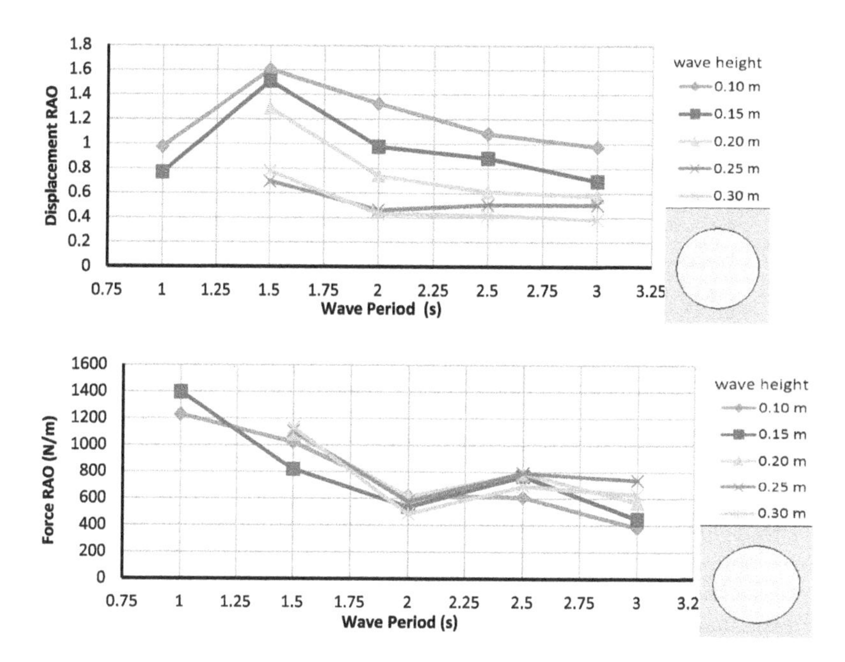

FIGURE 4.25 Performance of fully submerged cylindrical buoy

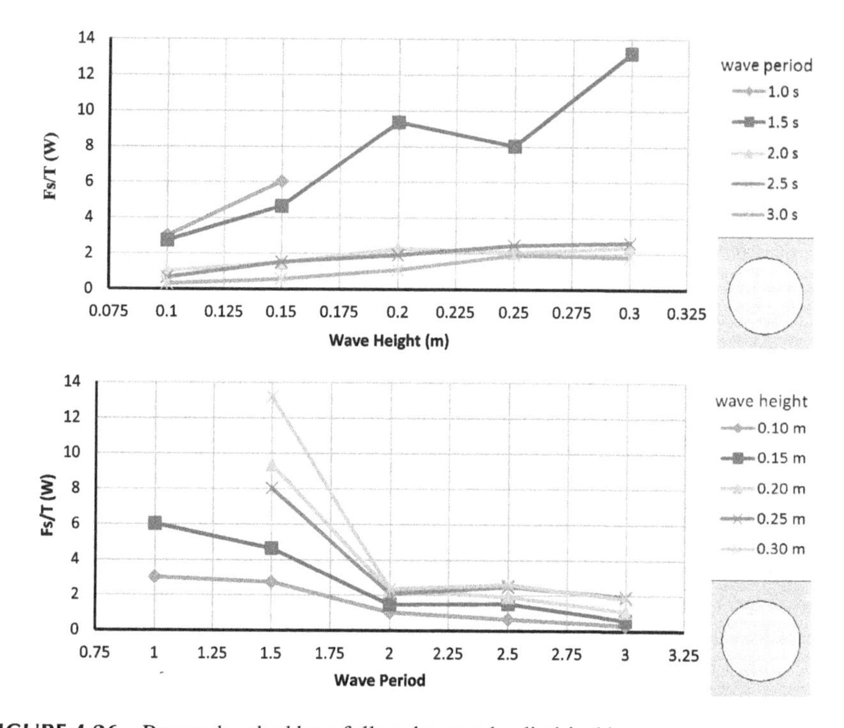

FIGURE 4.26 Power absorbed by a fully submerged cylindrical buoy

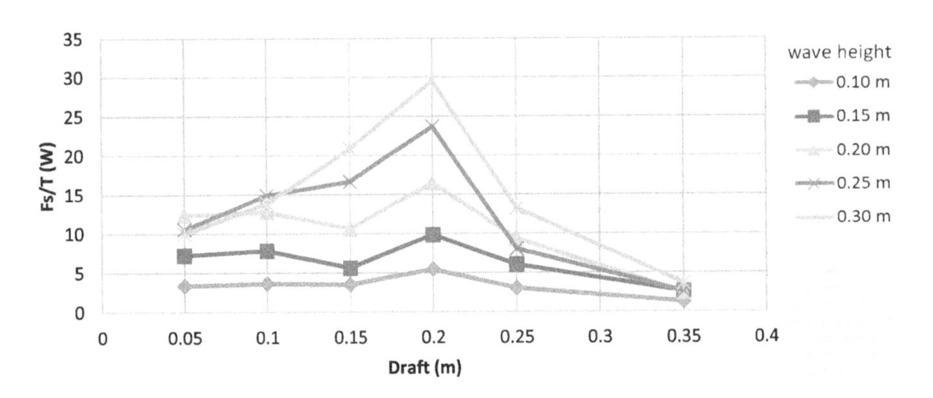

FIGURE 4.27 Power absorbed by a cylindrical buoy for different draft

Fig. 4.27 shows a comparison of the peak power absorbed by a cylindrical buoy for different drafts. It is seen that the maximum peak power is absorbed with a draft of 0.2 m. Power increases with wave height, and the increase is linear within the limitations of the experimental studies. For all the drafts, maximum power is observed for the period 1.5 s; there is a steep decrease in power absorption from 0.2 m draft to the fully submerged case.

4.8.2 Cylindrical Float Integrated with Fin

A fin is integrated into the cylindrical float to improve the power absorption capacity. Parametric variation, in addition to the draft of the buoy, is the angular position of the fin to the wave approach angle. The cylindrical float integrated with the fin is tilted placed to different angles, namely 24°, 54°, 219°, and 249° for the float diameters of (0.1, 0.15 m). Figs. 4.28 to 4.35 show the displacement and force RAOs under different wave heights and periods for different fin angles.

As seen in the above figures, displacement RAO is more sensitive to change in wave period than wave height for all fin angles; maximum displacement is seen at 1.5 s. Further, for the fin angle of 249°, displacement RAO is the least, but the force captured by the float is higher than other fin angles. By increasing the draft to 0.15 m, displacement response behavior is similar except that the force absorbed by the float at 249° is greater. Figs. 4.36 to 4.44 show the variation of power absorbed by the floats under the influence of wave height and periods for different fin angles and the draft of the floats.

It is seen from the above figures that the power absorbed by the cylindrical buoy with the integrated fin is significantly influenced by wave period and height. The peak power absorption is seen at a period of 1.5 s and 0.3 m wave height. For the higher angle of fins (say at 249°), the maximum peak is shifted to 2.0 s, but the influence of wave height on the absorbed power remains insignificant. By increasing the draft to 0.15 m, a similar trend is observed. For the fin's angle at 249°, maximum power is absorbed by the float at 0.3 wave height. These parameters help us choose the preliminary dimensions of the float and the fin angle to harness the maximum power.

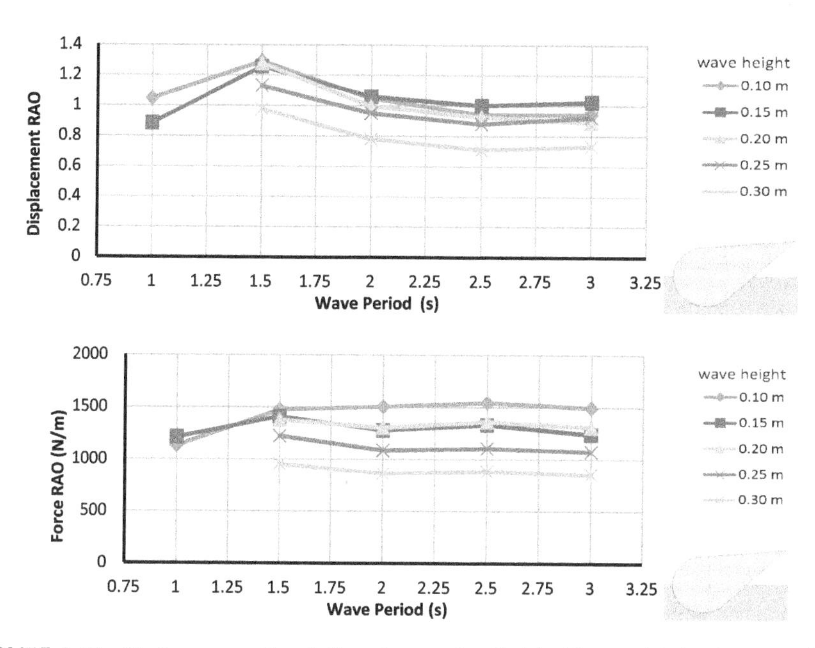

FIGURE 4.28 Performance of partially submerged cylindrical buoy with fin (0.10 m, 24^0)

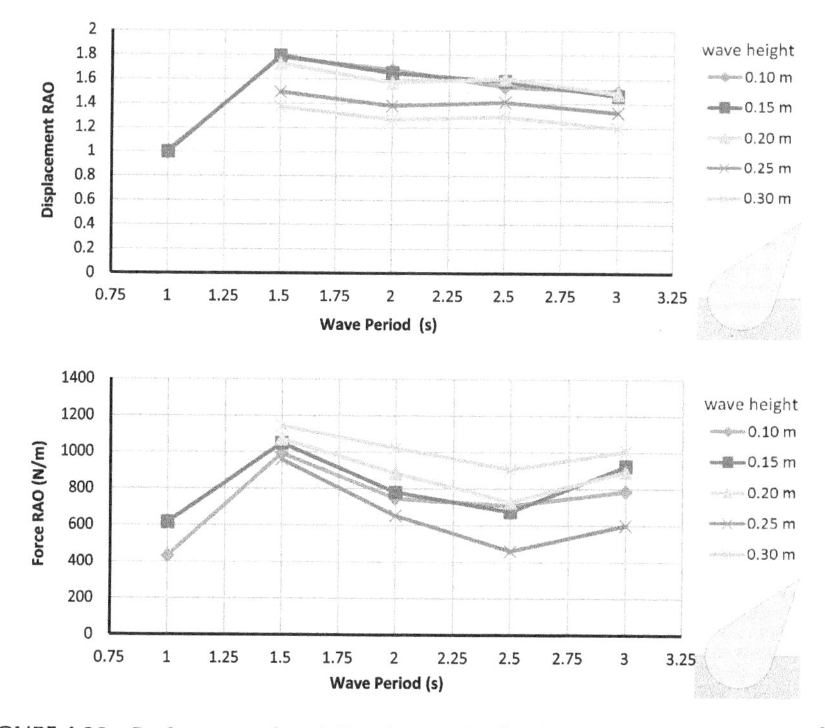

FIGURE 4.29 Performance of partially submerged cylindrical buoy with fin (0.10 m, 54^0)

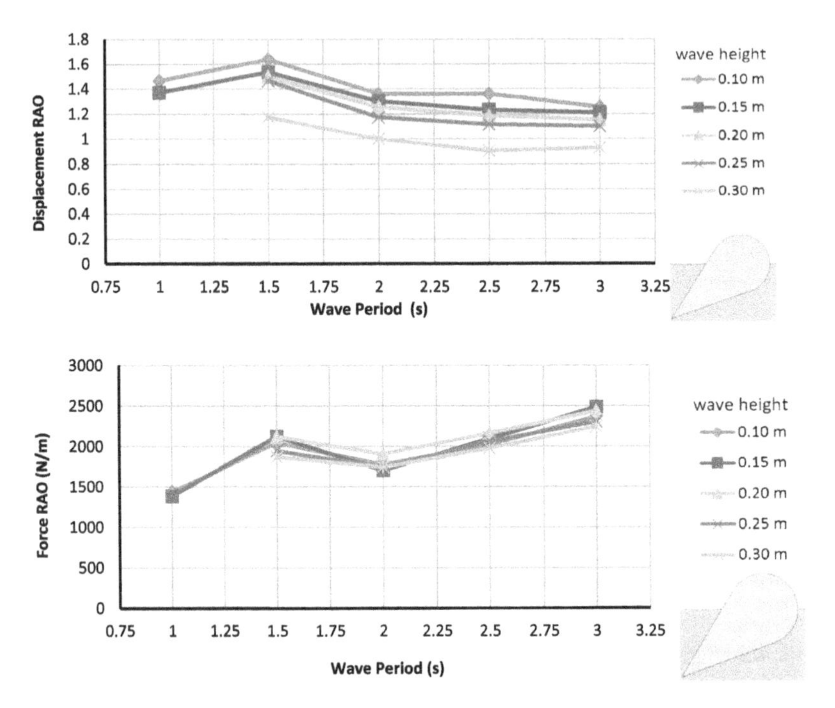

FIGURE 4.30 Performance of partially submerged cylindrical buoy with fin (0.10 m, 219^0)

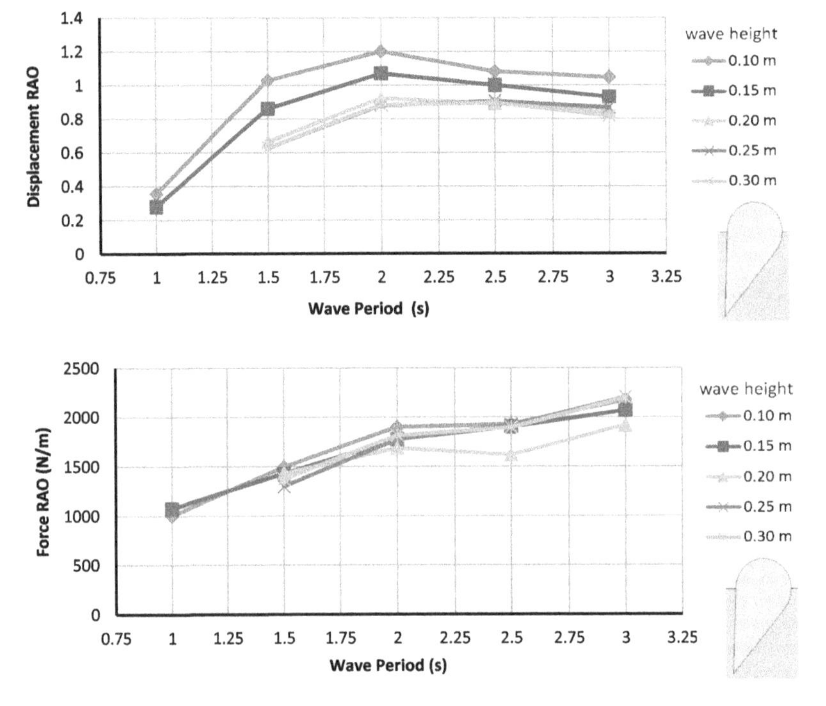

FIGURE 4.31 Performance of partially submerged cylindrical buoy with fin (0.10 m, 249^0)

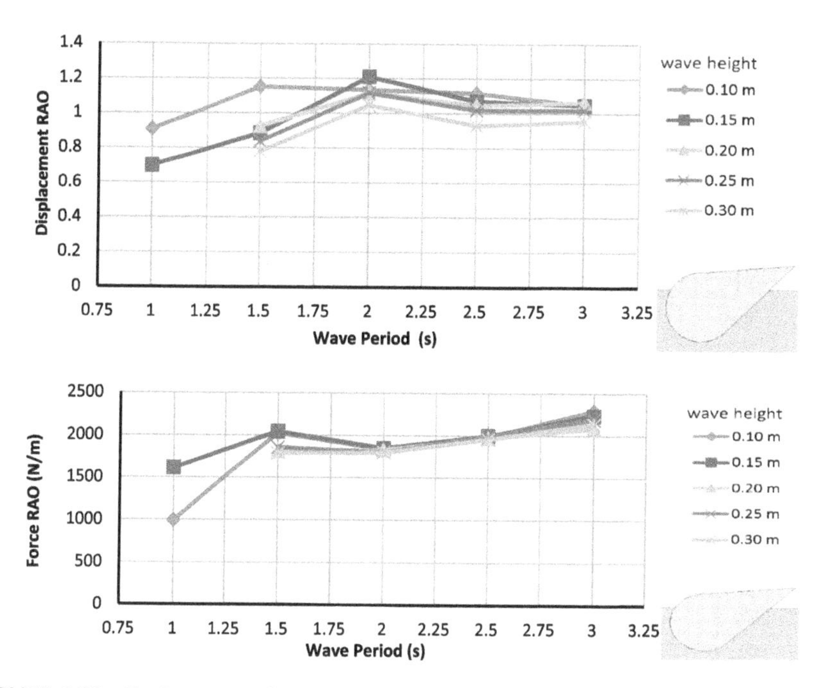

FIGURE 4.32 Performance of partially submerged cylindrical buoy with fin (0.15 m, 24^0)

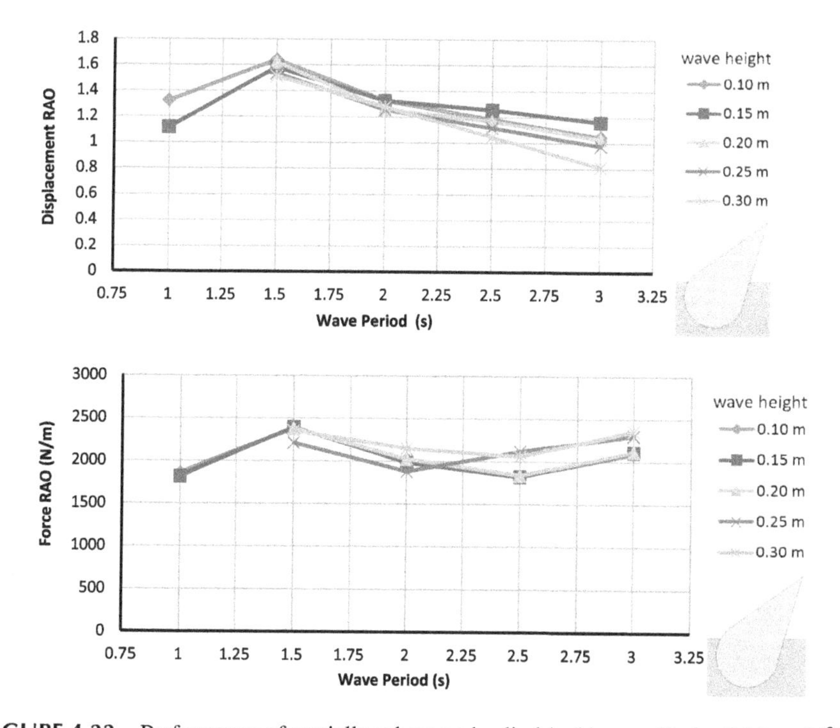

FIGURE 4.33 Performance of partially submerged cylindrical buoy with fin (0.15 m, 54^0)

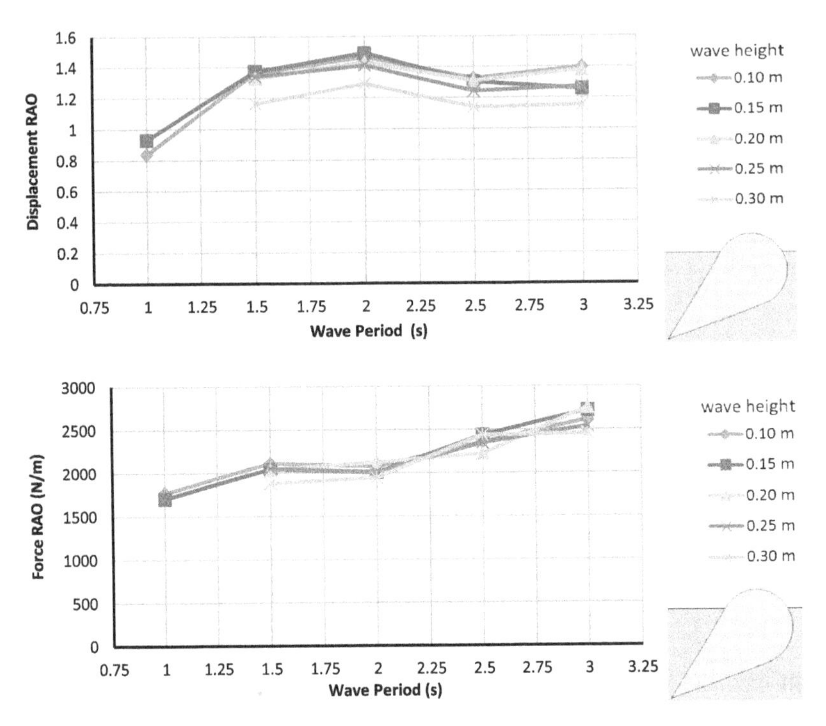

FIGURE 4.34 Performance of partially submerged cylindrical buoy with fin (0.15 m, 219^0)

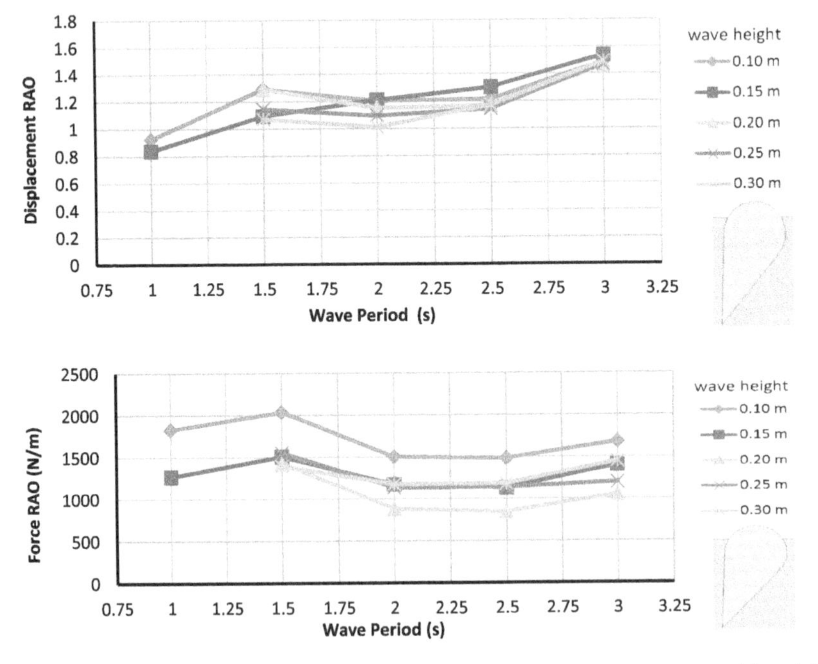

FIGURE 4.35 Performance of partially submerged cylindrical buoy with fin (0.15 m, 249^0)

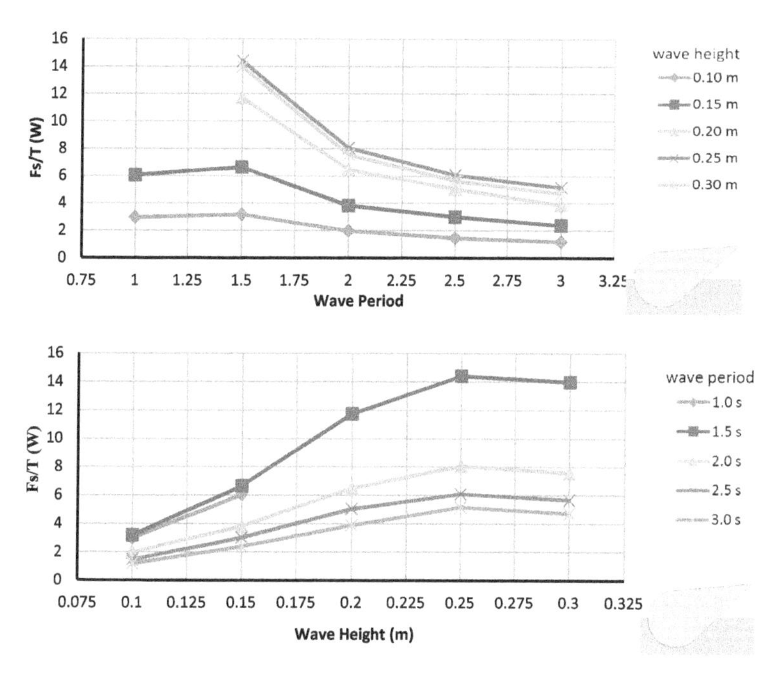

FIGURE 4.36 Power absorbed by a partially submerged cylindrical buoy with fin (0.10 m, 24^0)

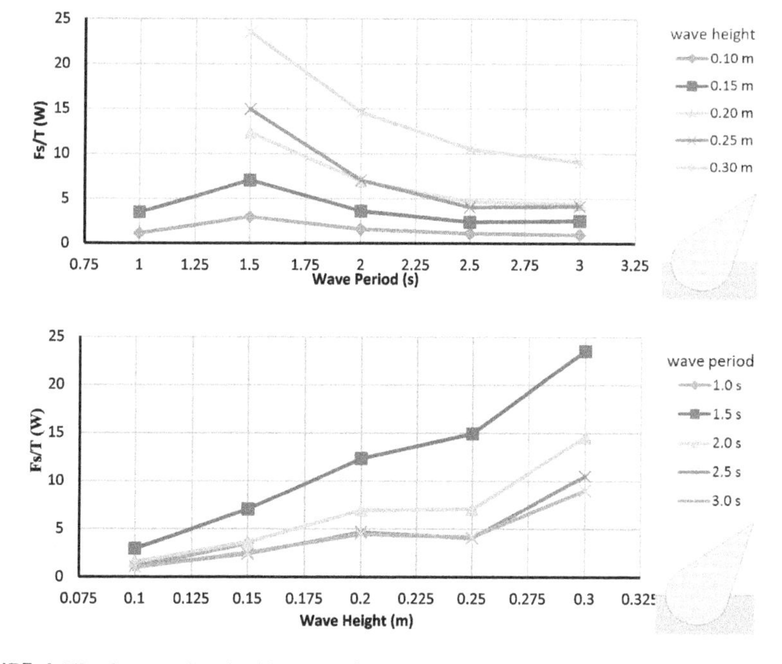

FIGURE 4.37 Power absorbed by a partially submerged cylindrical buoy with fin (0.10 m, 54^0)

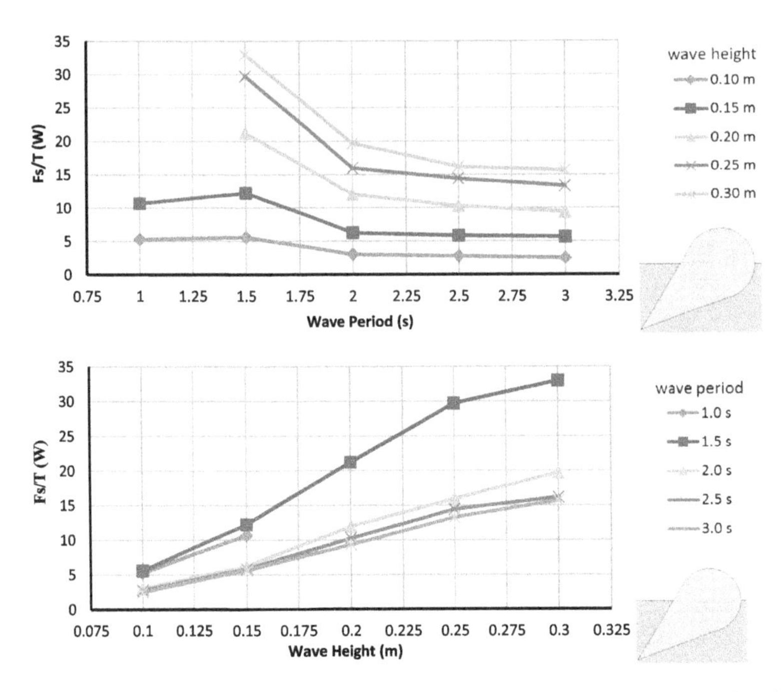

FIGURE 4.38 Power absorbed by a partially submerged cylindrical buoy with fin (0.10 m, 219^0)

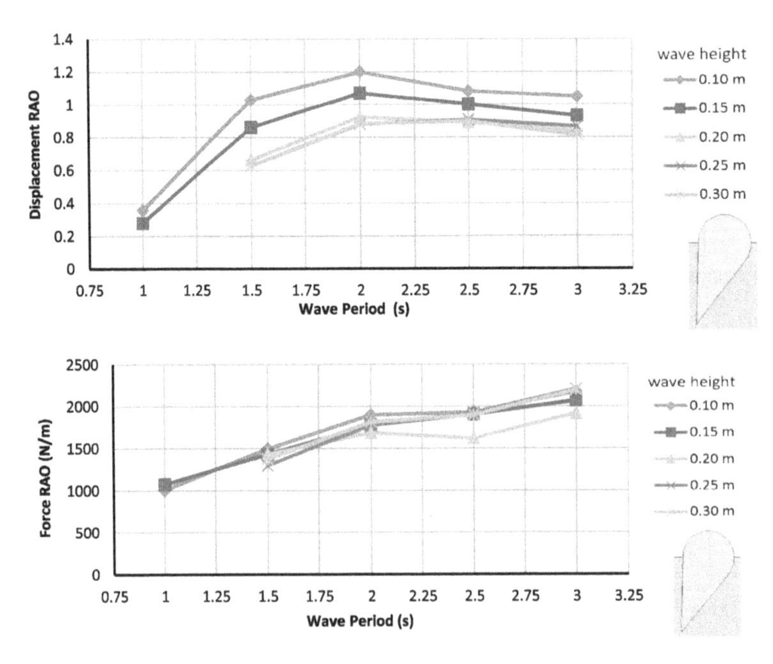

FIGURE 4.39 Power absorbed by a partially submerged cylindrical buoy with fin (0.10 m, 249^0)

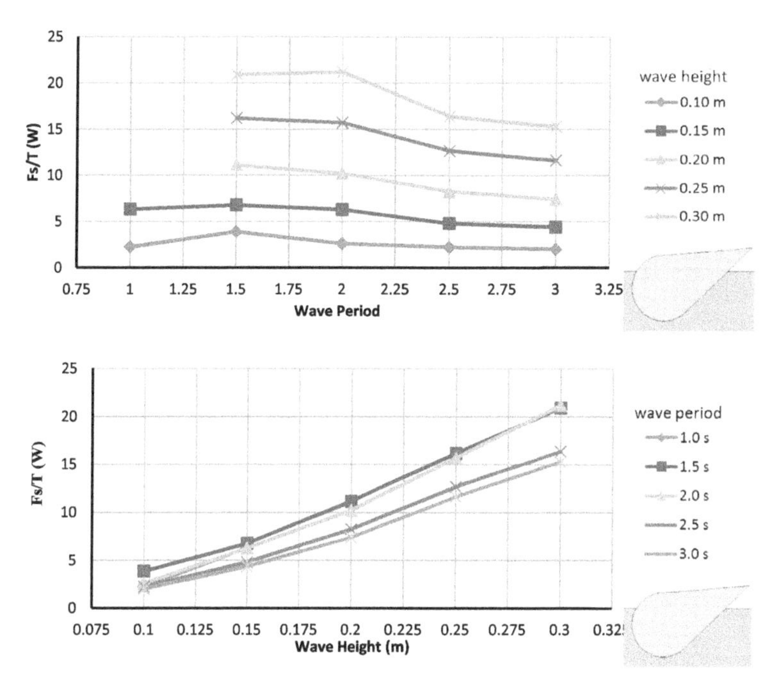

FIGURE 4.40 Power absorbed by a partially submerged cylindrical buoy with fin (0.15 m, 24⁰)

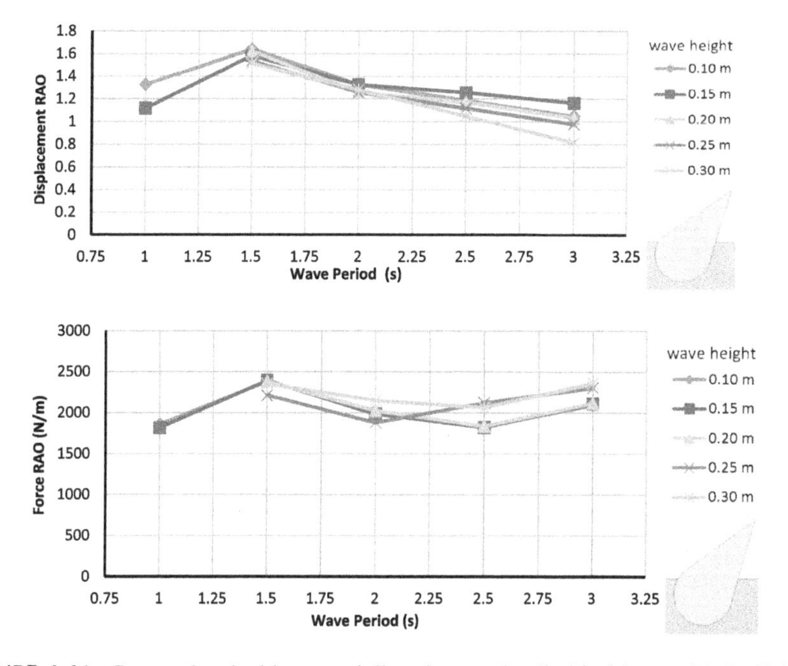

FIGURE 4.41 Power absorbed by a partially submerged cylindrical buoy with fin (0.15 m, 54⁰)

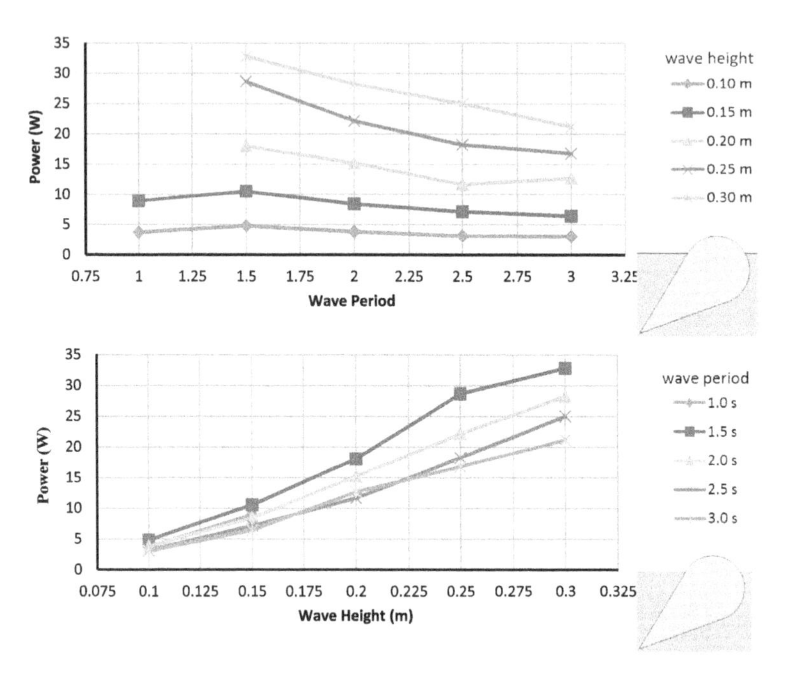

FIGURE 4.42 Power absorbed by a partially submerged cylindrical buoy with fin (0.15 m, 219^0)

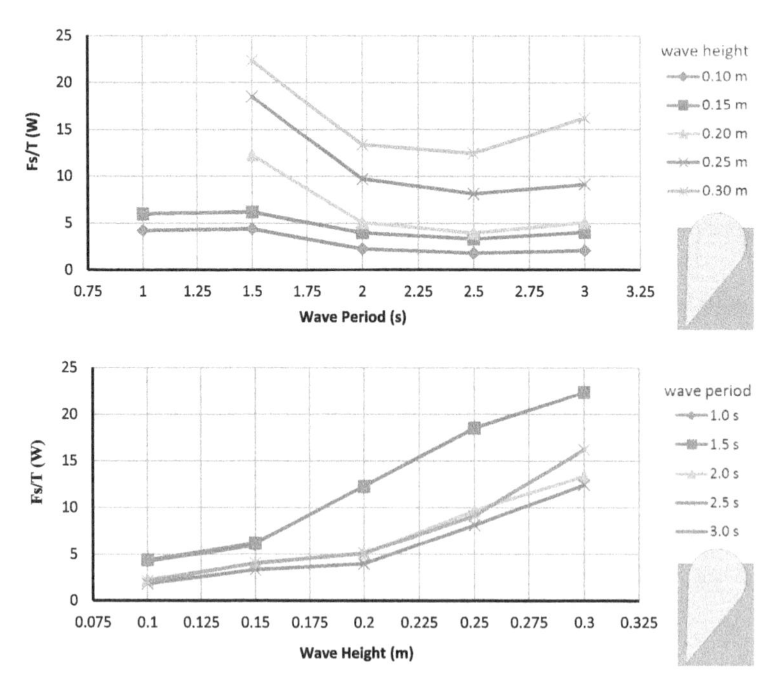

FIGURE 4.43 Power absorbed by a partially submerged cylindrical buoy with fin (0.15 m, 249^0)

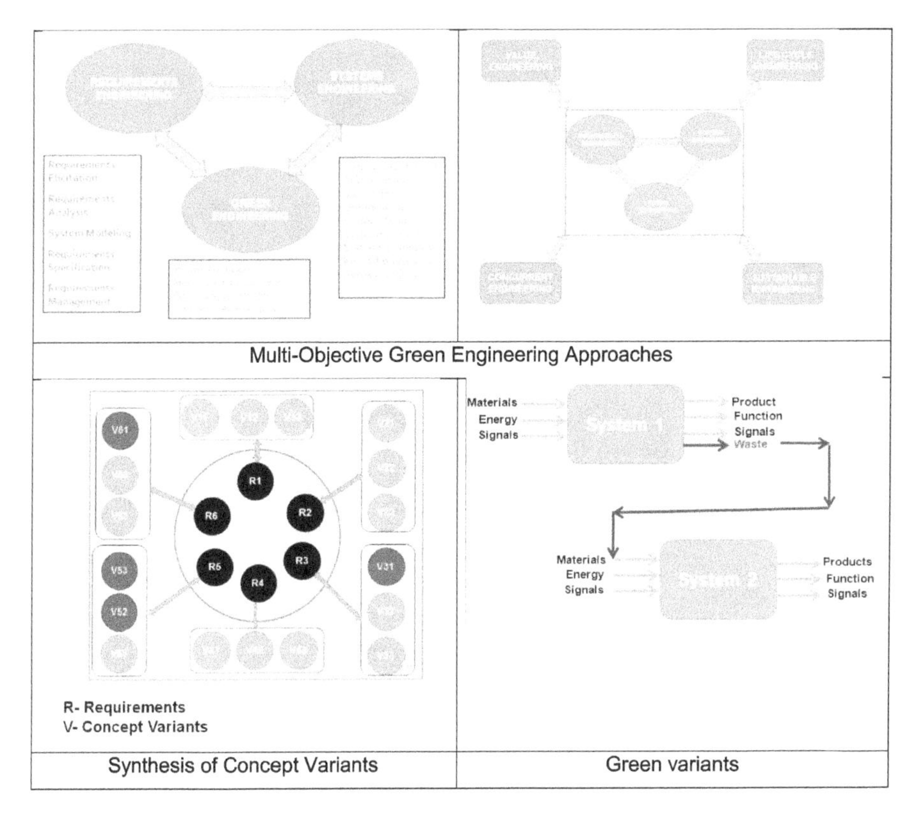

FIGURE 4.44 Multi-objective design synthesis approach for device development

4.9 MULTI-UTILITY DEVELOPMENT DEVICES

The advantages and disadvantages of wave energy have been discussed extensively by many policy makers in various countries, especially in comparison with solar energy. Some of the often-cited disadvantages have been higher costs and scalability issues. It is well established that most renewable energy systems are always economical when used as captive power plants providing a localized source of power to a specific energy user. When a few other utility areas are combined and design synthesis is carried out, considering engineering, safety, and aesthetics, wave energy devices could offer varying solutions in specific areas where other devices may fail.

Requirements engineering approaches used in the systems and software industry can be effectively used in ocean engineering to form the foundation for developing multi-utility development devices. Requirements elicitation, requirements analysis, requirements specification, requirements validation, requirements management are the common areas of requirements engineering. Requirements expansion for developing a closed ecosystem following green engineering principles of reuse, in analogy with nature, that 'Nature Wastes Nothing' will become a central area of focus in design synthesis when such an approach is adopted. The addition of principles of value engineering, life-cycle engineering, and sustainable engineering will add value,

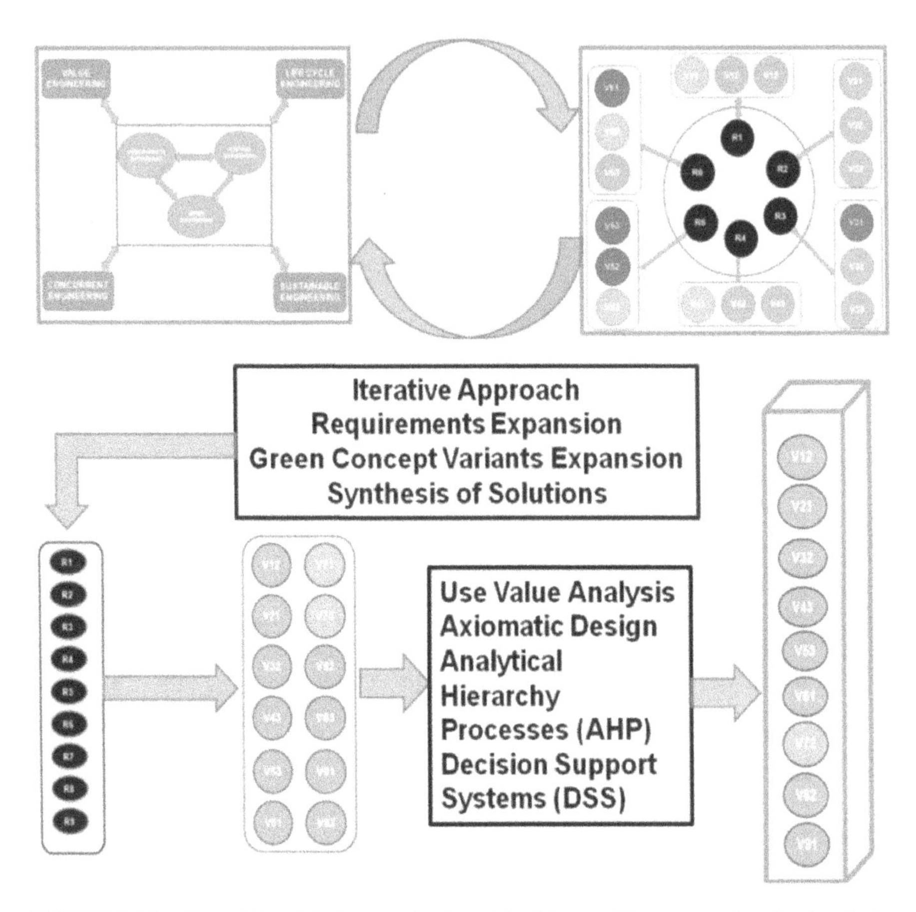

FIGURE 4.45 Consolidated design synthesis and decision-making approaches for developing multi-utility systems

life, and long-term sustainability. Concurrent engineering from principles of product development could be adopted and included to arrive at design variants for various requirements. After the arrival of design variants for various requirements validated for system synthesis, the variants that satisfy the individual requirements to develop a captive power system with minimum wastage must be selected as green variants. Green variants, when preferred over other variants where waste is present, may not be economical when individual requirements are addressed. However, when multiple utilities are considered in total, the system becomes economically viable. Fig. 4.44 shows the flow chart for adapting a multi-objective engineering approach, which is discussed in detail. Fig. 4.45 illustrates the consolidated approach used in the design synthesis and decision-making for developing such devices.

The commonly used engineering design synthesis tools and decision sciences tools like use value analysis (Gerhard et al., 2006), axiomatic design (Suh, 2001), analytical hierarchy processes (Saaty, 1990), systems engineering tools (INCOSE, 2015), and modern decision support system tools could be used to arrive at wave-energy systems that have multi-utility captive energy utilization devices and components. One such wave-energy system synthesized to serve a multi-utility area is discussed. The proposed system, TSUSUCA-DOLPHIN, is a tsunami flooding prevention, uninterrupted renewable energy conversion, and a captive consumption device. It has a wide application in desalination, ocean living simulation, protection of coasts, and the provision of health benefits from the ocean. TSUSUCA-DOLPHIN is designed as a spin-off benefit obtained from the research and system synthesis techniques developed for deep ocean mining (Raphel et al., 2004; 2007) and wave energy research.

A detailed state-of-art survey was carried out to identify the potential areas of designing the wave-energy devices to serve multi-functions, especially for controlling tsunamis flooding.

- Commonly discussed dangers due to tsunamis are in nuclear power plants like Kudankulam (India), Fukushima (Japan), low-lying populated areas experiencing tsunami inundation or flooding due to storms, and tourist beaches and resorts.
- A seawall is the currently available solution to stop tsunami waves, which can rise to 20 m in height on the shore. Seawalls can be quite expensive, and there are no additional uses. Further, seawalls cannot be easily dismantled if changes in the direction of orientation are required due to accretion, silting, etc.
- Twin barrier concepts have been proposed in the deep sea but are very expensive, require very long protection, and do not have any additional benefits.
- Several other concepts discussed in patents have the features of seawalls but do not have any other utility benefits.
- A device that could provide protection in the event of a tsunami and produce power from sea waves with high efficiency (in the wave breaker zone) combining with solar and other forms of energy and providing adequate protection eroding coasts is not present.
- There are many wave-energy devices, but they are either point absorber buoys, surge converters, or oscillating water columns where the efficiency is not very high. A combination of them could result in high efficiencies. These devices, however efficient, are not economical if power generation alone is considered. Captive utilization of power generated within the device would make the device more economically feasible, but such devices may not offer protection to tsunami flooding.
- A device that could act as a barrier to tsunami and storm surge flooding, generate power from waves and utilize it internally or within proximity, and provide coast protection could have a higher economic potential and protect major critical installations.

Figs. 4.46 to 4.49 show different views of TSUSUCA-DOLPHIN, namely schematic view, front view, plan, and a suggested location for the proposed device.

As seen in Fig. 4.46, TSUSUCA-DOLPHIN consists of the main hull shaped like a dolphin. It has a head (1), body (2), and tail (3). The bottom portion of the tail rests on

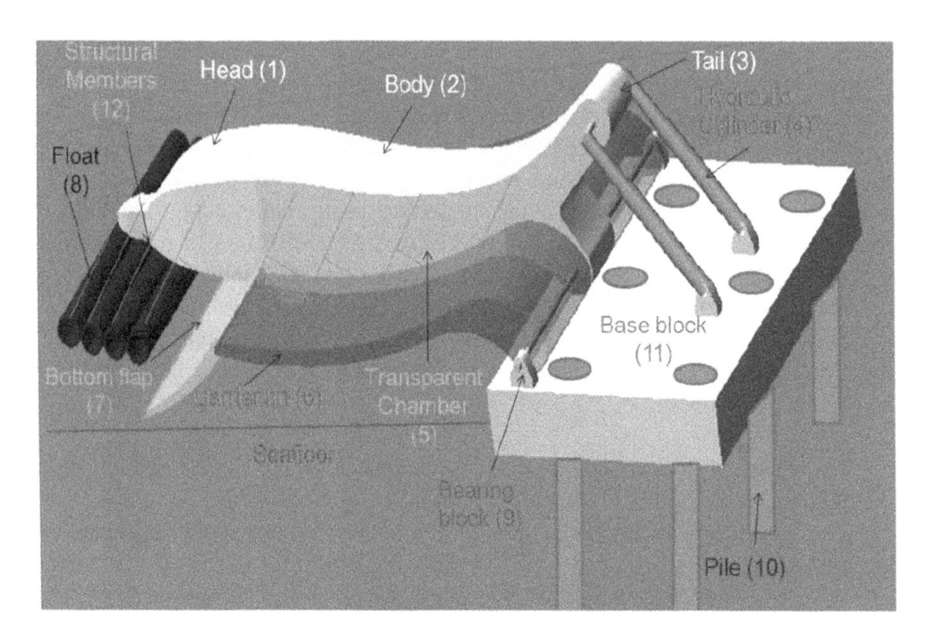

FIGURE 4.46 Plan of TSUSUCA-DOLPHIN

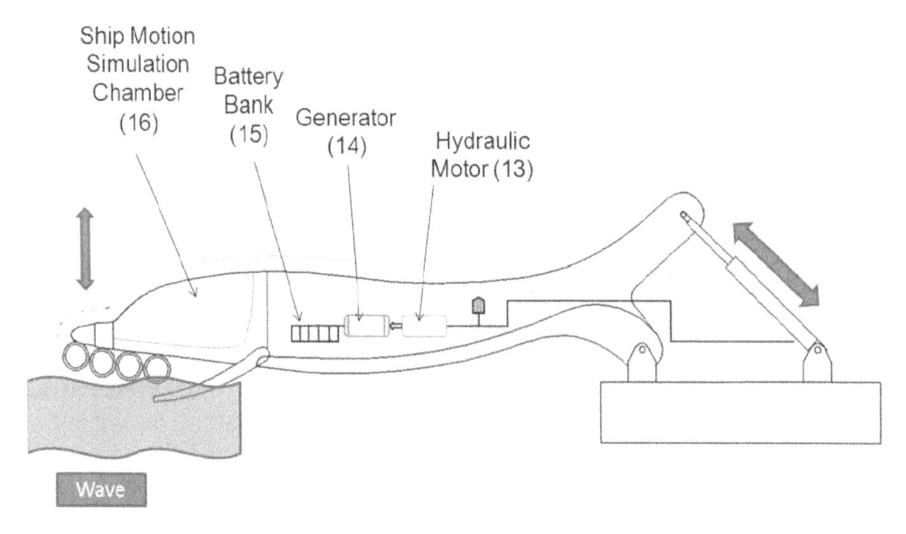

FIGURE 4.47 Front view of TSUSUCA-DOLPHIN

a base foundation block (11). The body is connected to two large barrier members (6), also called wings, on the sides. The tail is hinged at the bottom and can be connected to the power take-off unit mechanically through rack and pinion arrangement or a hydraulic cylinder (4). A shaft connected to the tail and barrier members is placed on the support blocks (9) to withstand the tsunami flooding load. It can be rotated

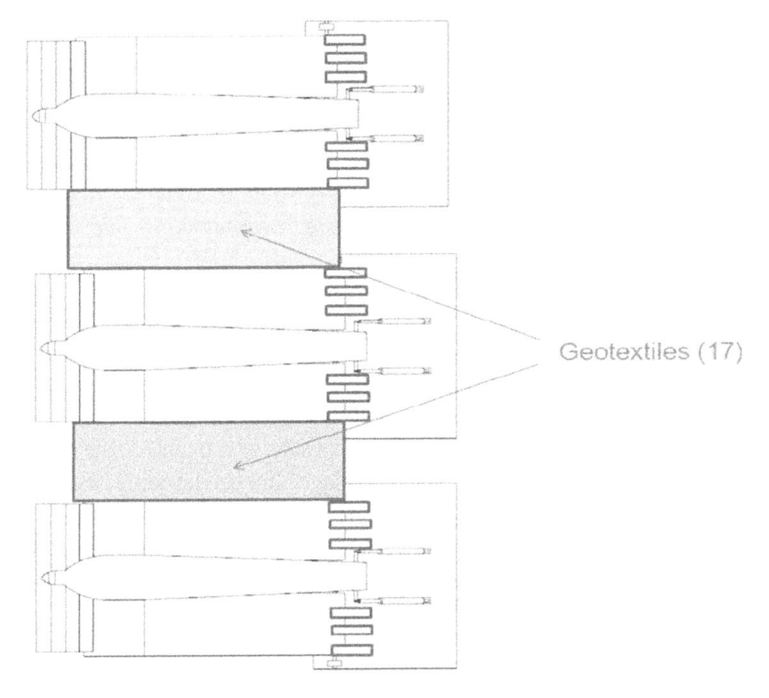

FIGURE 4.48 Plan view of TSUSUCA-DOLPHIN

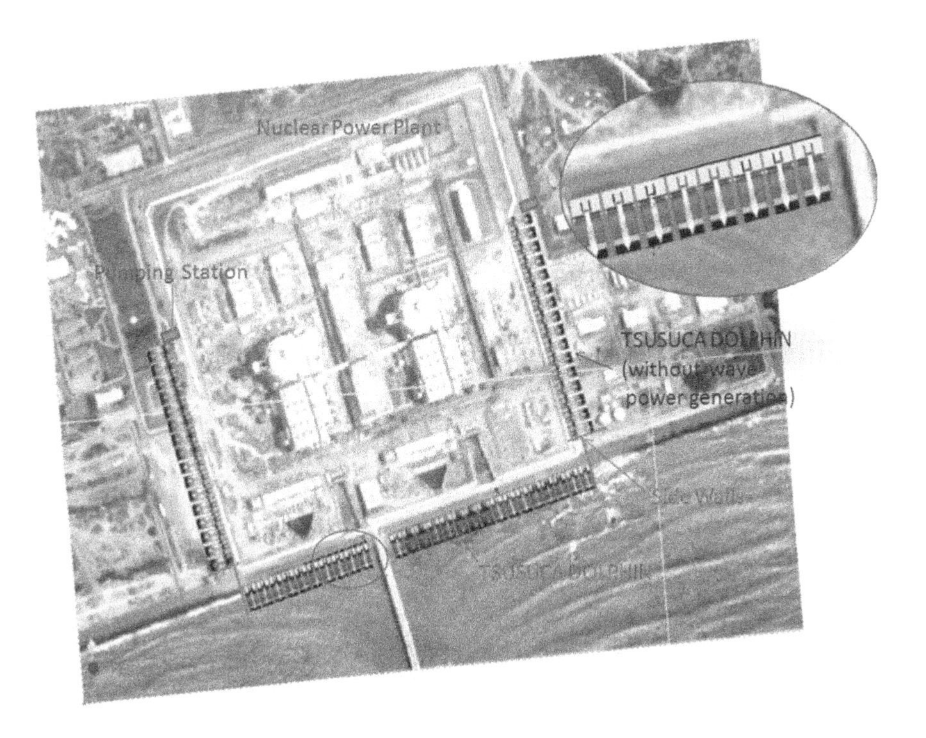

FIGURE 4.49 Proposed location of TSUSUCA-DOLPHIN

through a motor and gearbox and clutch arrangement to lift the chamber to block tsunami waves or storm surges. As the arrangement has to withstand tsunami waves of 20 m height, the structural arrangement is very rugged in construction. The head rests on buoyant members (8), which can support the heavy weight of the device, and are also essential when the wave height exceeds 5 m to help break the wave from the coast. The tail can be present before the wave breaker zone, which is closer to the shore or beyond the wave breaker zone, to cause wave breaking near the mouth of the device.

The location of the device can be decided based on sediment movement patterns. The pins of the hydraulic cylinders (4) can be released by electric or hydraulic actuators or can also be manually released. The actuation can be through automated arrangement or manual override to facilitate lifting the main structure during tsunamis. The pins also behave like shear pins in the event of failure of structure to lift by automated or manual means and during unpredicted tsunamis. The body houses the power generation unit and a desalination plant. The barrier wings are made up of a flat bottom and have heavy truss members to provide the structural strength required to resist tsunami and wave loads. Solar panels can be mounted on the wings, and solar films can be wrapped around the body to generate solar energy. A part of the same portion where solar panels are mounted can also be used as sunrooms for sunbathing and wave motion to promote beach tourism.

A bottom flap (7) is also connected to the barrier member and body through hinged joints. The profile of the bottom flap is configured with a breaker to cause wave breaking within the device and increases the height of motion of TSUSUCA-DOLPHIN, thereby capturing most of the energy from the waves to result in higher power generation. The bottom flap closely follows the profile of the seafloor. The bottom flap (7) also prevents erosion of sediments below the foundation block (11). Fig. 4.47 shows the transparent inner view of TSUSUCA-DOLPHIN. A ship motion simulation chamber (16) with transparent view ports is present in the head section. The chamber will have seats in which people can sit and experience the wave motions. Seats can also be designed to be supported on ball joints with small electromagnetic actuators and swivels programmed to actuate through accelerometers through embedded electronics and can have transverse motions also to simulate roll motions. This chamber can be used for recreation purposes, tourism, etc. It can also be used as a platform for simulating ship motions (16) for trainee sailors (sea sickness/ motion sickness training).

The base foundation block (11) can be of multiple steel structure units bolted together and filled with small concrete blocks arranged one over the other for handling in wave breaker zones. The steel structure units can be towed to the site by floating using air bags or buoyancy floats and installed in a location by releasing the floats. The steel structures have central pipes through which screw piles or helical piles (10) can be driven. Alternatively, the base blocks can be steel structure units filled with quick setting underwater concrete or suitable RCC pumped through concrete pumps. The whole base block could also be designed as a floating unit ballasted by filling sea water at the site. Multiple units of TSUSUCA-DOLPHIN can be lined along the coast interconnected by strong geotextiles (17) supported by ropes or such materials with adequate strength, which will allow differential motion between two units to a considerable extent as well as withstand pressure and minimize water flow in the gaps during tsunamis. In addition to geotextiles, a rubber-lined rotating wedge block hinged on one unit and rotated to cause sealing on the connected units in the upright position will block the water flow between two units. The units without power generation features will be lined perpendicular to the coast to prevent water from entering the sides. The corner gaps can be suitably sealed against a local L-shaped high wall.

Fig. 4.50 shows the power generation setup connected to TSUSUCA-DOLPHIN. Working concepts of the proposed multi-utility device are explained in Figs. 4.51 and 4.52. The action of the device during a tsunami can be seen as a conceptual idea conceived from the design. Further, solar panels installed on the top of the hull serve as multi-utility power generation units.

In the event of an unpredicted tsunami that occurs if the earthquake's epicenter is close to a nuclear plant (like the Fukushima tsunami flooding), there may not be sufficient time to switch on the motor manually. During such an event, the TSUSUCA-DOLPHIN will rise due to buoyancy. The limit switch on the hydraulic cylinder will cause an electric actuator to disconnect the pins and allow free movement of one end of the cylinder connected to the foundation. Another switch in the sequence gets operated, causing the clutch to become engaged, and the motor and gear box unit is switched on to lift TSUSUCA-DOLPHIN against the tsunami waves. If the actuator fails, the base pin will shear off to cause the hydraulic cylinders to slide away and lift TSUSUCA-DOLPHIN to the height of the water swell caused by the tsunami wave. In an unpredicted tsunami, people can also stay safe inside the DOLPHIN as the entire system is buoyant and comes out when the tsunami flooding subsides.

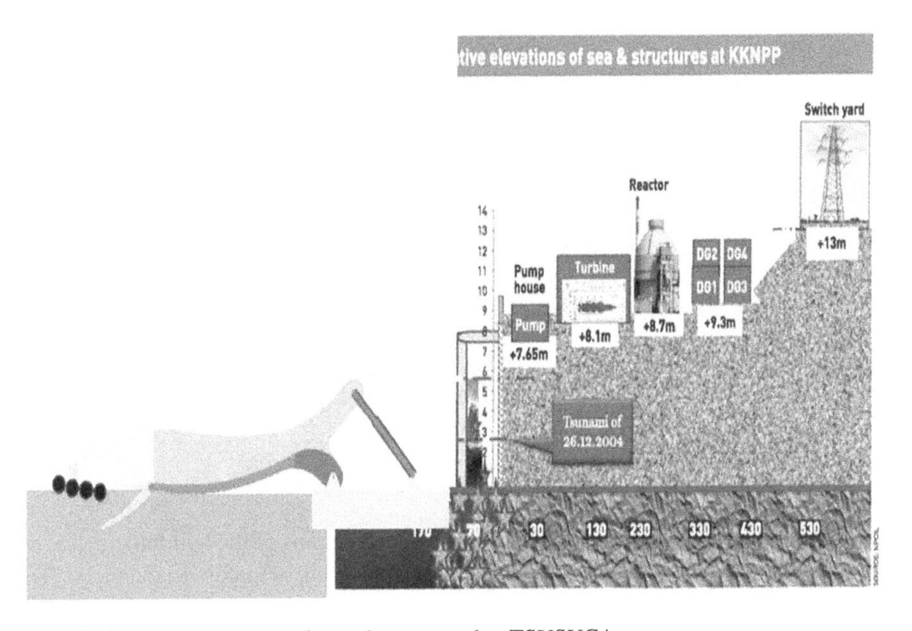

FIGURE 4.50 Power generation unit connected to TSUSUCA

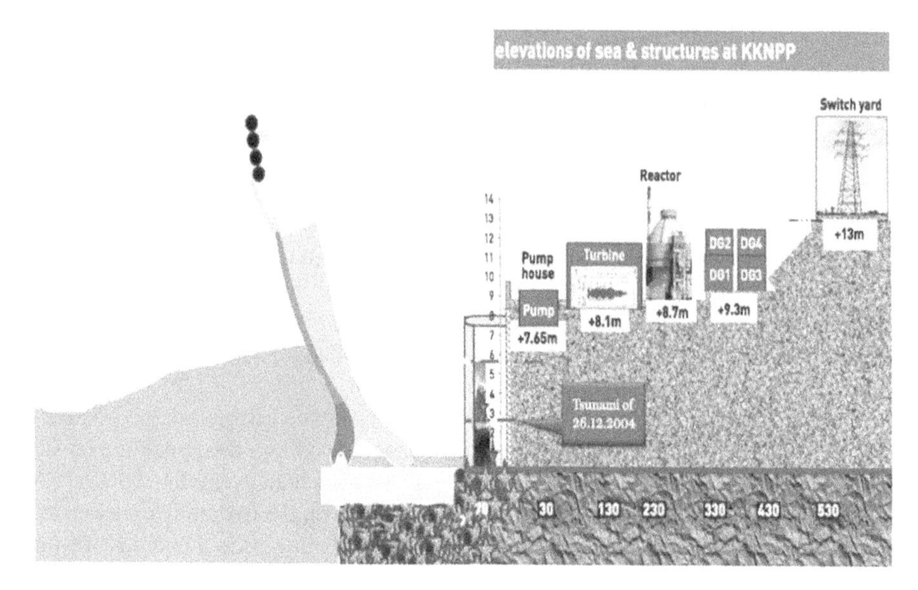

FIGURE 4.51 TSUSUCA during tsunami

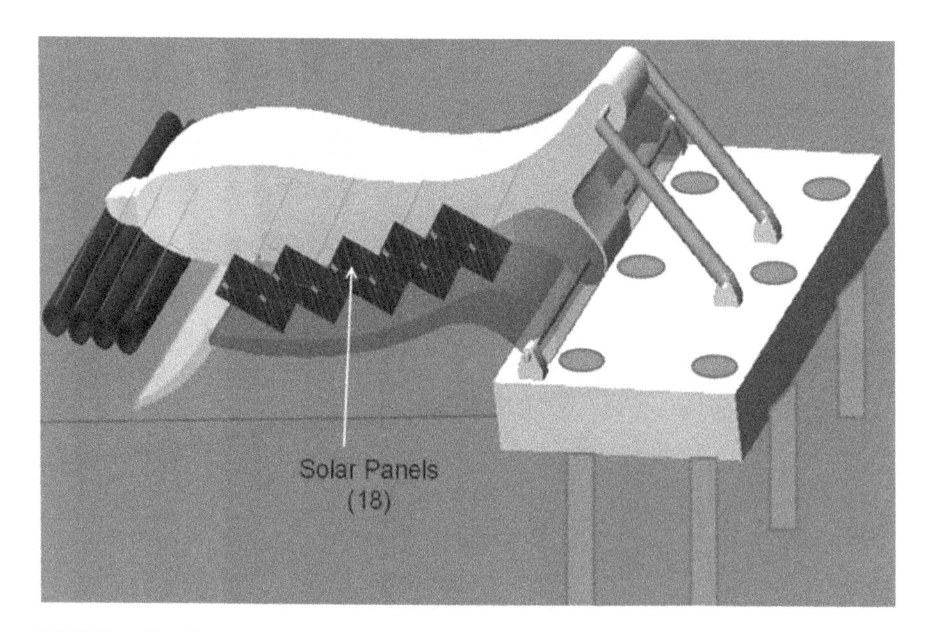

FIGURE 4.52 Solar panels installed on the hull of TSUSUCA

Under the predicted tsunamic and storm surges, the pins of the hydraulic cylinders (4) of each TSUSUCA-DOLPHIN can be removed using actuators or manually. The clutch, which is normally disengaged, will be engaged to connect the motor and the gear box unit. The motor can be operated to lift the TSUSUCA-DOLPHIN. Multiple units can be lifted to form a large wall to protect coastal structures like a nuclear power plant. More than 90% of the flow can be blocked. However, about 10% may flow through the bottom gaps and gaps between two TSUSUCA-DOLPHINs, which can be drained through channels to a different salt field if close by or pumped out using pumps safely. People can be asked to leave the area as there is a time gap (usually a few hours) between tsunami flooding and the earthquake.

During normal operations, the clutch is disengaged, and as waves are incident on the buoyant member (8), the device moves up and down, the amplitude determined by the energy of waves. The movement is further increased by wave breaking occurring in the bottom flap (7). As the device moves, the hydraulic piston reciprocates within the cylinder, and the working fluid causes a hydraulic motor (13) to rotate, in turn, rotates an electric generator (14) connected to it. Alternatively, a rack and pinion mechanism can move a toothed wheel instead of a hydraulic cylinder. The speed of the wheel can be increased through a gear box to drive the generator. The generator can be connected to a battery bank (15) with capacitors in parallel to reduce fluctuations in voltage generated. The generated power can be used for self-contained applications internally. A hydraulic accumulator connected in the circuit dampens the sudden increase in pressure caused in the hydraulic system when wave breaking occurs in the device. A flywheel performs the dampening action instead of the accumulator in the mechanical rack and pinion-toothed wheel arrangement. Solar panels and solar films can be mounted on the TSUSUCA-DOLPHIN and connected to the battery pack and solar power generated. Thus, the proposed device is seen as a combined solar and wave power generator, with multi-utilities as explained graphically in Figs. 4.53 to 4.59.

FIGURE 4.53 TSUSUCA-DOLPHIN as a coastal protection device

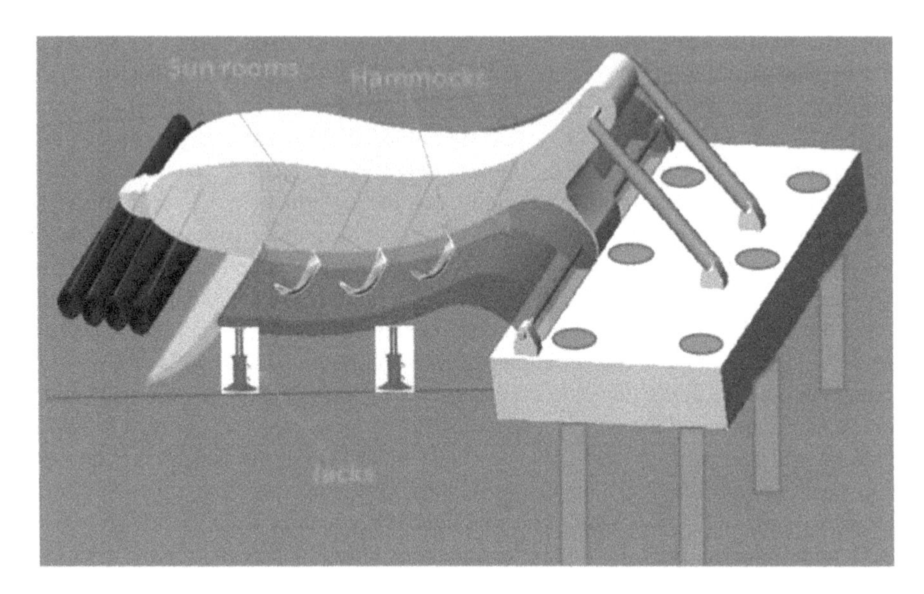

FIGURE 4.54 TSUSUCA-DOLPHIN with sunrooms, hammocks, jacked-up for stationary mode

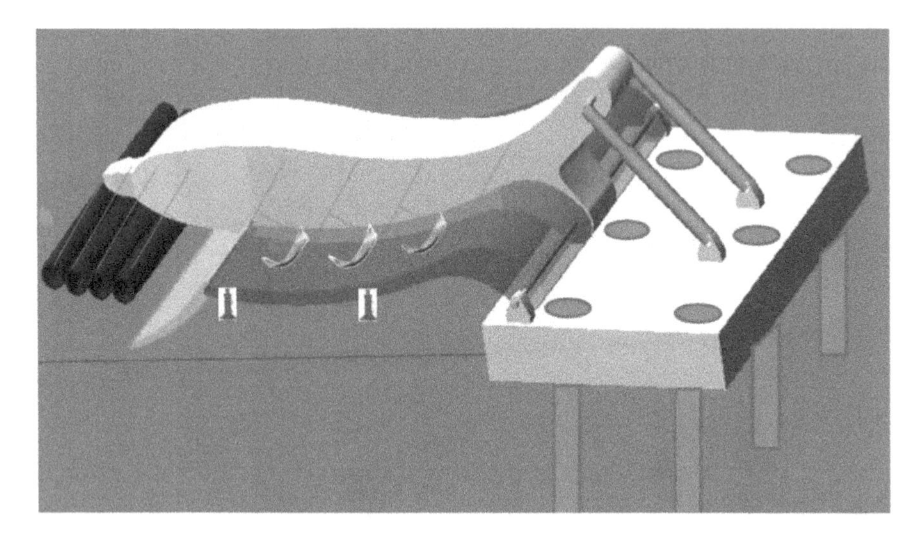

FIGURE 4.55 TSUSUCA-DOLPHIN with sunrooms, hammocks with jacks lifted for wave-riding experience

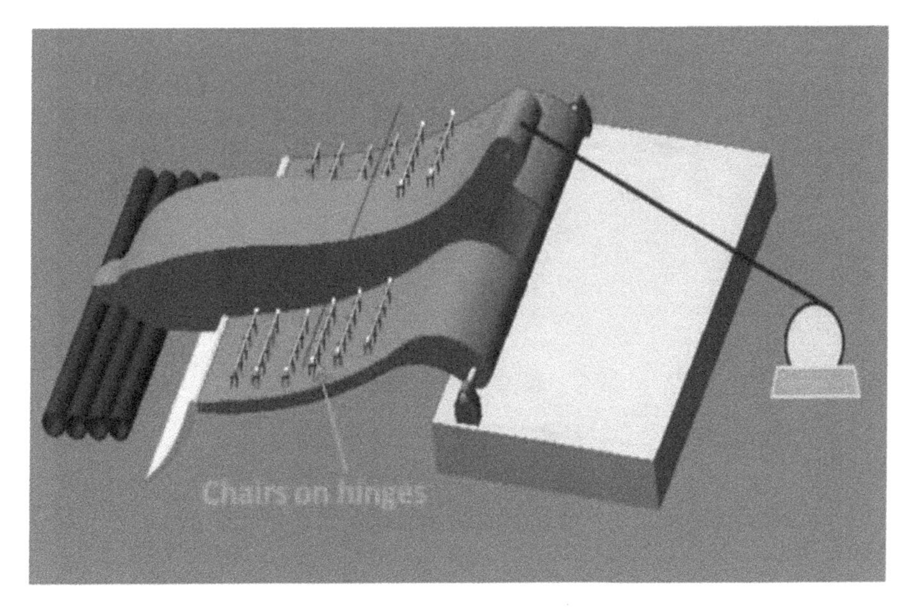

FIGURE 4.56 TSUSUCA-DOLPHIN with chairs for seating people for entertainment tourism on beaches

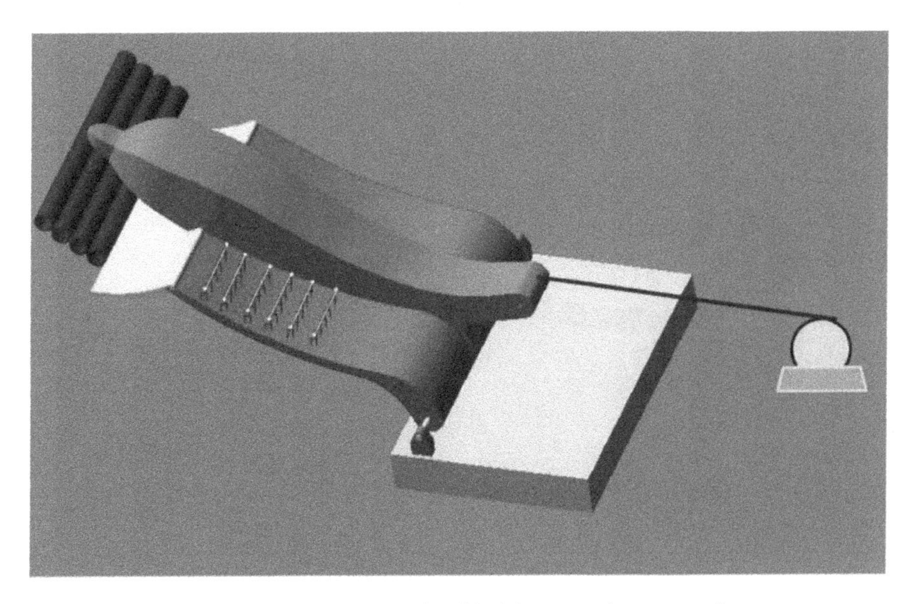

FIGURE 4.57 TSUSUCA-DOLPHIN being lifted for entertainment tourism

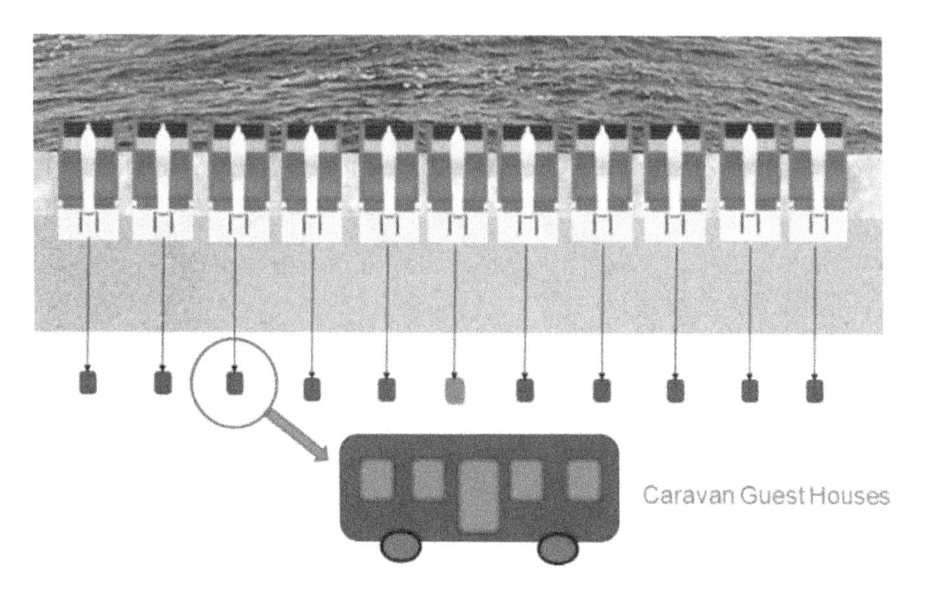

FIGURE 4.58 TSUSUCA-DOLPHIN powered caravan guest houses for beach tourism

FIGURE 4.59 TSUSUCA-DOLPHIN as a ship motion simulator for trainee sailors (for overcoming sea sickness)

4.9.1 Applications of TSUSUCA-DOLPHIN

Desalination: A reverse osmosis plant has been accommodated in the body of TSUSUCA-DOLPHIN. The potable water produced is pumped to a centralized storage tank for further distribution through water tankers or nearby guest houses on beaches.

Coastal erosion and accretion: The energy from the waves is used to lift the floats and the bottom flaps, which further prevents the forward motion of water to a considerable extent. As wave energy is converted to electrical energy, the impact of waves on the coasts is minimal leading to coastal protection. Beach accretion can also take place behind TSUSUCA-DOLPHINs, which is usually a desirable one. Sometimes depending on the flow pattern, erosion can also take place in front of TSUSUCA-DOLPHIN. To prevent tilting of the arrangement in such cases, the foundation base blocks could be jacked up at specified locations, and a barrier plate lowered in front of the foundation to prevent further erosion. A sand pump provided within the chamber can fill sediments if erosion occurs below the foundation. The same pump could be used for flushing away sediments if accretion occurs outside the chamber. The area subjected to accretion can be fluidized using jets by operating the pump, and the solid-liquid mixture can be pumped periodically to a tank. The sand can be washed with fresh water from the desalination plant and used to fill low-lying areas or for other useful purposes.

Health benefits from the ocean–solar cubicles/sunrooms: The space between the barrier members and the support trusses (12) can be covered using transparent tops (5) to form sunrooms. The power generated from TSUSUCA-DOLPHIN can be used to air-condition the sunrooms. These sunrooms can have necessary features for people to enjoy air-conditioned sunbathing, thereby providing health benefits due to vitamin D. These sunrooms can promote beach tourism. Initially, the base block made of multiple units is installed, followed by direct installation of the TSUSUCA-DOLPHIN in case the area is easily accessible and large mobile cranes movement is possible. Alternatively, the TSUSUCA-DOLPHIN can be built from the base block as the entire steel structure is of prefabricated and bolted construction, floated and ballasted at the site either as a single unit or multiple part units.

4.9.2 Workable Alternatives

1. TSUSUCA- DOLPHIN can also be made without any power generation features as buoyant side protection for flooding prevention.
2. TSUSUCA-DOLPHIN can be used in various ways to promote beach tourism and serve as a tsunami barrier. A winch can be used to simulate and show how the device will work in the event of a tsunami. People can also be present inside the device during such simulation
3. TSUSUCA-DOLPHIN can be made in any other shape. The present shape is like that of a DOLPHIN. However, later it can be made in any other shape.
4. Power conversion can be through a rotary gearbox and also through a motor/generator with a flywheel arrangement.
5. Stay lines can also be provided from the front side (facing the ocean) anchored to the seabed of TSUSUCA-DOLPHIN so that reaction loads due to flooding can be minimized on the foundation.

Typical specifications for TSUSUCA-DOLPHIN for a few applications are listed in Tables 4.9 to 4.11, namely nuclear power plants, coastal protection systems, and health tourism.

TABLE 4.9
TSUSUCA-DOLPHIN for Nuclear Plants

Dimensions of 1 module
Length (overall including buoyancy region): 30 m
Length (tsunami blocking part, excluding nose): 21 m
Width: 20 m
Tsunami flooding blocking capability: 20 m
Wave power:20 kW – 200 kW (depending on sea state)
Solar power: 40 kW
Captive desalination plant:10000 lpd
Battery storage bank:100 kWh
Cost of one module: USD 0.40 Million

TABLE 4.10
TSUSUCA-DOLPHIN for Desalination and Coastal Protection Modules

Dimensions of 1 module
Length (overall including buoyancy region): 30 m
Length (tsunami blocking part, excluding nose): 10–21 m
(Structural design can be done with lighter members depending on the tsunami flooding in an area)
Width: 20 m
Tsunami flooding blocking capability:20 m
Wave power:20 kW – 200 kW (depending on sea state)
Solar power: 40 kW
Captive desalination plant: 10000 lpd
Battery storage bank: 100 kWh (during nights)
Cost of 1 module: USD 0.2–0.4 million

TABLE 4.11
TSUSUCA-DOLPHIN for Beaches (Health Tourism and Coastal Protection Modules)

Dimensions of 1 module
Length (overall including buoyancy region): 30 m
Length (tsunami blocking part, excluding nose): 10–21 m
(Structural design can be done with lighter members depending on the tsunami flooding in an area)
Width: 20 m
Tsunami flooding blocking capability: 20 m
Wave power:20 kW – 200 kW (depending on sea state)
Battery storage bank: 10 kWh (during nights)
Cost of 1 module: USD 0.15–0.3 million

4.9.3 Cost Benefits

The presence of a tsunami barrier like TSUSUCA could have protected the coastal Fukushima Daiichi nuclear power plant. The cost of construction of the TSUSUCA barrier would have been about USD 60–75 million (for a 3 km barrier with 500 m side protection). The plant had a power generation capacity of 4700 MW. The cost of the plant would have been about USD 12.5 billion. TSUSUCA for such a location could have resulted in wave power generation of 4.5 to 30 MW (for the 3 km barrier, depending on the sea conditions). A combination of solar power of 40 kW per module of 20 m width and 30 m length could generate an additional 6 MW of power. Though the capital investment would be high, about USD 6.5 million per MW for TSUSUCA, the revenue generated, if combined with income generated through its captive utilities like desalination (15 MLD), could generate an income of USD 20,000 per day for RO water. A few airconditioned sunrooms modules could be added for sunbathing with ocean wave movements, which could also generate tourism revenue (if permitted near nuclear power plants, for which special permissions would be needed). In such cases, the investment in TSUSUCA System would have rendered the investment attractive in addition to protecting the Fukushima nuclear power plant. The cost of construction of a 20 m high wall for 3 km length to act as a dam against tsunami flooding would result in an expenditure of more than USD 150 million with no other income benefits from the arrangement. Japan has already constructed 400 km of seawalls of 13 m height, and it is reported that the people feel that it looks like a prison, and the sea view is also blocked. However, in the case of TSUSUCA-DOLPHIN, the sea view is blocked only during a tsunami. It is also shaped like a dolphin, and it can be designed to provide an aesthetic appearance to beautify beaches.

Other tsunami devices and wave-energy devices can alleviate tsunamis but may not be sufficient to provide full protection to a nuclear plant. TSUSUCA could be considered for the recently discussed Kudankulam or Kalpakkam nuclear plants in India or other coastal nuclear plants. Similarly, TSUSUCA could be used for combating coastal erosion for a 10 km length between Chennai port and Ennore port and could have very high cost benefits considering the high value of land in a city like Chennai. It could also provide desalinated water, health benefits, and tourism income as many configurations could be designed considering the 10 km length of the coastline. Further, as green engineering concepts are used in the proposed design, there shall be no waste generation in TSUSUCA-DOLPHIN. Table 4.12 shows the patent citations and non-patent citations, based on which the concept is conceived.

Non-patent Citations

Twin wing tsunami barrier: www.globalfloodds.com/, www.noort-innovations.nl/TWTB_Documentation.pdf
Offshore tsunami walls: www.aktsunami.com/draft/lessons/58/unit8/atep_58_StructuralCountermeasures_VA.pdf

TABLE 4.12
Patent Citations

Internet References	Patent Details
WO2012023650	Retaining Wall Construction System for Preventing Tsunami and Flood Damages and Construction Method thereof
US6050745	Wave Breaker Steps for waterfront bulkheads, seawalls, and seacoast
US4856935	Beachfront wave energy dissipation structure,
GB987271A	Breakwater for coast protection,
FR2729981A1	Wave Protection Barrier,
CN1804224A	Tsunami preventing method,
WO 2011128524 A1	Wall for protection against tsunamis
WO 1999024675 A1	Protective elements, devices comprising said elements and methods for protecting a zone against floods and avalanches,
US 20070253784 A1	Hurricane pulse and tsunami duty protective seawalls
WO 2013030810 A1	Structure and method for protection against tsunami waves and high sea waves caused by storms
WO 2013029007 A1	Pile arrangement for wave barriers and methods,
US 4006598 A	Breakwater system

SUMMARY

Unlike the other forms of matured technology like wind energy, wave-energy systems are still at several stages of development, competing against each other. In the last two decades, most of the research and development activity in wave energy has been focused in the developed countries mainly due to the financial support and coordination provided by the investment policies. However, in the last few years, scientific interest in wave-energy exploration has grown rapidly in other parts of the world. In general, developing a wave-energy device from the concept to the commercial stage is a slow and expensive process. Yet still, significant progress is made in the theoretical and numerical modeling of wave-energy converters by various researchers worldwide. While model testing in a wave tank is time-consuming and considerably expensive, but still essential to validate the proof-of-concept, a final stage is to test the prototype under real sea conditions. It is important to note that the geometry of the proposed devices is constrained to the wavelength possible under the experimental setup while aiming at the energy absorption under the resonance period. For these reasons, even if pilot plants are to be tested in the open ocean, they must be large structures. However, high costs of fabricating, deploying, maintaining, and testing these prototype devices pose serious limitations to the advanced development of wave-energy systems. Most of the successful inventions are backed by liberal financial assistance extended by the local government.

The research methods deliberating the design and development of a novel device discussed in this chapter is a sincere attempt made by the authors to intuit more novel designs soon. The presented case studies demonstrate the potential of electrical power generation using the proposed devices. Research methodology and the results of the experimental works are truly an alternative to traditional fossil fuels. Further, the preliminary studies, as discussed, are found to be suitable for a wide range of wave periods; they are also cost-effective to manufacture and maintain over a while. Experimental investigations on a novel shape of a deep-water wave energy converter showed the motion response and force characteristics of a cylindrical wave float at different drafts. The influence of the change of shape of the float is presented by attaching a fin to the cylinder. Based on the experimental studies conducted, a draft of 0.2 m of the float generated maximum power. The performance of the buoy varied greatly with approach angle. For some approach angles of the fin, the device's performance is better than that of a buoy without fin. In general, the addition of fin decreased displacement and increased force. For all conditions, peak power is observed at a 1.5 s period and an approach angle of 180° which showed the highest power capture.

Mechanical wave energy converters (MWEC) have other advantages in deep ocean mining to provide part of the energy requirements. They also can be designed effectively to satisfy multiple utilities. One such wave device that can be used for tsunami flooding prevention and for captive utilization of the power generated internally (TSUSUCA-DOLPHIN) has been described in detail. It has potential for wide application in coastal desalination, ocean living simulation, coastal protection, and providing health benefits from the ocean, with all machinery present internally within the device. Such devices can become highly cost-effective as well as protect installations like nuclear plants.

EXERCISES

1. What makes ocean energy a higher potential source of green energy?
2. List the differences between wind and wave energy. Compare benefits of wave energy with other renewable energy sources.
3. Draw a neat sketch of an MWEC as described by the author and explain its working principle.
4. What is the use of the RPM multiplier in the MWEC, as discussed by the author?
5. Write the equation of motion of a floating buoy, restricted in heave motion, and explain its components.
6. What are the factors that govern the power take-off design?
7. Explain the design principles of rack and pinion, as used by the author in MWEC.
8. What is called the power transmission ratio?
9. How is a double rack MWEC superior to MWEC? Explain with a neat sketch.
10. How is failure assessment done for mechanical components of the device in the design stage?
11. Compare design FMEA with process FMEA
12. What do you understand by earlier and late failure mode? Explain your answer.
13. What are the objectives of carrying out design FMEA?
14. Explain the conceptual design of deep ocean wave energy converters.

15. What is slapping and slamming forces?
16. How do you define a response amplitude operator?
17. Explain multi-objective green engineering approaches
18. Draw a neat sketch of TSUSUCA-DOLPHIN and explain its working principle.
19. What are various applications of TSUSUCA-DOLPHIN, as discussed by the author?
20. How are multi-objective design modules cost-effective? Explain this in the context of TSUSUCA-DOLPHIN

5 Offshore Wind Turbines

5.1 INTRODUCTION

Non-renewable energy resources such as oil, nuclear power, coal, and natural gas are the primary energy sources for most geographies. Fossil fuels, the primary source of these resources, are limited and subject to market price fluctuations. Further, the extraction of oil and gas also poses a threat to the environment. Successful generation of nuclear power brings with it the possibility of a nuclear catastrophe; radioactive waste storage and disposal are serious challenges (Chandrasekaran and Sricharan, 2021). Wind energy is one of the most promising options of energy extraction, because of its renewable nature and pollution-free extraction process (Musial et al., 2006; 2004; Musial and Butterfield, 2004; Butterfield et al., 2004; 2005; 2007). European countries are ahead in offshore wind energy extraction. The offshore wind energy plant of about 900 MW capacity, commissioned in the Baltic and the North Sea, is a classic example. Wind energy is one of the evolved and cheapest technology compared to its counterparts. Environmental problems and human discomfort are major concerns of onshore wind farms, overcome in offshore wind energy plants (Taboada, 2015; Sclavounos, 2008; Tong, 1998). Consistency and strength of wind are attributed as the prime reason for the greater potential of offshore wind energy (Diaz and Guedes, 2020; DNV, 2018). The following are a few advantages and disadvantages (Henderson and Patel, 2003; Henderson and Morgan, 2003; Henderson and Witcher, 2010).

- Turbulence intensity is lesser in the sea than land, and shear at land is more than at sea.
- Constraints are not imposed by road or rail on the size of an offshore wind turbine installation, unlike onshore plants.
- Noise and visual disturbances are avoided if they are installed a sufficient distance from shore.
- Vast expanses of the uninterrupted open sea are available, and the installations will not occupy land, interfering with other land uses.

Disadvantages are as follows:

- CAPEX cost is very high due to the turbine and components, support system, and its foundation.
- Accessibility is highly compromised, which increases the downtime of the plant in case of maintenance and repair.
- Under the aerodynamic loads, the design of offshore wind turbines is complex.

DOI: 10.1201/9781003281429-5

The predominant availability of offshore wind resources is seen in water depths greater than 30 m along with Norway, China, United States, and Japan. Fixed-bottom support structures are widespread in Europe due to shallow water installations at depths lesser than 20 m; however, these technologies are not economically feasible in deeper waters (Musial and Butterfield, 2004). Floating platforms prove a more viable option (Butterfield et al., 2004). These platforms possess excellent response characteristics as they alleviate laterals loads by their geometric design, not by strength. FORM-dominance is the key factor (Chandrasekaran, 2015; 2016b; 2017; 2019b; 2020). Butterfield et al. (2004; 2005) presented the economic potential of the floating platforms, highlighting a wide range of platform configurations for offshore wind turbines. The factors considered are the combinations of mooring systems and ballast methods to pre-install the platform. The design classifications are based on the ability to achieve static stability. One of the classic and common examples of such design is a SPAR buoy whose stability is achieved by lowering the mass center below the center of buoyancy.

5.1.1 Support Systems for Wind Turbines

A single point anchor reservoir platform consists of a deep-drafted floating caisson, a cylindrical hollow structure similar to a huge buoy (Moo, 2013). The deep draft of SPAR platforms provides favorable motion characteristics and offers a stable support system for offshore wind turbines. Due to deep-draft geometry, SPAR is position-restrained by a catenary mooring system and remains stable even when the mooring lines are disconnected (Hang et al., 2003; Islam et al., 2017). SPAR platforms have a low pitch and heave motion compared to other platforms, making them safe and operable up to a water depth of about 3000 m. Fig. 5.1 shows the schematic view of a SPAR platform.

Triceratops is one of the recently conceived new-generation offshore platforms. It consists of a deck and three buoyant legs, which are position-restrained by a set of taut-moored tethers, as shown in Fig. 5.2. Triceratops's most innovative component different from other platforms is the ball joint (Chandrasekaran and Jamshed, 2017; Chandrasekaran and Kiran, 2018; Chandrasekaran and Senger, 2017). Ball joints connect the deck and buoyant legs and restrain the transfer of rotational motion between them; only the translational motion between the deck and buoyant legs is transferred (Chandrasekaran and Madhuri, 2012a; 2012b; 2012c). As a support system for offshore wind turbines, offshore triceratops help to isolate the deck from the legs using ball joints. This improves the deck's stability to support the wind turbine mast (Chandrasekaran, 2017; Chandrasekaran et al., 2013a; 2010; 2011). Triceratops is stiff in the vertical plane and compliant horizontally, similar to TLPs (Chandrasekaran and Nagavinothini, 2017; 2018; Chandrasekaran and Jamshed, 2015; Chandrasekaran et al., 2015b). The buoyant legs of triceratops resemble the hull of the SPAR platform. Thus, triceratops derives advantages from both TLP and SPAR (Chandrasekaran and Vishruth, 2013). Besides, triceratops attracts less force due to reduced waterplane area.

Offshore triceratops has a triangular deck, which is supported by deep-draft buoyant legs. These legs are taut-moored to the seabed using pre-tensioned tethers.

FIGURE 5.1 SPAR platform (Courtesy: Chandrasekaran and Jain, 2016)

While compliancy offered in the horizontal plane is similar to that of a TLP, both taut-moored tethers and deep-draft buoyant legs offer stiffness in the vertical plane. Offshore triceratops is, therefore, a hybrid combination of a TLP and SPAR. Motion characteristics of triceratops resemble that of a TLP and SPAR in the horizontal and vertical planes, respectively. Triceratops exhibits rigid body motion in translational degrees of freedom while remaining flexible in rotational degrees of freedom. Ball joints placed between the deck and buoyant legs partially isolate the deck by restraining rotation transfer from buoyant legs to the deck and vice-versa. Under lateral loads, rotational responses of the deck are significantly less than that of the buoyant legs. On the other hand, as the waterplane area of buoyant legs is greater due to deep-draft conditions, the possibility of corrosion is higher.

Alternate support systems for offshore wind turbines are TLP, whose stability is regulated by the tether tension variation. Both TLP and SPAR are designed to remain flexible in the horizontal plane and stiff in the vertical plane. In particular, TLPs have considerable periods (about 100 to 120 s) in surge, sway and yaw motion while short periods in heave, pitch, and roll motion (about 2–5 s). This implies a hybrid property,

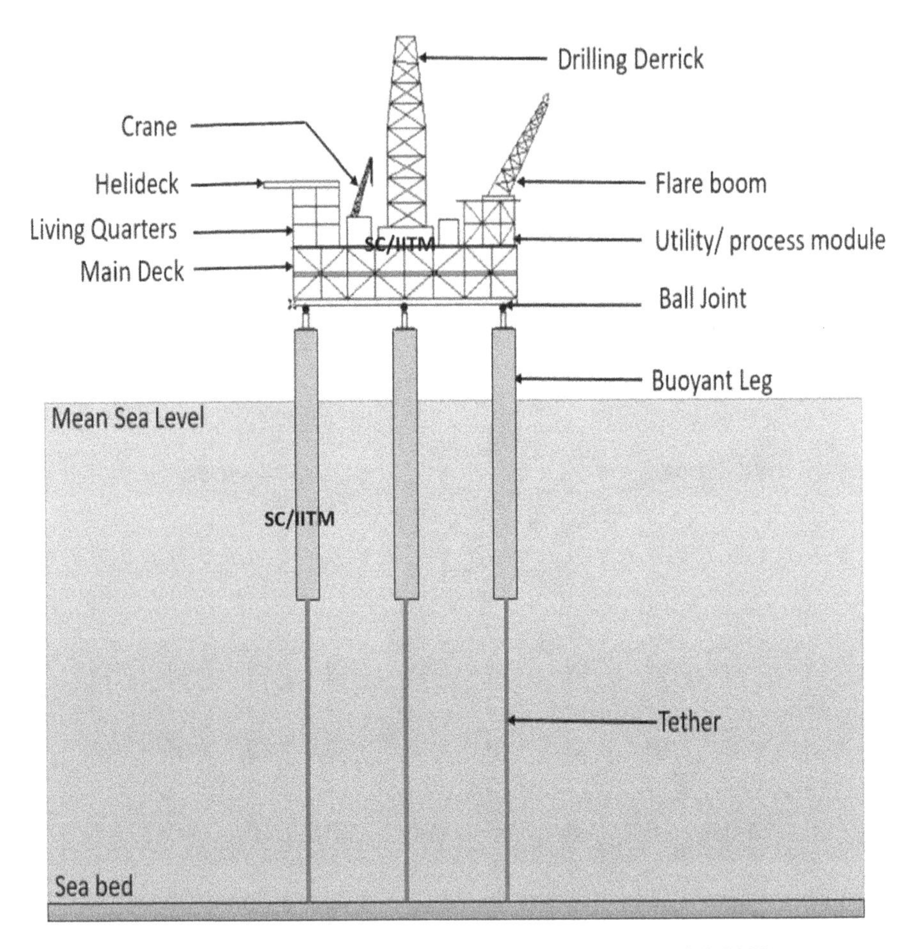

FIGURE 5.2 Triceratops (Courtesy: Chandrasekaran and Nagavinothini, 2020)

named compliant platforms (Chandrasekaran and Jain, 2002a; 2000b). Floating offshore platforms are designed for long-term survivability for oil and gas exploration and are competent to withstand severe wave and wind loads (Henderson and Patel, 2003; Bae and Kim, 2011). Therefore, using them as a base support for offshore wind turbines is a feasible idea, but the design of the base becomes more complex due to big moments and payload that arise from the mast of the wind turbine (Fulton et al., 2006; Nielsen et al., 2006; Joseph et al., 2009). Design requirements for onshore wind turbines are well defined and insist a coupled dynamic analysis of the wind turbine and the base support before the machine certification (IEC-61400, 2005; IEC-61400–3, 2006; Wayman et al., 2006; Bryne et al., 2002).

5.2 WIND POWER

Many offshore wind farms have been commissioned in the shallow waters in Denmark, United Kingdom, and the Netherlands in the recent past. In terms of installed power, the main projects are namely Lynn and Inner Dowsing, U.K. (194 MW), the Kentish

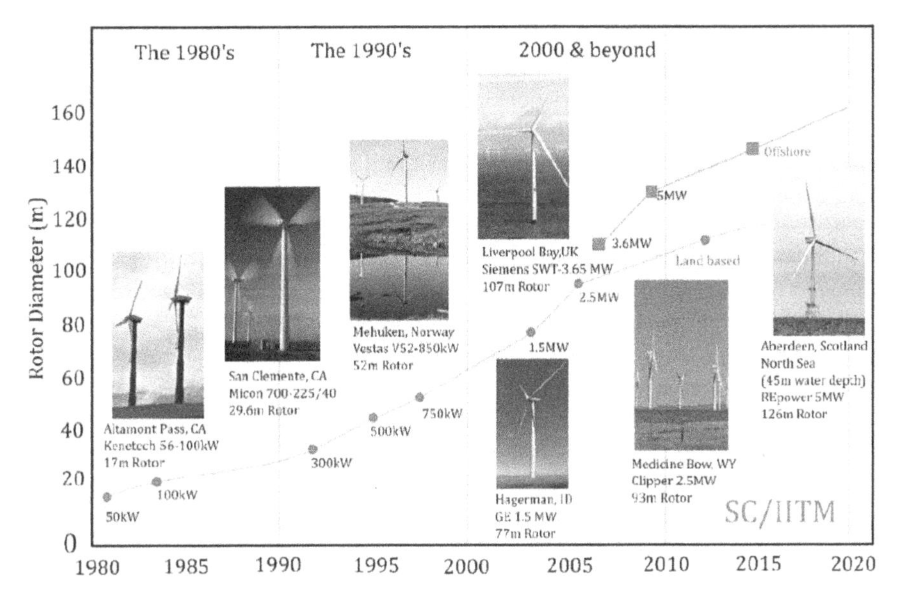

FIGURE 5.3 Evolution of wind turbine technology with time

Flats project, U.K. (90 MW), the Burbo Banks project, U.K. (90 MW), the Q7 project, the Netherlands (120 MW), the Nested Offshore Windfarm, Denmark (165 MW) and the Horns Rev project, Denmark (160 MW). As the water depth is less than 25 m for these wind farms, it proved economical to use concrete gravity structures or steel monopiles as base support systems (Henderson and Morgan, 2003; Karimirad, 2010; 2014). Recently, in Portugal, a 2 MW prototype offshore wind turbine is supported by a floating platform, WindFloat (Roddier et al., 2009; 2010; Aubaalt et al., 2009; Cervelli et al., 2009). In the early and mid-1980s, the typical wind turbine size was less than 100 kW. By the late 1980s and early 1990s, the size had increased up to 500 kW, further improving up to 1000 kW. Now turbines are available with capacities above 5MW (Fig. 5.3) (Vivek and Dennis, 2016). Offshore floating wind farms are becoming highly attractive. They can minimize the scenery disturbance, avoid noise generated by wind-driven blades, provide high wind speed by low surface roughness, and use extremely abundant deep-water wind resources.

5.3 NUMERICAL TOOLS

Numerical analysis is an inherent part of offshore wind turbine aerodynamic analysis and must respond to its response behavior under different sea states. Ansys AQWA can simulate linearized hydrodynamic fluid wave loading on floating or fixed rigid bodies. This is accomplished by employing three-dimensional radiation/diffraction theory and Morison's equation in regular waves in the frequency domain. Unidirectional or multiple directional, second-order drift forces are evaluated by the far-field, near field solution, or a full, quadratic transfer function (QTF) matrix. Free-floating hydrostatic and hydrodynamic analyses in the frequency domain can also be performed to assess the response behavior of the turbine.

Ansys AQWA can estimate the platform's equilibrium characteristics and dynamic stability, coupled with the mast under the steady-state environmental loads. It can perform frequency-domain statistical analysis of the coupled responses of the platform under irregular waves. The linearized drag due to Morison elements (tube, disc), wind, and dynamic cables can also be simulated in Ansys AQWA (Zambrano et al., 2006; Larsen and Hanson, 2007).

The real-time motion of a floating body under regular or irregular waves can be simulated, in which non-linear Froude-Krylov and hydrostatic forces are estimated under instantaneous incident wave surface. Additionally, the real-time motion of the floating body while operating in multi-directional irregular waves can be simulated using the first- and second-order wave excitations. Wind and current loading can also be applied to the bodies as external forces at each time step through a user-defined dynamic-link library. For multi-body dynamics, coupling effects between the bodies can also be simulated. The convolution approach is used to account for the memory effect of the radiation forces. Wave loads on the platform are estimated using the radiation/diffraction approach (Lefebvre and Collu, 2011; Agarwal and Manuel, 2011).

The FAST (Fatigue, Aerodynamics, Structures, and Turbulence) code is used for simulations, employing a combined modal and multi-body dynamics formulation. FAST code can model different configurations of wind turbine configurations, namely three-bladed turbines with a rigid hub, two-bladed turbines with a rigid or teetering hub, turbines with gearboxes or direct drives, turbines with induction generators or variable-speed controllers, turbines with active blade-pitch regulation or passive stall regulation, turbines with active or passive nacelle-yaw control, and turbines with passive rotor or tail furling (Wayman et al., 2006; Butterfield et L., 2005). The FAST code was developed by National Renewable Energy Laboratory (NREL), and it is a fully coupled dynamic analysis simulator that could be used for floating wind turbine concepts. FAST code uses two subroutines such as AeroDyn and HydroDyn; wind and wave data are vital data inputs for the analysis. AeroDyn calculates the wind load along blades using blade element theory. TurbSim (Turbulence modeling scaling) generates wind data for the FAST code, and HydroDyn calculates wave load on the supporting platform. The FAST code simulates stochastic time-domain turbine response and computes hydrodynamic loads using Morison's equation.

5.4 OFFSHORE WIND TURBINE CLASSIFICATIONS

Offshore wind turbines are classified into three major types based on water depth: shallow water, transitional and floating wind turbines. Shallow water wind turbines are commissioned up to 30 m water depth while the transitional type is installed between 30 to 60 m. The floating type is preferred for water depths greater than 60 m. Wind turbines in shallow waters rest on mono-pile, gravity-base, or suction-bucket structures. In contrast, the transitional ones are supported by different foundation systems: tripod tower, guyed mono-pile, full-height jacket, submerged jacket, or enhanced suction-type structures. Floating wind turbines rest on SPAR, TLP, or a triceratops. With the recent insight of offshore wind power projects proposed in deep-water to capture higher velocities, floating wind turbines resting on compliant type offshore platforms are under comprehensive exploration.

5.5 OFFSHORE FLOATING WIND TURBINE: COMPONENTS

There are two types of wind turbines, namely the horizontal axis (HAWT) and the vertical axis wind turbine (VAWT). The major sub-systems of a wind turbine are blades, nacelle, controller, generator, rotor, and the tower (also called the mast). Rotor houses the blades, which in turn are attached to a hub of the turbine. The number of blades, their geometric profile, and their length govern the performance of a wind turbine. The upwind or downwind design approach is used in designing the rotor, while the design of the blades is governed by the method of controlling the pitch motion. Nacelle comprises the generator, controller, gearbox, and shafts and acts as a protective cover. The yaw-drive system controls the nacelle alignment, which connects the nacelle to the top of the tower. The gyroscopic moment is generated by the yaw of a wind turbine and is proportional to the rotational moment of inertia of the rotor. To avoid excess moments, the rate of yaw motion should be controlled. Rotor diameter and the nature of loading conditions determine the height of a tower at a given site. Dynamic coupling between the tower and rotor can lead to reduced life of the machine due to fatigue. The tower or the mast is about 60–80 m high and consists of three telescopic sections. The tower supports the wind turbine nacelle and the rotor. A generator is used to convert the mechanical output of the wind turbine to electrical output. Induction of synchronous generators is deployed, whose relative rotational speed between the rotor and the rotating magnetic field in the stator generates power. The difference between these respective speeds is called slip, which governs the wind turbine's power output.

The control system monitors the changes in the blade pitch angle, generator loading, and nacelle yaw of a wind turbine. A continuous modification in the alignment of the nacelle to the wind direction is required to maximize the power output. Controlling the pitch of the turbine blades helps reduce the torque of the rotating shaft and, in turn, minimizes fatigue damage. As the electric output from the offshore wind turbines is fluctuating, the direct integration to the electric grid is not feasible. Typically, offshore wind turbines transmit power through cables buried in the seabed; alternatively, the generated power can be stored in batteries and used later. Fig. 5.4 shows the components of a wind turbine.

5.6 OFFSHORE FLOATING WIND TURBINES

Currently, there are several offshore floating wind turbines under various stages of development. The main concern of such wind turbines is to study the effect of greater water depth on their performance to improve their power generation capabilities. The support systems for floating wind turbines in deep water fall into five main categories: SPAR type, TLP type, pontoon (barge) type, semi-submersible type, and triceratops type. In general terms, the SPAR type have superior heave performance than semi-submersibles due to their relatively deep draft and reduced vertical wave exciting forces. But more pitch and roll motions reduce their stability as the

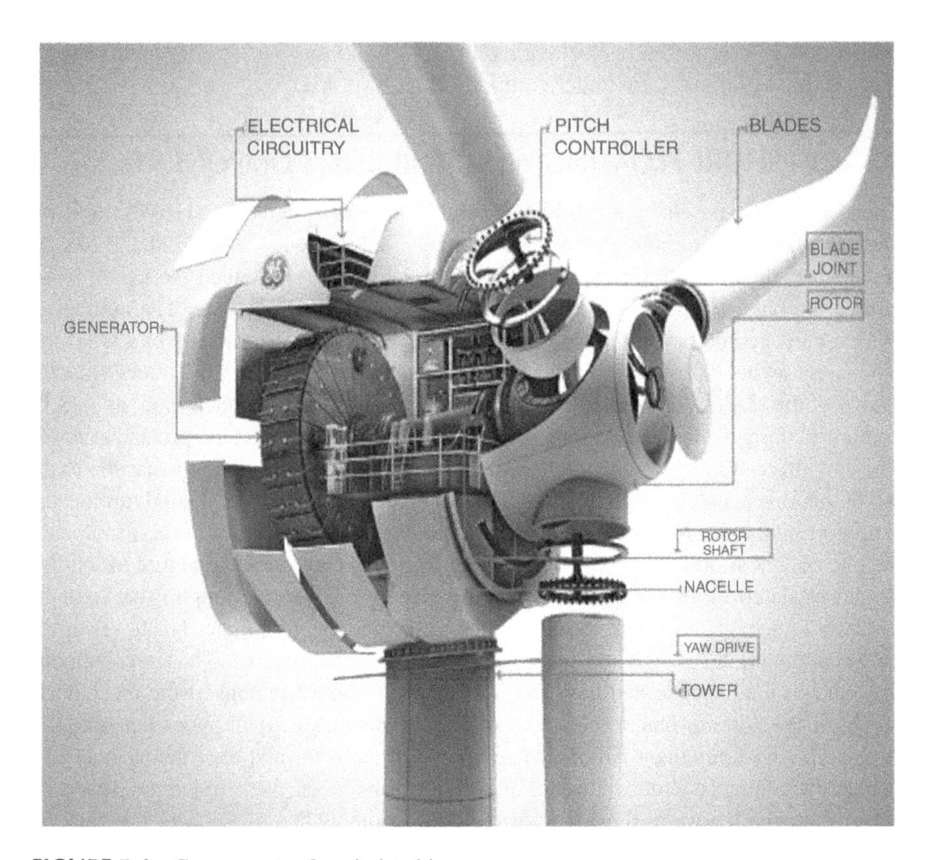

FIGURE 5.4 Components of a wind turbine

waterplane area contribution becomes limited. TLP-types exhibit a controlled heave and pitch motions mode, but some factors govern their choice as a base support for wind turbines. They are, namely, complexity and cost of the mooring installation, change in tendon tension due to tidal variations, and frequency coupling between the mast and the mooring system. Semi-submersible types are also useful as they possess better operational stability with a shallow draft. It is equally convenient to tow, install and commission a semi-submersible-type wind turbine in shallow-draft conditions. Henderson presents more details of offshore wind energy in Europe (Henderson and Morgan, 2003; Henderson and Patel, 2003), while detailed studies on the ocean, wind, and wave energy utilization can be seen in Nielsen et al. (2006) and Skaare et al. (2007, 2015). The development of offshore wind energy in the United States in the recent past showed a promising contribution and advanced research on floating wind turbines; more details of various types of offshore wind turbines are discussed in the following section.

5.6.1 Single Point Anchor Reservoir Type

SPAR-supported wind turbines consist of a supporting foundation known as the floater, the tower, and the rotor nacelle assembly (RNA). Towing is done under calm

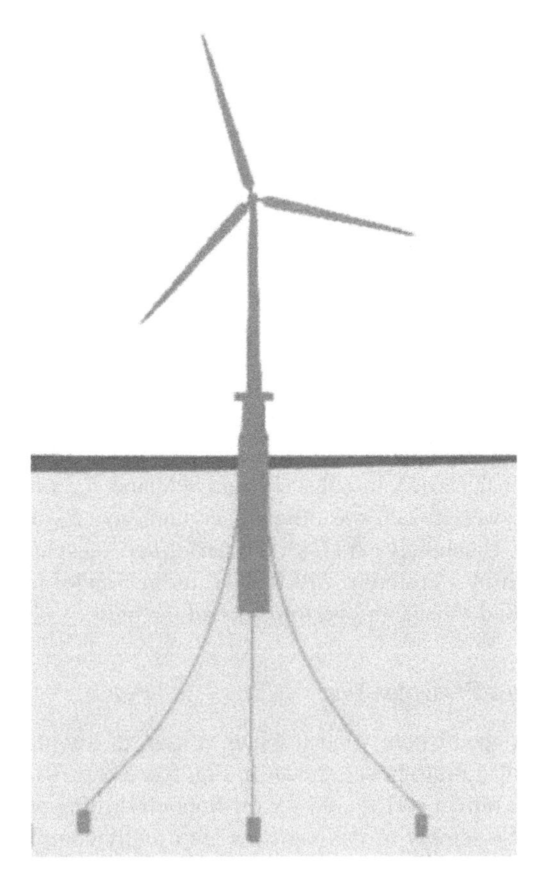

FIGURE 5.5 Schematic view of SPAR-type offshore wind turbine

water conditions, closer to the prospective deployment site. It is then followed by upending and stabilizing the support system. Later, a derrick barge is employed to mount the tower and RNA and then towed to the deployment site to connect with the mooring. Fig. 5.5 shows the schematic view of the SPAR-type wind turbine (Utsunomia et al., 2009).

Supporting foundation consists of cylindrical steel sections, in-filled with gravel and water to maintain a higher center of gravity but below the center of buoyancy. It ensures a stable floating condition (or a stay-upright condition) with a large, righting moment. It also enables greater inertial resistance to pitch and roll motion. Though SPAR-type wind turbines are recommended for greater water depths, a sufficient keel-to-seabed vertical clearance shall be ensured for the mooring system to be effective. SPAR-type wind turbines are position-restrained by the catenary mooring systems. They consist of anchor chains, steel cables, or synthetic fiber ropes. One of the common mooring configurations is a single, vertical tendon which is position-restrained at the base by a swivel connection. It allows a free rotation of the turbine mast and is adaptable to the change in wind direction. SPAR-type wind turbines are advantageous as they show better power optimization; however, under fatigue loads, a single vertical tendon may fail to prevent unrestrained drifting of the floater (Karimirad, 2010; 2014).

SPAR-type wind turbines use a large single-cylinder structure constructed of steel or concrete and ballasted by water and soil (Athanasia and Genachte, 2013). SPAR-supported wind turbines pose a higher initial cost of the foundation because of their large size. Due to its deep draft, the SPAR platform has a superior heave performance compared with semi-submersibles but has higher pitch and roll movements. The monolithic structures of SPAR are better compared to TLPs and semi-submersibles. One of the main concerns is that SPARs produce considerable wake effects in surface water than semi-submersibles or TLPs with smaller floating features. The first demonstration of SPAR-type platforms is Hywind (Atcheson et al., 2016) and SPAR at NRMI (Utsunomiya et al., 2009). The full-scale prototype of Hywind was installed in 2009 with a 2.3-MW offshore wind turbine in a water depth of about 200 m. The numerical analysis for the dynamic behavior of Hywind is presented in Nielsen et al. (2006) and Skaare et al. (2015). In 2017, Hywind Scotland, the world's first commercial floating wind power project, was officially commissioned with a 30-MW power generation capacity. Utsunomiya et al. (2009) performed a test at a 1:22.5 scale and analyzed the motion of a prototype SPAR wind turbine under regular and random waves. Studies showed satisfactory performance of the unit.

5.6.2 Tension-Leg Platform Type

The TLP-type floating offshore wind turbines consist of a typical platform structure to support the wind turbine, as shown in Fig. 5.6. These wind turbines consist of a smaller hull form than a TLP used for offshore oil drilling and production. The commissioning and assembly of the pontoons are usually completed onshore, thus

FIGURE 5.6 Schematic view of TLP-type offshore wind turbine

reducing the difficulties encountered in commissioning. Vertical tendons are axially pre-tensioned using the transfer of excessive buoyancy of the assembly. Tendons are anchored to the seabed using a template foundation, suction caissons, or piles. TLP-supported wind turbines offer better stability against the overturning moment due to a large compliancy in the horizontal plane. As buoyancy controls the design in such wind turbine supports, righting stability is achieved through the pretensioned tethers. It exhibits a lesser dynamic response under wave excitation (Bae and Kim, 2011).

The TLP is tethered to the seafloor using a taut-leg mooring system known as vertical tendons, which restrain the heave motion of the platform (Atcheson et al., 2016; Henderson and Witcher, 2010). The primary benefit of the TLP is that the substructure is smaller and lighter than other types of floating platforms, resulting in cheaper material costs. Its ability to be commissioned at the deeper water depth depends on the seabed state and geology. A TLP-supported wind turbine has a relatively less dynamic response to wave excitation than other floating structures, such as barges, SPARs, and semi-submersibles. TLPs possess desirable heave and pitch/roll motion but have an expensive and complex mooring system installation. The GICON-SOF Pilot by GICON and Blue H, a prototype TLP platform with a small wind turbine installed in a water depth of 113 m, are two experimental studies that used a TLP structure (Kolios et al., 2016; Musial et al., 2004).

5.6.3 Pontoon (Barge) Type

Barge-type floating wind turbines comprise a large pontoon base to support the wind turbine superstructure. The concepts of distributed buoyancy and weighted waterplane area are employed to achieve the required stability and righting moment. Barge-type floating wind turbines are moored by catenary anchor chains, offering a primary disadvantage of high roll and pitch motions. Hence, its suitability is limited to calm seas such as those of a lagoon or a harbor. Fig. 5.7 shows a barge-type offshore wind turbine. Of all the floating foundation types, barges have the shallowest draft, and this

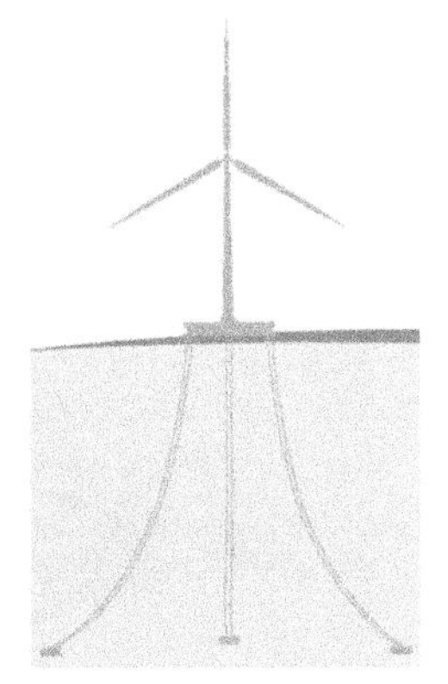

a) Schematic view

b) Conceptual view

FIGURE 5.7 Barge-type offshore wind turbine

characteristic makes them a good option when the turbine is installed near a wharf (Ikoma et al., 2021; 2019). However, the design will be subjected to more significant wave movements, necessitating more robust mooring devices. The barge is normally moored with catenary anchor chains (Ma, 2019). A moonpool is used in specific barge designs to reduce wave-induced loading (James et al., 2018). Ikoma et al. (2019) suggested that the hydrodynamic performance would be improved by applying multiple moonpools instead of a single moonpool. The downside of the barge type of FOWT is its weakness in roll and pitch motions (Chandrasekaran, 2020). The barge platform has been consented to in FLOATGEN in France by IDEOL with a 2-MW and 3-MW wind turbine (Liu et al., 2016).

5.6.4 Semi-Submersible Type

Semi-submersible-type offshore wind turbines consist of large column tubes linked to each other through tubular members. Semi-submersibles offer a high degree of compliancy and a stable vertical plane configuration (Chandrasekaran, 2020). Two prime possibilities that exist in the design are, namely, i) either a wind turbine may be placed on one of the column tubes; or ii) there could be a cluster of wind turbines supported by the columns. Column tubes are instrumental in providing the ballast as they are filled with water partially. The waterplane area of the columns provides stability to the system under floating conditions, while shallow draft expands its site employability. Fig. 5.8 shows a schematic view of a semi-submersible-type offshore wind turbine.

A semi-submersible floater is a structure with a shallow draft. It is buoyancy-stabilized at the surface. A semi-submersible-supported wind turbine can be categorized as a flexible type due to its adaptability to various site circumstances. However, the construction and installation of this type of platform are complex due to the numerous welded connections and relatively high steel mass. The structure is moored with three centenary-based mooring lines, similar to that of SPAR-supported

FIGURE 5.8 Schematic view of a semi-submersible-type offshore wind turbine

FIGURE 5.9 Semi-submersible platform supporting wind turbine

wind turbines. Column members are braced to improve strength under lateral loads (Karimirad, 2014; 2010). The wind turbine rests either on one of the tubes or all the columns, as shown in the figure (Chandrasekaran et al., 2020). In terms of met-ocean conditions, a semi-submersible may demonstrate cost-effectiveness in benign sea conditions but with a larger seabed footprint (Mast et al., 2015). Fig. 5.9 illustrates the conceptual view of a semi-submersible platform supporting a wind turbine mast. The first demonstrations of semi-submersible concepts are the Principal Power (WindFloat) and Fukushima FORWARD. A detailed design, development, and aerodynamic analysis of the WindFloat semi-submersible-type floating wind turbine was performed by Roddier et al. (2009; 2010), which showed satisfactory performance of these types.

5.6.5 Triceratops

In the recent past, there has been an evolution in the geometric forms of offshore platforms in terms of innovativeness and motion characteristics under deep and ultradeep waters. Triceratops has desirable characteristics under deep and ultra-deep-water conditions. A conceptual view of triceratops, useful for oil and gas exploration, is shown in Fig. 5.10. Novelty in the design is the deployment of ball joints between the deck and the buoyant legs. The triangular deck of the platform is supported on three buoyant legs through the ball joints, which partially isolate the deck from the legs. The ball joints isolate rotational degrees of freedom, whereas the translational

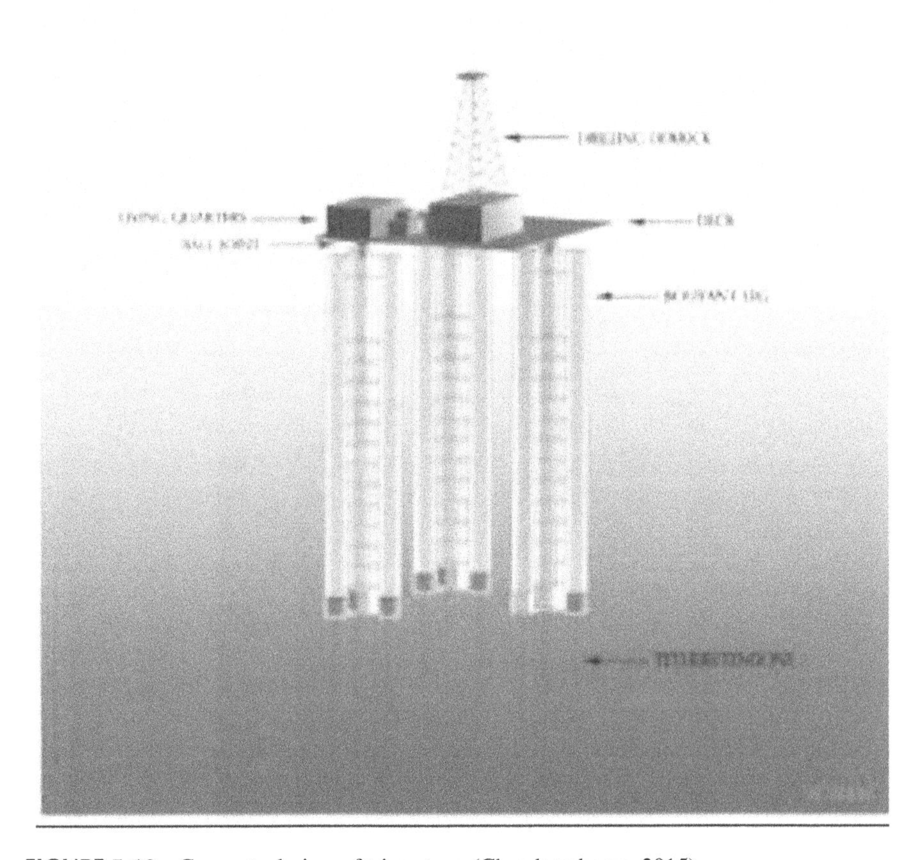

FIGURE 5.10 Conceptual view of triceratops (Chandrasekaran, 2015)

degrees of motion are transferred from the legs to the deck. The buoyant legs are subsequently anchored to the seabed using taut-moored tethers.

Triceratops is a hybrid combination of a TLP and SPAR. Motion characteristics of triceratops resemble that of a TLP in the horizontal plane and SPAR in the vertical plane. Triceratops exhibits rigid body motion in translational degrees of freedom while flexible in rotational degrees of freedom. It is markedly different from conventional platforms, where rigid body connections between all the structural units make it behave as one rigid unit (Chandrasekaran and Nagavinothini,2020). Rotational responses of the buoyant legs differ from that of the deck; buoyant legs are partially isolated from the deck, which helps reduce large displacements in the horizontal plane. It may be a useful and desirable characteristic to support an offshore wind turbine. In addition, the derived geometric form has a few salient advantages, namely i) considerable decrement in forces exerted on the platform due to minimize exposure of the structure near the free surface; ii) risers are protected from the influence of lateral loads as they are placed inside the buoyant legs; iii) dynamics and motion characteristics are superior to that of TLP and SPAR;

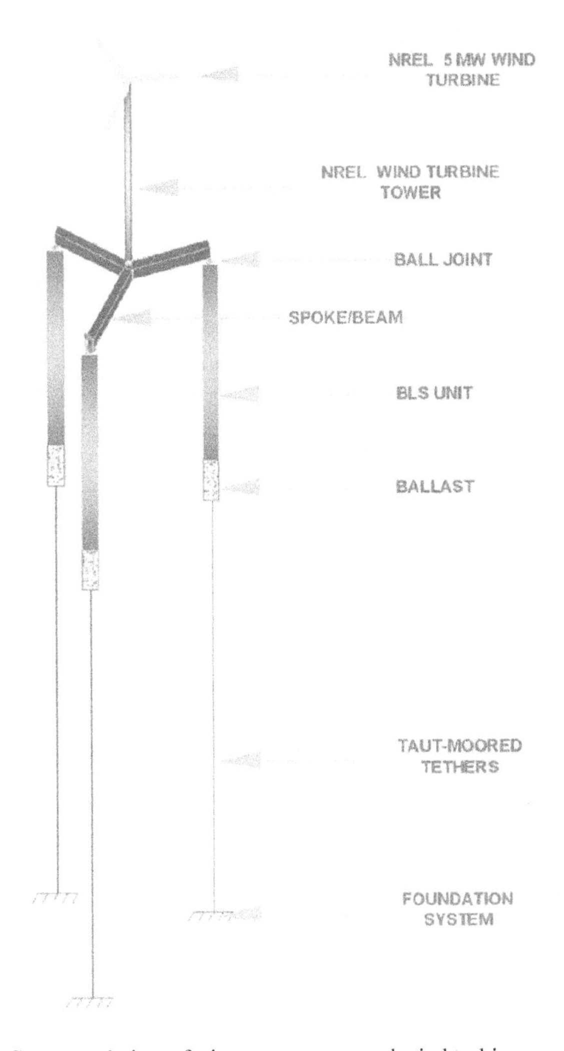

FIGURE 5.11 Conceptual view of triceratops-supported wind turbine

iv) installation and decommissioning are relatively simple; v) a high strength tether system is no more an indispensable requirement unlike that of a TLP; vi) simple station keeping characteristics; vii) a stable configuration; and ix) convenient to relocate and retrace.

Fig. 5.11 shows a schematic view of triceratops deployed for offshore oil exploration in ultra-deep waters. Even though several advantages make triceratops look far superior to other alternatives, there are some precautions one needs to have in mind. As the waterplane area of the buoyant legs is higher, there is a higher possibility of corrosion. However, these can be monitored by various means such as (i) employing sacrificial anodes; (ii) cathodic protection techniques; and (iii) anti-corrosive coatings (Chandrasekaran, 2015).

Triceratops is a recent innovative structural form of offshore structures for deep-water exploration. In ultra-deep waters, triceratops showed the ability to tolerate wave, wind, current, and ice loads, particularly in deep waters (Chandrasekaran et al., 2010; 2012a; 2012b; Chandrasekaran and Sengur, 2015; 2017, Chandrasekaran and Nagavinothini, 2018b; Chandrasekaran and Nagavinothini, 2020; Chandrasekaran and Nagavinothini, 2019a). Offshore triceratops consists of a deck and three buoyant legs (BLS). Ball joints connect the buoyant legs to the deck, connected to the restoring system comprising taut-moored tethers. This arrangement of position-keeping is similar to TLPs. Offshore triceratops differs from the other classic offshore structures because of their ball joints. These ball joints partially isolate the deck from the legs. They transfer heave, surge, and sway motion but restrain rotations between them (Chandrasekaran et al., 2013). As a result, the response of the deck under waves is reduced, which is desirable in compliant structures (Nagavinothini and Chandrasekaran, 2020). A series of taut-moored tethers hold the buoyant legs in place, making it rigid in the vertical plane while providing compliancy in the horizontal plane. Researchers examined this arrangement to its suitability for FOWT and found it satisfactory even at depths beyond 200 m (Ma, 2019). Fig. 5.12 is a conceptual view of a triceratops platform mounted with a wind turbine (Ittyavirah and Philip, 2016).

Compared to other types of foundations, the triceratops foundation is a simple installation, and the decommissioning procedure is less complex in deep water (Chandrasekaran et al., 2019). The platform can be installed in part or as a whole structure. Due to the ball joints, the platform can operate effectively, even in

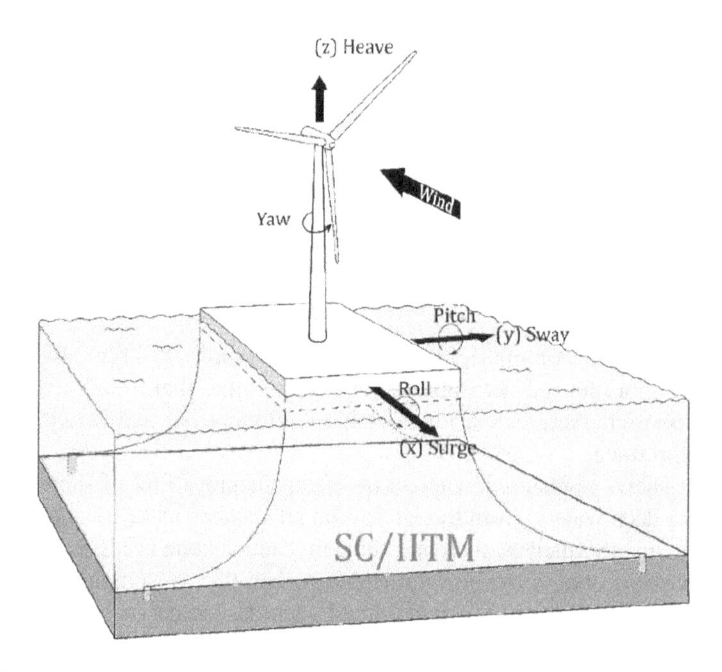

FIGURE 5.12 Degrees of freedom of the supporting platform

ultra-deep-water conditions, by preventing rotational motion from the buoyant legs to the deck (Chandrasekaran et al., 2013). It has better motion characteristics and better dynamics than TLPs and sparse platforms (Chandrasekaran, 2015). It has a simple restraining system. It is cost-effective, especially in deep water (Chandrasekaran, 2015). The deck has less pitch response, even under harsh environments (Chandrasekaran et al., 2020). A failure in tether will not result in the complete collapse of the structure (Chandrasekaran et al., 2020).

5.6.6 Triceratops-Supported Wind Turbines

As the triceratops is a new platform, a few early studies on triceratops were previously conducted concerning wind energy. These experiments confirmed the effectiveness of ball joints by preventing the transfer of the deck's rotation to the buoyant legs under the influence of wind. Ittyavirah and Philp (2016) investigated the response of offshore triceratops supporting a wind turbine under hydro-aerodynamic stresses. The non-linear analysis highlighted the response of triceratops, supporting a 5-MW NREL offshore wind turbine in 314 m water depth; FAST is used to analyze the aerodynamics. It is observed that the response amplitude operators (RAO) changed gradually with gentle peaks, showing that the structure was stable under wave loading. The sway motion remained unaffected by wind speeds, indicating that the turbine's dynamic characteristics did not affect sway. At the same time, the dynamic characteristics of the turbine have little impact on the yaw response. However, other DOFs are significantly affected by wind speed.

As triceratops is still under geometric upgradation, extensive evaluations of this type of platform with various structural adjustments may help to improve the platform's operational advantages. There are two modified types: stiffened triceratops and triceratops with elliptical buoyant legs (Chandrasekaran and Sengur, 2015). In the stiffened triceratops, buoyant legs are interconnected by additional stiffeners, making the unit a monolithic platform. Under such stiffened conditions, responses of the deck were found to be reduced but resulted in a significant increase in the tether tension of the taut-moorings (Chandrasekaran and Jamshed, 2017). Chandrasekaran et al. (2015b) presented an experimental study of a scaled model of a stiffened triceratops under regular waves. It was found that the natural periods of triceratops with stiffened buoyant legs are reduced in surge and sway motion. The stability increase observed is not just advantageous during installation but also under operational sea states as well. In addition, it was also observed that the wave direction has no effect on the surge and pitch motion of the triceratops model, which makes it suitable for deep and ultradeep waters.

In the present scenario, elliptical cross-sections for large structures are being considered, especially in offshore oil platforms. The complex hydrodynamics of an array of elliptical cylinders affect the overall response of the structure. Studies on the triceratops with buoyant legs of elliptical cross-sections (Nagavinothini and Chandrasekaran, 2020). The cross-section of buoyant legs has been shown to influence the natural frequency in stiff degrees of freedom (DOF) in free oscillation tests. It was found that even under unidirectional waves, diffraction and the Froude-Krylov forces are active; sway response increases with the increase in the eccentricity of elliptical sections. Triceratops with the ellipsoidal buoyant leg of eccentricity 2.0 performed better than other cross-sections. In different DOF, a reduction in the deck response was observed while there was an increase in the stability; a lesser tether tension variation contributes to its benefit.

5.7 EXPERIMENTAL AND NUMERICAL ANALYSES

While designing sub-structures for a floating wind turbine, one of the key considerations is to minimize the motion to achieve maximum stability. Fig. 5.13 shows the six degrees of freedom for a floating wind turbine. Environmental loads encountered by the wind turbine can be grouped into the above six DOF; waves, wind cause most of these loads, and tidal motions (Butterfield et al., 2007), as shown in Fig. 5.14.

Numerical studies are essential to quantify the response variations of an offshore platform, supporting a wind turbine on its superstructure. In recent years, several software codes have been found more appropriate and useful: FAST (Fatigue, Aerodynamics, Structures, and Turbulence) developed by NREL (Jonkman and Buhl, 2005); ADAM (Automatic Dynamic Analysis of Mechanical Systems); 3DFloat, which is used for non-linear, coupled time-domain simulations of offshore structures

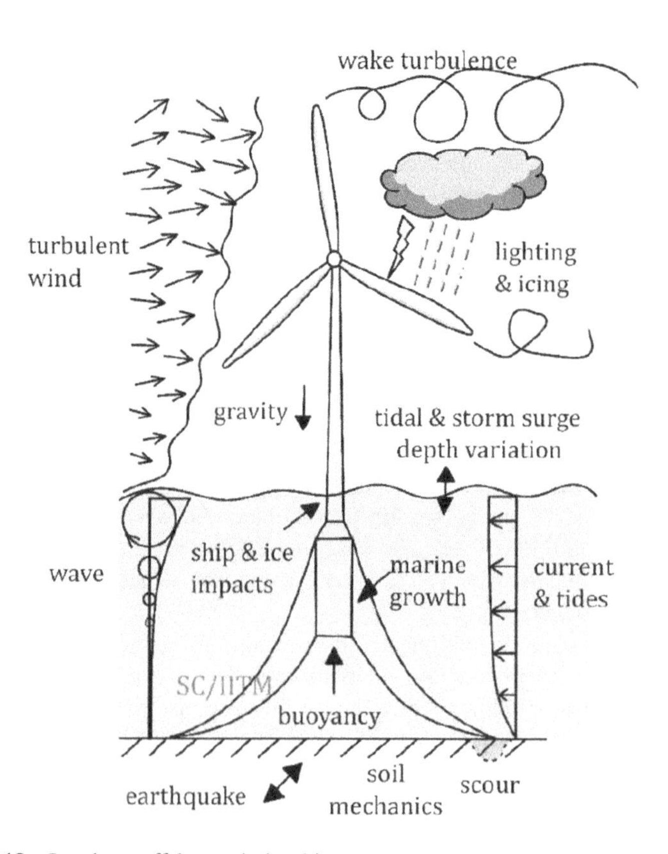

FIGURE 5.13 Loads on offshore wind turbines

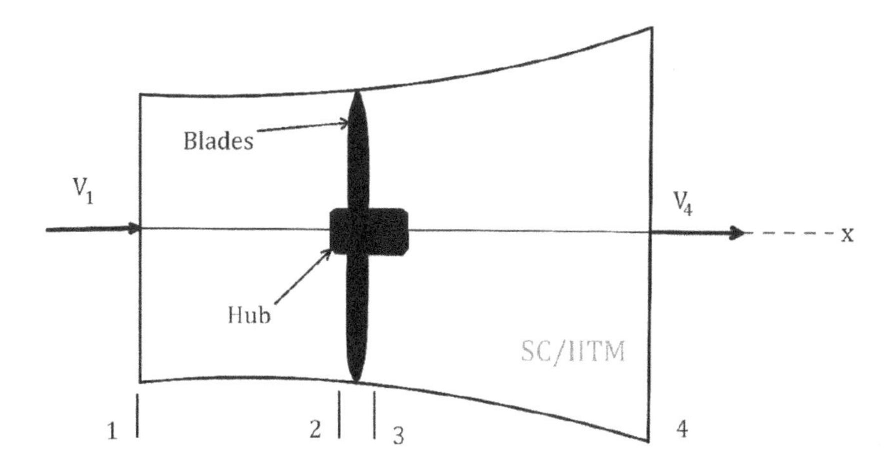

FIGURE 5.14 Schematic section of wind turbine with critical sections

(Nygaard et al., 2016); and Ansys AQWA, which is an integrated system for undertaking hydrodynamic and mooring analyses. They are used for modeling the dynamic environmental loadings and structural responses. To investigate the aerodynamic response of an offshore wind turbine under the combined action of wind and wave loads, a detailed mathematical formulation has been developed by Chandrasekaran et al. (2013). According to the study carried out on triceratops, a significant reduction in the triceratops deck response with no rotation transfer from the buoyant legs to the deck; deck remains horizontal for the entire duration of the encounter. Even at more significant wave heights, the deck's response in the pitch and yaw DOF is lower than that of the buoyant legs.

The generation of regular and irregular waves is essential to simulate the responses of the platform supporting a wind turbine. Tethers are key parts of the floating offshore platforms. Under the excitations of current, waves, and winds, failure of one or more tethers may affect the stability and functioning of the platform. An increase in the roughness of the sea state may result in the dynamic tension variation in tethers, but the effect of wind is insignificant (Chandrasekaran and Nagavinothini, 2020). The stress created in each wire of the mooring system is less when the wind and waves act together than when the waves act alone. Even at very high sea states, the maximum tension in the tethers was found to be less than the yield stress, ensuring no failure (Chandrasekaran and Nagavinothini, 2018c). The integrated wind, wave, and current impacts still cause very little pitch response, supporting the effectiveness of this new design platform.

In addition to the environmental forces, wind turbines are also sensitive to special loads, which can jeopardize the strength and stability of the entire structure. Chandrasekaran et al. (2019) studied the ice-induced response of triceratops to analyze the platform's suitability under ice crushing for different sea states. The influence of ice thickness, velocity, and crushing strength on the triceratops' response was explored. It was noted that ice loads in all DOF produce a shift in the mean response position. The analysis also showed that no rotating reaction could be transferred from the leg onto the deck. The response of the offshore structures under the extreme natural hazards that occur unexpectedly should also be considered. Although FOWTs have a lower risk to human life and the environment than oil and gas platforms (Atcheson et al., 2016), extreme events may threaten wind turbines' reliability and performance.

Earthquakes produce large tether tension fluctuations in taut-moored structures. Response analyses of the triceratops to seismic activity in the presence of waves were examined by Chandrasekaran et al. (2014c). Results showed that the heave and pitch responses of the buoyant legs and the deck increase due to the induced seismic activities. This study showed the advantages of using ball joints in triceratops platforms under seismic action. Until now, no full-scale triceratops-based wind turbine has been launched because of the high cost and complexity of constructing a full-scale offshore wind turbine. However, some triceratops physical models have been studied to examine offshore triceratops' stability and dynamic behavior (Chandrasekaran, 2015; Chandrasekaran and Madhuri, 2015a; Chandrasekaran and Madhuri, 2012a). All studies confirm the necessity to take additional care during installation and decommissioning due to the heave period of the free-floating buoyant leg, which lies within the wave period zone.

5.8 MATHEMATICAL BACKGROUND

The mathematical formulation of floating wind turbine motions under complex wind and wave interaction loading includes aerodynamics of rotor blades, multi-body dynamics of a wind turbine, and the relevant hydrodynamic theories. Some of the concepts discussed below are useful to compute forces and moments, both on the wind turbine and floater system. The following expression gives the basic form of rigid body motion for any dynamic system:

$$[M]\{\ddot{x}\}+[C]\{\dot{x}\}+[K]\{x\}=\{F(t,x,\dot{x})\} \tag{5.1}$$

The blade momentum theory is useful to estimate the induced velocities on the wind turbine blades in both the axial and tangential directions; it is an appendage of the actuator disk theory. It assumes that the airflow through the rotor of the blade creates a momentum (pressure drop) in the rotor plane and is used to determine the induced velocities on wind turbine blades. The aerodynamic forces are calculated individually and then integrated along the blade span to calculate the total forces and moments on the turbine. Flow in the rotor plane is disturbed due to these induced velocities, which leads to the change in forces obtained by the blade element theory.

5.8.1 Blade Momentum Theory

The law of conservation of momentum gives the amount of force acting on a body. In the case of a wind turbine, the one-dimensional blade momentum theory assumes an ideal rotor; extension can be made by incorporating the rotation of the rotor. Fig. 5.15

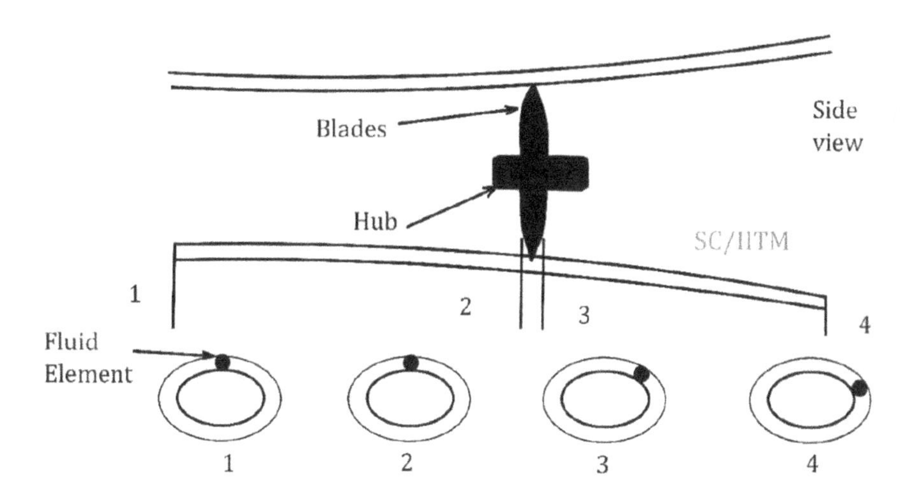

FIGURE 5.15 Side view of wind flow

shows the schematic drawing of wind turbine marked with different critical sections on the upstream and the downstream sides. Fig. 5.16 shows the side view of the wind flow.

Considering a homogeneous, steady-state, incompressible flow, the following condition is valid:

$$p_1 = p_4 \tag{5.2}$$

Hence, from the continuity equation, one can state the following:

$$v_2 = v_3 \tag{5.3}$$

Considering v_1 as uniform at Section-1, axial force and torque can be estimated using the momentum theory. By applying Bernoulli's equation to the flow between sections (1–2) and (3–4), the following equations are derived:

$$\left(p_1 - p_2\right) = \frac{1}{2}\rho\left(v_2^2 - v_1^2\right) \tag{5.4}$$

$$\left(p_3 - p_4\right) = \frac{1}{2}\rho\left(v_3^2 - v_4^2\right) \tag{5.5}$$

where $(p_{i,}\ i = 1\,to\,4)$ are the static pressures, ρ is the air density and $(v_{i,}\ i = 1\,to\,4)$ are the flow velocities at the corresponding sections. Substituting Eqns. (5.2 and 5.3) in Eqns. (5.4 and 5.5) and simplifying, we get:

$$\left(p_2 - p_3\right) = \frac{1}{2}\rho\left(v_1^2 - v_4^2\right) \tag{5.6}$$

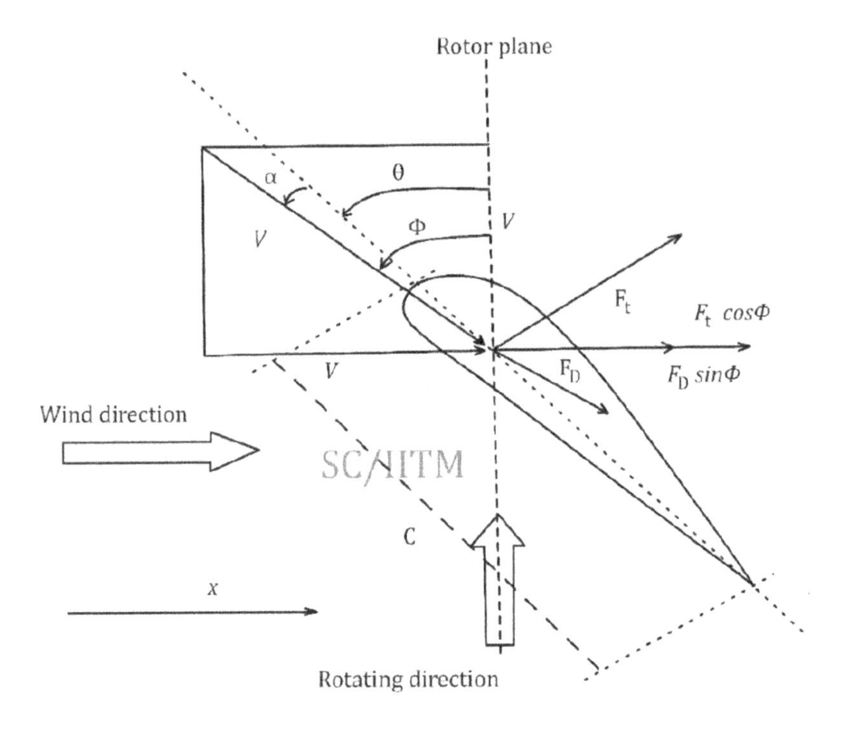

FIGURE 5.16 Lift and drag forces on wind turbine elements

An axial force is given by:

$$dF_x = (p_2 - p_3)\ dA \tag{5.7}$$

$$dF_x = \frac{1}{2}\rho\left(v_1^2 - v_4^2\right)dA \tag{5.8}$$

$$dF_x = \dot{m}\left(v_1 - v_4\right) \tag{5.9}$$

Considering the axial induction factor (a) as the ratio of the velocities of flow due to the presence of blades to that of the original wind velocities in the upstream, the following expressions hold good:

$$a = \frac{(v_1 - v_2)}{v_1} = 1 - \left(\frac{v_2}{v_1}\right) \tag{5.10}$$

$$v_2 = v_1(1 - a) \tag{5.11}$$

$$v_4 = v_1(1 - 2a) \tag{5.12}$$

Hence, the net axial force on the blade is given by:

$$dF_{x=} \frac{1}{2}\rho\left\{v_1^2 - \left(v_1\left(1-2a\right)\right)^2\right\}2\pi r\,dr \quad (5.13)$$

where (r, dr) is the radius and differential radius of the annular ring at a section. When the blades rotate, flow across the blade is not linear, and rotor rotation generates angular momentum, rotor torque. With reference to Fig. 5.16, sections (1, 2) of the wind element (marked by the dot) are unchanged, while the sections (3, 4) are influenced by the rotational effect of the moving blades on the wind element. It is characterized as the wake rotational speed (ω). The angular induction factor ($\acute{a}$) and torque are given by the following relationships:

$$\acute{a} = \frac{\omega}{2\Omega} \quad (5.14)$$

$$T = \frac{d(I\omega)}{dt} = \frac{d(mr^2)\omega}{dt} = \frac{dm}{dt}r^2\omega \quad (5.15)$$

Where Ω is the blade's rotational speed, and I is the moment of inertia of the blade element. The following relationships hold:

$$dT = d\dot{m}\,\omega r^2 \quad (5.16)$$

$$d\dot{m} = \rho dA v_2 = \rho 2\pi r\,dr\,v_2 \quad (5.17)$$

$$dT = \rho 2\pi r\,dr\,v_2\,\omega r^2 \quad (5.18)$$

Substituting, we get:

$$dT = 4\acute{a}\left(1-a\right)\rho\pi\Omega v_1 r^3 dr \quad (5.19)$$

Eq. (5.13) and Eq. (5.19) can be used to compute the net axial force and the torque on the blade, respectively.

5.8.2 Blade Element Theory

The blade element theory is based on the aerofoil theory; forces on the blade depend on the flow conditions and the two-dimensional, sectional wing data. The axial and rotational forces are calculated using the lift and drag of the blade section. Flow over an aero-foil leads to pressure distribution over its surface. The lift force caused by the unequal pressure distribution on the upper and lower surfaces of the aerofoil remains perpendicular to the incoming airflow direction while the drag is parallel. Drag shall exist due to the viscous and friction forces. At the same time, separation at the trailing edge creates unequal pressure distribution on the aerofoil surfaces, one facing towards and the other facing away from the incoming flow. Aerofoil characteristic data are required to calculate the lift and drag forces over the blade element and

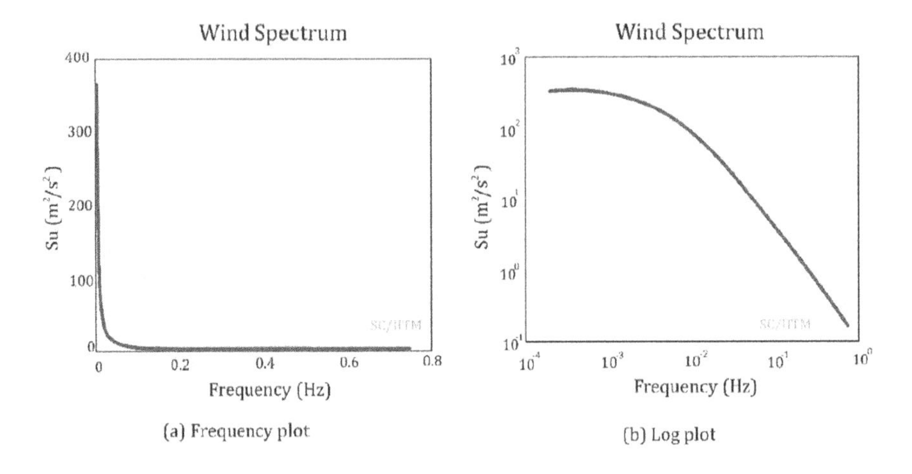

FIGURE 5.17 Frequency and log plot for the IECKAI model

function the relative velocity over the aerofoil. The velocity of a rotor blade element has two components: one along with the flow and the other tangential. The resultant flow velocity, acting over the aerofoil, is the sum of these two velocity components. Fig. 5.17 shows the lift and drag components of an aerofoil.

The average tangential speed of flow at the radial location on the blade is given by the following relationship:

$$v_t = \Omega r^2 + \frac{1}{2}\omega r = \Omega r\left(1 + a'\right) \tag{5.20}$$

where $(\omega/2)$ is the average rotational flow over the rotor plane due to wake, and Ω is the blade rotation speed. It is also important to note that the axial flow, just before the rotor is given by:

$$v_2 = v\left(1 - a'\right) \tag{5.21}$$

The following relationship gives the angle of attack (α):

$$\tan\left(\alpha\right) = \frac{\Omega r\left(1 + a'\right)}{v_1\left(1 - a'\right)} = \lambda_r \frac{\left(1 + a'\right)}{\left(1 - a'\right)} \tag{5.22}$$

Where λ_r is the tip speed ratio and defined as below:

$$\lambda_r = \frac{\Omega r}{v_1} \tag{5.23}$$

The lift and drag forces can be obtained using the following relationships:

$$dL = C_L \frac{1}{2} \rho W^2 cdr \quad (5.24)$$

$$dD = C_D \frac{1}{2} \rho W^2 cdr \quad (5.25)$$

where C_L is the lift coefficient, and C_D is the drag coefficient of the aerofoil.

The axial and tangential forces are given by the following expressions:

$$dF_x = dL\sin\alpha + dD\cos\alpha \quad (5.26)$$

$$dF_\theta = dL\cos\alpha - dD\sin\alpha \quad (5.27)$$

Combining the above equations, the following expressions are valid:

$$dF_x == B\frac{1}{2}\rho W^2 (C_l \sin\alpha + C_l \cos\alpha) cdr \quad (5.28)$$

$$dF_\theta = B\frac{1}{2}\rho W^2 (C_l \sin\alpha + C_l \cos\alpha) cdr \quad (5.29)$$

where B is the number of blades. The following expression gives torque:

$$dT == B\frac{1}{2}\rho W^2 (C_l \cos\alpha - C_D \sin\alpha) cdr \quad (5.30)$$

5.9 AERO-ELASTIC MODEL

The aero-elastic model shall be created conveniently using FAST, which is useful for two-and three-bladed HAWT. It combines different numerical models, namely FAST2 and FAST3; the former is for two blades, and the latter is for three blades, while AeroDyn is a specific subroutine for HAWT. One of the primary advantages of this numerical tool is that both Lagrange and Newton-Euler methods of multi-body dynamics are combined and used; Hence, there is no necessity to define the interactive and constraint forces between bodies, which is substituted using the generalized force approach. Furthermore, energy functions are avoided, and algorithms based on vector products are used in the code. The following equation holds:

$$M_{ij}(q,u,t)\ddot{q}_j + f_i(q,q,\dot{u},t) = 0 \quad (5.31)$$

where M_{ij} is the component of the inertia mass matrix, f_i is the forcing function, r_i is the set of wind turbine control inputs. This method for a system of bodies uses generalized coordinates to define the motion. The forces are balanced in an inertial reference frame by the following equation.

$$F_r^* + F_r = 0 \; (r = 1,2,\ldots P) \tag{5.32}$$

where P is the number of DOF and F_r^* is the generalized inertia forces, and F_r is the generalized active forces except inertia forces. In FAST, the generalized inertia forces are given as the summation of tower force, nacelle force, hub force, and the blade force, as given below:

$$F_r^* = F_r^{*(Tower)} + F_r^{*(Nacelle)} + F_r^{*(Hub)} + F_r^{*(Blades)} \tag{5.32}$$

FAST models a wind turbine with nine rigid bodies and four flexible bodies (Jonkman and Bhul, 2005; Wayman et al., 2006). The rigid bodies include the earth, support platform, base plate, nacelle, armature, gears, hub, tail, and structure furling with a rotor and the flexible bodies are tower, driveshaft, blades (in-plane), and blades (flapwise). Distributed stiffness, mass properties, and mode shapes govern the analysis. It is important to note that FAST has limitations on mode shape. In a blade, two flap-wise and one edgewise bending modes are only allowed in the analysis. For a three-bladed HAWT model, FAST considers 24 degrees of freedom, namely translation, and rotation (6 DOFs), tower flexibility (4 DOFs), nacelle yaw (1 DOF), variable generator and rotor speeds (2 DOFs), blade flap-wise (6 DOFs), blade displacement (3 DOFs), rotor furl (1 DOF), and tail furls (1 DOF).

As sub-systems are interrelated and influence the response, the reference frames for different sub-systems are handled using the coordinate system. Writing an equation of motion for a specific sub-system is much easier when compared to that of the whole system. Such equations, written using a local coordinate system, can then be related to the global coordinate system through coordinate transformations. It is at the different coordinate systems or reference frames that the Kane method defines each DOF or generalized coordinates. Derivatives of these coordinates define the partial velocities and accelerations. With the obtained velocities and accelerations, inertia forces can be calculated and balanced with generalized active forces. FAST uses nine different coordinate systems to define input and output parameters in the FAST user guide.

5.9.1 Kane Method

The following are the salient steps involved in the Kane method. The model is defined with the pre-identified rigid bodies; reference frames and points of interest are successively obtained.

(i) DOFs are defined, and generalized coordinates are obtained.
(ii) Speed, partial velocity, and acceleration are defined.
(iii) Mass and inertia of the bodies are defined.
(iv) Generalized active forces are defined.
(v) Finally, the $F_r^* + F_r = 0$ is evaluated.

5.9.2 Wind Load

An offshore wind turbine is designed to withstand extreme wind conditions safely. These wind conditions for safety and load-based considerations can be divided into normal and extreme wind conditions. Normal wind conditions frequently occur, which is referred to more than once in a year when the turbine is under operation. The extreme conditions have a recurrence period of (1–50) years. It is to be noted that the wind speed has a cubic relation with the available energy but is reduced in the presence of obstructions such as trees or buildings. Due to the effect of turbulence, the actual wind speed varies about its mean value with time and direction at any location. Also, the force exerted by the Earth's surface on the moving air retards the flow, and the region of influence constitutes the boundary layer. The power law and logarithmic methods are widely used mathematical models to define the vertical variation of mean wind speed over different regions. The following relationship gives mean wind speed using the power law:

$$V(z) = V_{\text{Hub}}\left(\frac{z}{z_{Hub}}\right)^{a} \tag{5.33}$$

Where $V(z)$ is the average wind speed at height (z) above the still water level, V_{Hub} is the wind speed at hub height, and (a) is the power-law exponent; for offshore locations, this is taken as 0.14 (IEC 61400-3, 2006). The wind conditions involve a constant mean flow combined with either varying gust profile or turbulence. Turbulence refers to the random wild velocity variations acquired from the average of 10 min records. It includes the effect of change in wind speed and direction. There are three salient components of turbulent wind velocity, namely, lateral, longitudinal, and upward.

5.9.3 Wind Shear Effect

The design of a wind turbine is dependent on the maximum load and the structural excitations, fatigue prediction, control system operation, and power quality (Bae and Kim, 2011). For an optimal design, it is important to identify the fluctuations related to turbulence. Wind shear is caused by the boundary layer above the ocean surface. It is characterized by an increase in the wind speed with altitude. Further, blades pointing upwards experience a higher wind speed compared to those pointing downwards. Hence, the wind velocity profile becomes non-linear under the influence of wind shear. Loads are generated by the wind shear on the spinning rotor blades change the rotor frequency periodically. Constant off-axis winds also cause a similar load on the turbine and the blades. The tower of a wind turbine obstructs the incoming airflow, forming a boundary layer. A stagnation point in front of the tower beyond which the velocity is reduced. Every time a blade passes the tower, the aerodynamic forces on that passing blade drop on each occasion. This oscillation excites the blade at the rotor frequency and both the tower and nacelle at thrice of the rotor frequency. Offshore wind turbines are designed so that the dominant natural frequencies of the tower and platform are well above (usually 10%) or below the rotor frequencies. It is necessary to avoid resonance, which in turn reduces the service life of the turbine.

5.10 NORMAL TURBULENCE MODEL

Turbulence adds severity to the fatigue loads and affects many major wind turbine components. Hence, knowledge of the turbulence at a site is of utmost importance.

For the normal turbulence model (NTM, the standard deviation of the turbulence (σ_1) is given by the 90% quantile of the design wind speed at a particular hub height (IEC 61400–2005):

$$\sigma_1 = I_{ref}(0.75\ V_{Hub} + b);\ b = 5.6\ m/s \tag{5.34}$$

Where I_{ref} is the expected value of turbulence intensity at 15m/s. Turbulence intensity (TI) quantifies wind speed variation within an observation period of about 10 minutes and is given as follows:

$$TI = \frac{\sigma_1}{V_{Hub}} \tag{5.35}$$

5.11 WIND SPECTRUM

Fluctuations of wind can be assumed as a combination of mean wind and sinusoidally-varying wind. These variations, including their magnitude and distribution to the frequency, can be defined using the wind spectrum. Several spectral models are available, namely, the IECKAI (IEC Kaimal model), the IECVKM (IEC Von Karman Isotropic Model), the Riso smooth-terrain model, and several NREL site-specific models (NWTCUP, GP_LLJ, WF_UPW, WF_O7D, and WF14D). Standard deviations are computed by integrating the velocity spectra (S) as given below:

$$\sigma^2 = \int_0^\infty S(f)\,df \tag{5.36}$$

For example, the PSD function of the IEC Kaimal Model spectrum, in a non-dimensional form, is given by the following relationship:

$$\frac{f\,S_k(f)}{\sigma_k^2} = \frac{4fL_k / V_{Hub}}{\left(1 + 6f\,L_k V_{Hub}\right)^{5.3}} \tag{5.37}$$

Where f is the frequency in Hz, k is the velocity component direction index (where 1 - longitudinal, 2 - lateral, and 3 - upward), S_k is the single-sided velocity component spectrum, σ_k is the standard deviation, L_k is the integral scale parameter. Fig. 5.18 shows the probability density function variation in the longitudinal direction of the IECKAI model for a wind speed of 11.4 m/s.

5.12 NUMERICAL ANALYSIS OF TRICERATOPS-SUPPORTED WIND TURBINE

Wind turbine mounted on triceratops at 600 m water depth is modeled; details of the platform are given in Table 5.1. The deck is modeled as a plate element, while the buoyant legs are modeled as cylindrical members. Meshing is done using quadrilateral plate elements with a total number of nodes and elements are 9406 and 9398,

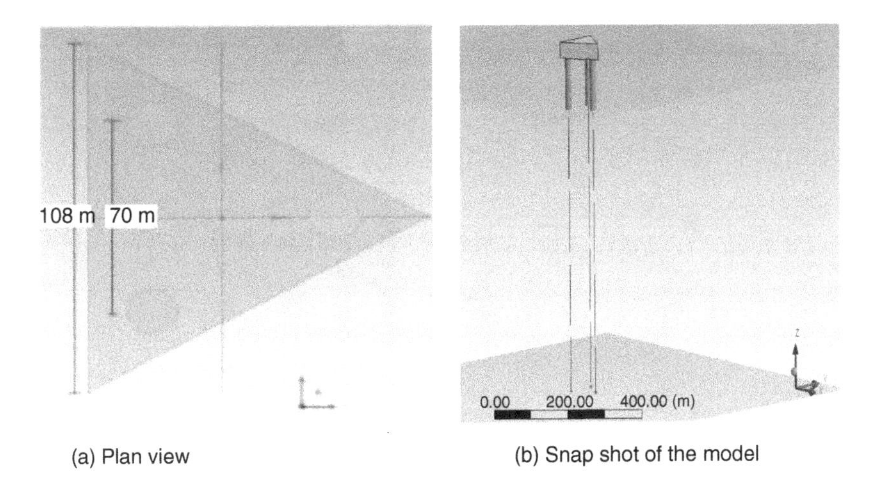

(a) Plan view (b) Snap shot of the model

FIGURE 5.18 Triceratops for supporting wind turbine

TABLE 5.1
Details of the Platform and the Wind Turbine

Description	Units	Values
Water depth	m	600
Material density	kg/m3)	7850
C/C of legs	m	70
Diameter of each leg	m	17
$I_{xx,}I_{yy}$ of the legs	Ton-m^2	26016366
I_{ZZ} of the legs	Ton-m^2	688795.11
$r_{xx,}r_{yy}$ of the legs	m	51.46
r_{zz} of the legs	m	8.37
$I_{XX,}$ I_{YY} of the deck	Ton-m^2	3558415.8
I_{ZZ} of the deck	Ton-m^2	5894093.2
$r_{xx,}r_{yy}$ of the deck	m	29.6
r_{zz} of the deck	m	38.1
Area of deck	m^2	1732.4
Freeboard	m	25
Draft	m	76.7
Metacentric height	m	12.24633
Axial stiffness of tether	kN/m	84000
Buoyancy force	kN	86740
Self-weight of platform	kN	39818.79
Weight of wind tower mast	kN	6842
Initial tether tension	kN	46921.21

FIGURE 5.19 Snapshot of wind turbine on triceratops

respectively. Wind turbine mast is considered as point load and applied on the CG of the platform. Schematic view and a snapshot of the numerical model are shown in Fig. 5.19. A snapshot of the platform supporting the wind turbine is shown in Fig. 5.20.

After verifying the stability check, the model is subjected to wave loads for sea state-3 with a wave heading of 0^0. The response amplitude operators (RAO) in the active degrees of freedom are obtained. The power spectral density plots of the free decay response are shown in Fig. 5.21 for the active degrees of freedom; PSD of the input wave spectrum is also over-laid for comparison. Rotor frequency is also marked in the figures for clarity. As seen in the plots, peaks occur at 0.4375 Hz and 0.2387 Hz for heave and roll motion; for surge and yaw, they occur at 0.0159 Hz and 0.02475 Hz, respectively. These peaks occur at the natural frequencies of the respective degree of freedom. Due to a symmetric geometric form, (roll, pitch) and (surge, sway) responses are similar under the 0^0 wave heading; hence would share the same natural frequencies.

It can be noted that the power spectral density (PSD) plots of the responses shift towards the right of the dominant frequency of the chosen wave spectrum in stiff degrees of freedom (heave, roll). In the case of the flexible degrees-of-freedom, this shift is towards the left, ensuring no-resonance condition. Further, as the peaks of

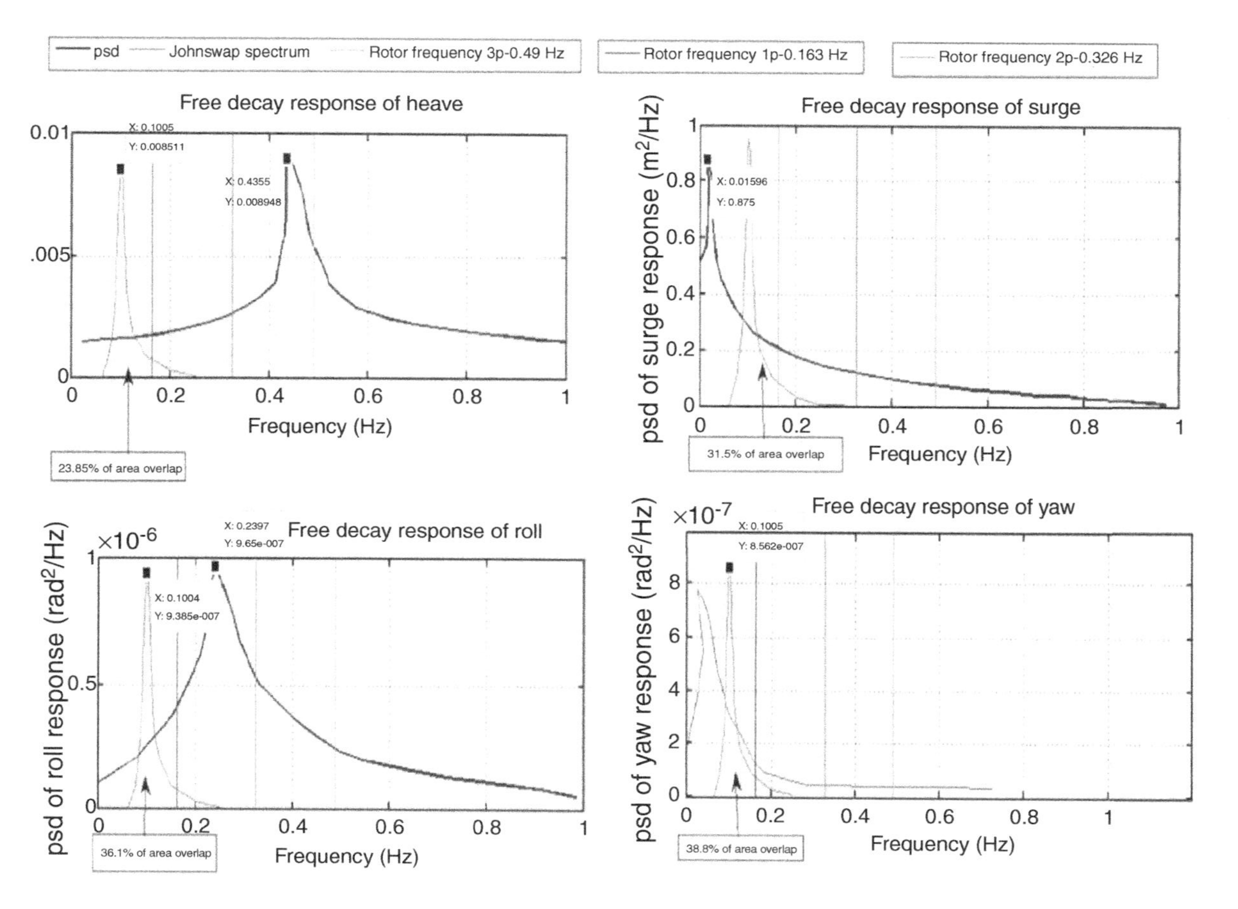

FIGURE 5.20 Power spectral density plots of free decay responses

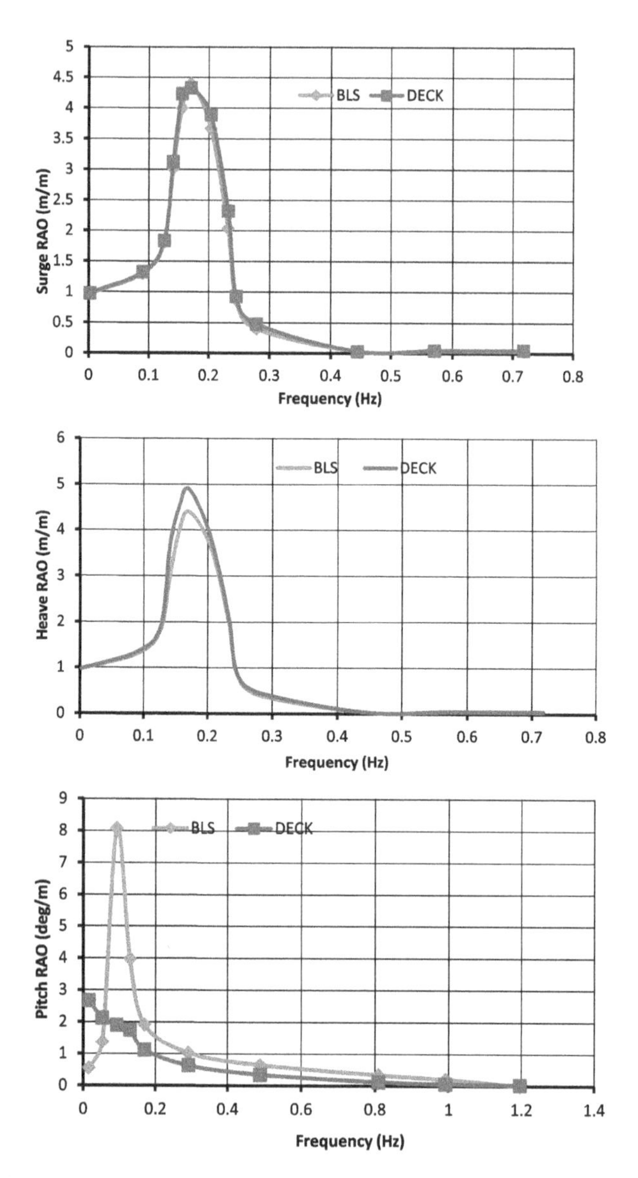

FIGURE 5.21 RAO plots in various degrees of freedom for deck and buoyant leg

the wave spectrum and rotor frequency are well-separated from the peaks of the responses, a safe margin is established. It ensures a safe working condition of the wind turbine as the excitations of wave and wind-driven rotor fail to excite the system in the degrees-of-freedom. Thus, the wind turbine is dynamically safe and stable. Although the distinct differences in the peak frequencies, as shown in the plots, verify this statement, there are areas of overlap. The wave and response spectrum overlap is about 24% for heave, 32% for surge, 36% for roll, and 39% for yaw. It re-iterates the fact that the influence of wave energy is maximum in yaw motion, which is followed

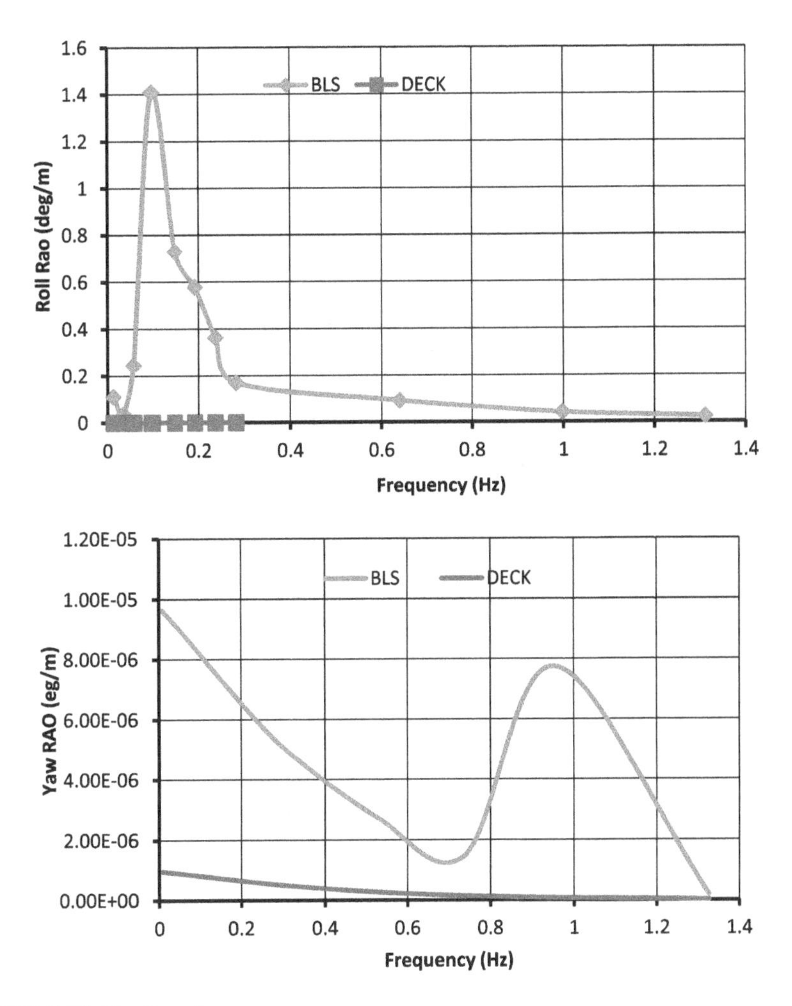

FIGURE 5.21 Continued

by the roll and surge; it is minimum for heave, which is desirable. Although the heave natural frequency is closer to the rotor frequency, heave motion is not excited under the chosen operating conditions. Hence the safety of the deck is not compromised.

Fig. 5.21 shows the RAO plots of the surge, heave, pitch, roll, and yaw motion of the deck and one of the buoyant legs located in the front side of the wave heading direction. It is seen from the figures that deck response in all the rotational degrees of freedom is insignificant in comparison to that of the buoyant legs due to the presence of ball joints. The ball joints restrain motion transfer from the legs to the deck. However, the marginal response of the deck in the rotational degrees of freedom is due to the differential heave encountered by the platform. Differential heave is caused due to the unequal tension variations in the tethers, although initial pre-tension in all the sets of tethers is identical.

5.13 RESPONSES UNDER OPERABLE LOADS AND PARKED CONDITIONS

Fig. 5.22 shows the response of the wind turbine platform in all six degrees of freedom under operable and parked conditions. It is seen that the heave response has a similar magnitude under both the conditions at the same frequency; the response spectrum is broader under operable conditions compared to the parked ones. Due to the design objective of restraining the operating mass of the wind turbine up to 17% of that of the total platform mass, heave motion is not excited. Roll and pitch responses show a distinct shift in the frequencies between the operable and parked states of the wind turbine, with a reduction in this shift at higher frequencies. The distinct shift is attributed to the coupling of the tower and the platform response. There are no distinct shifts in the flexible degrees of freedom (surge, sway, and yaw) frequencies for operable and parked conditions, confirming a monolithic behavior of the platform and the tower under operational loads. Multiple peaks can be seen in all responses due to multiple excitations from the loading conditions.

Table 5.2 summarizes the RAO variations to the wave heading angle. Considering the direction of the wind to be opposite to that of the wave heading at 0^0, surge

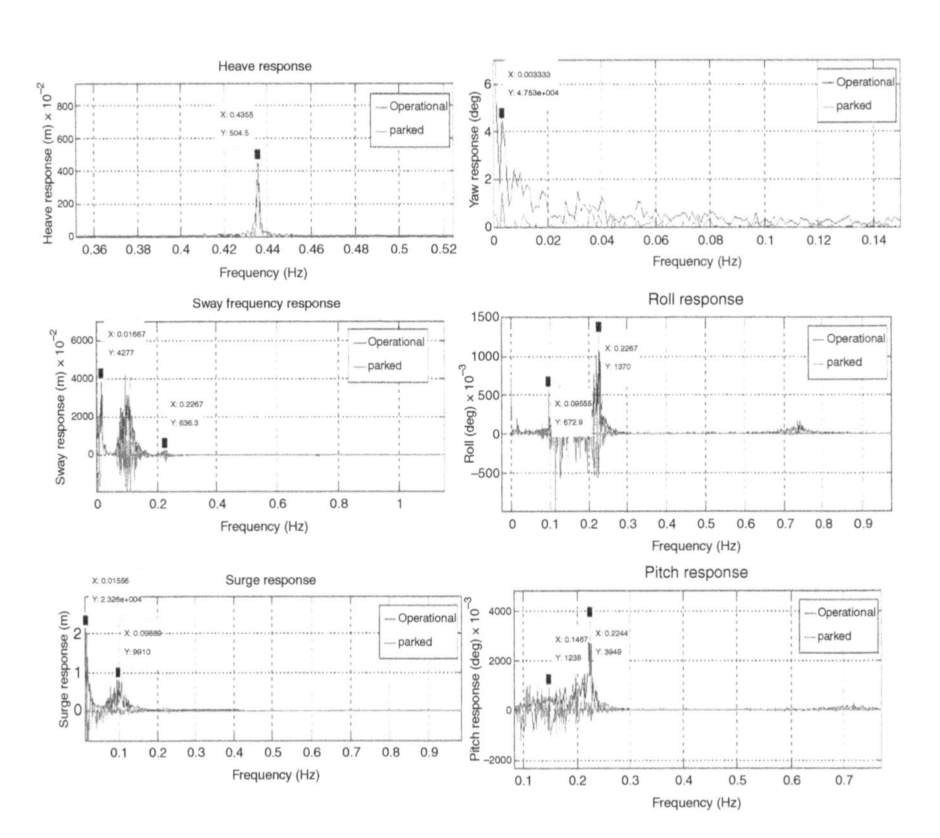

FIGURE 5.22 Responses under operable and parked conditions

TABLE 5.2
Variation of RAO with Change in Wave Heading Angle

Wave Heading	Surge (m)	Sway (m)	Heave (m)	Roll (deg/m)	Pitch (deg/m)	Yaw (deg/m)
0	0.857	0.0118	5.03E-03	0.0287	0.0361	7.71E-03
30	0.634	0.272	1.54E-02	0.0302	0.0338	7.36E-03
45	0.496	0.502	1.72E-02	0.0315	0.0318	3.25E-02
60	0.254	0.711	3.06E-02	0.0323	0.03015	5.51E-02
90	0.0123	0.852	3.67E-02	0.0345	0.0270	6.85E-02

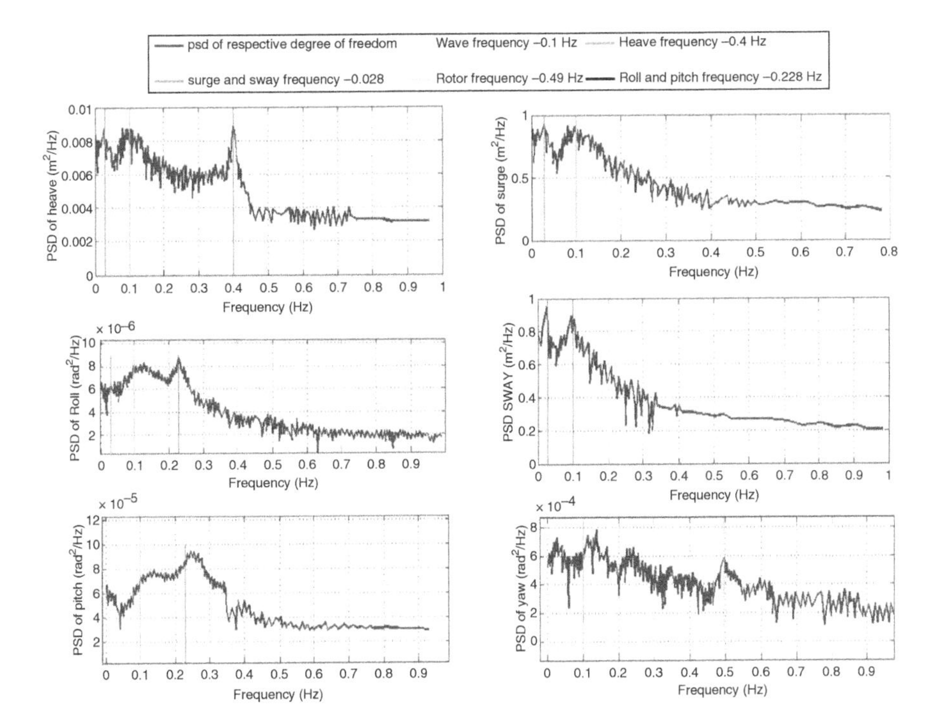

FIGURE 5.23 Power spectral density plots under operating condition

response increases with the wave heading angle, while this trend is reversed in sway response. It can be attributed to the change in the relative velocities between the wave and wind. Further, surge at 0^0 is not the same in magnitude to sway at 90^0 due to the variation in the aerodynamic damping between this degrees-of-freedom. Roll response follows the trend of sway response, while pitch follows the surge trend, indicating a strong coupling between these pairs. Though yaw is a flexible degree of freedom, lesser RAO in all wave heading angles is due to the novel geometry of Triceratops as a support system. The maximum yaw response observed at 90^0 is due

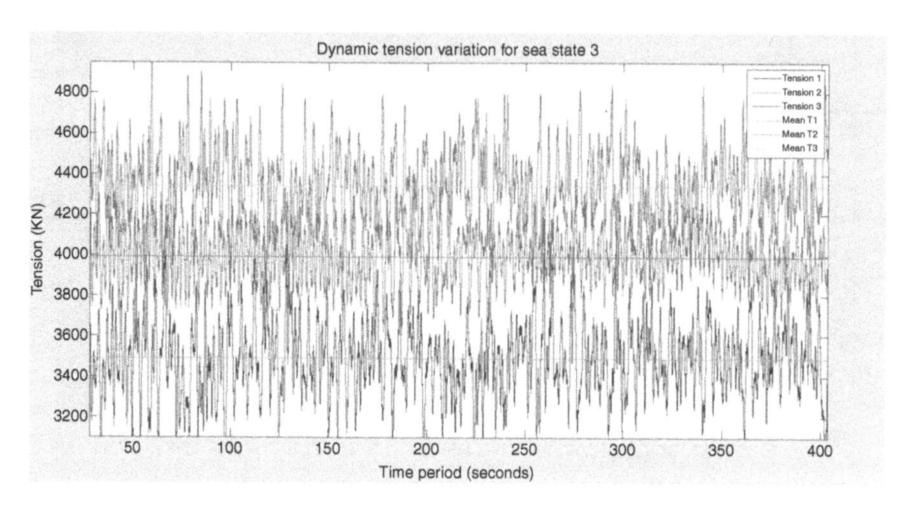

FIGURE 5.24 Dynamic tether tension variation

to the combined roll and sway along the aerodynamic thrust. Heave follows the trend of the surge, indicating a strong coupling between them.

Power spectral density plots are shown in Fig. 5.23 under the combined wave and wind excitations. Heave PSD shows peaks at dominant wave frequency and also at surge and sway frequencies, respectively. It confirms a strong coupling between surge-heave, which is due to the set-down effect in compliant systems. Roll and pitch exhibit peaks at frequencies close to their natural frequencies. The turbulence causes low-frequency excitations by the wind. It is observed that the standard deviation of the sway response is about 22% lesser than that of surge due to the predominant wave direction along with surge; also, the rotor thrust is absent in the sway direction. The yaw PSD demonstrates the effect of all other degrees of freedom on the yaw response. It can be realized from the smaller peaks at the natural frequencies of other degrees of freedom. The rotor load excitation is seen as a peak at 0.49 Hz, suggesting the influence of rotor loads on the yaw response. Yaw PSD is highly noisy compared to all other responses. Several peaks can be seen at frequencies of all other degrees of freedom, showing their influence on the yaw response. Surge/sway and roll/pitch peaks are comparable to the wave frequency peaks. A peak is observed at the rotor's 3P excitation, indicating the effect of rotor loads on the yaw response.

5.14 DYNAMIC TENSION VARIATION OF TETHERS

Damage of the compliant offshore platforms can be quantified through the damage of tethers. Fig. 5.24 shows the tension variation in all the three legs of triceratops under the Douglas Sea state-3. Dynamic tension variation is attributed to the nature of the wave excitation force. It can be observed from the figure that a peak in tether tension in one of the tethers does not necessarily mirror the other two tethers. It is because there is a phase lag between the wave approach on each of the tethers. When one tether gets taut moored, the other gets slackened and vice versa. Table 5.3 shows the

TABLE 5.3
Normalized Tension Variation in Tethers

Tether	Max Tension (KN)	Min Tension (KN)	Mean Tension (KN)	$(T_{max} - T_{min})/T_{mean}$ (%)
Tension1	3988	2969	3490	29.2%
Tension2	4337	3662	3990	16.8%
Tension3	4874	4001	4340	20.1%

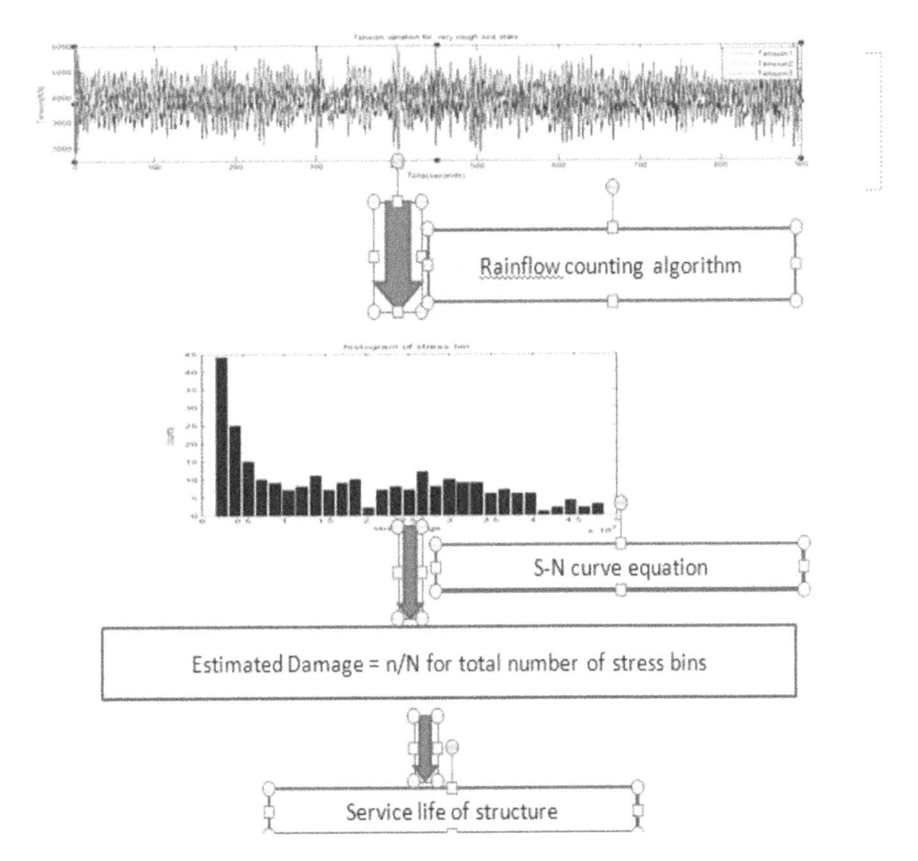

FIGURE 5.25 Methodology for fatigue life estimate

normalized tension variations in all the groups of tethers. These variations are within the permissible limits, ensuring the operational safety of the platform with the wind turbine.

The dynamic tension variations in the tethers can result in fatigue failure and hence to be assessed. Fig. 5.25 describes the flowchart used for fatigue life estimate. Firstly, the tether tension variation is simulated for a rough sea state from which the stress time history is obtained. After that, the rain flow counting algorithm is used to get the stress histogram. This histogram results in the output of the stress ranges (S) in the x-axis and the number of counts (n) in the y-axis. Using the S-N curve approach and the guidelines suggested in the code, one can estimate the value of cycles (N) (DNV-RP-C203).

Further, one can estimate the damage for a given simulation time (say, 900 s) using the Palmgren-Miners rule and sum it up for the total number of stress bins. This damage value is then extrapolated to damage =1 to derive the structure's service life. Table 5.4 summarizes the damage estimate parameters.

TABLE 5.4
Damage Estimate Parameters

Stress Range (Pa)	Stress Range (Mpa)	N (Cycles)	n (Counts)	$D = n/N$ (Damage)
8112758.333	8.112758333	2.62117E+12	3500	1.33528E-09
24329275	24.329275	97187967077	44	4.52731E-10
40545791.67	40.54579167	20997260987	25	1.19063E-09
56762308.33	56.76230833	7652791033	15	1.96007E-09
72978825	72.978825	3600886229	10	2.77709E-09
89195341.67	89.19534167	1972302293	9	4.5632E-09
105411858.3	105.4118583	1194900082	7	5.85823E-09
121628375	121.628375	777848980.2	8	1.02848E-08
137844891.7	137.8448917	534352654	11	2.05857E-08
154061408.3	154.0614083	382752828.9	7	1.82886E-08
170277925	170.277925	283481650.2	9	3.17481E-08
186494441.7	186.4944417	215775594.8	10	4.63444E-08
202710958.3	202.7109583	168022839.4	2	1.19031E-08
218927475	218.927475	133382605.1	7	5.24806E-08
235143991.7	235.1439917	107646110.2	8	7.43176E-08
251360508.3	251.3605083	88126974.8	7	7.94308E-08
267577025	267.577025	73055604.31	12	1.64258E-07
283793541.7	283.7935417	61233978.34	8	1.30646E-07
300010058.3	300.0100583	51831353.96	10	1.92933E-07
316226575	316.226575	44259337.32	9	2.03347E-07
332443091.7	332.4430917	38093253.11	9	2.36262E-07
348659608.3	348.6596083	33021369.9	6	1.81701E-07
364876125	364.876125	28811353.36	7	2.4296E-07
381092641.7	381.0926417	25287640.58	6	2.3727E-07
397309158.3	397.3091583	22315893.08	6	2.68867E-07
413525675	413.525675	19792131.25	1	5.05251E-08
429742191.7	429.7421917	17635022.45	2	1.13411E-07
445958708.3	445.9587083	15780328.99	4	2.5348E-07
462175225	462.175225	14176858.87	2	1.41075E-07
478391741.7	478.3917417	12783476.52	3	2.34678E-07

TABLE 5.5
Service Life Estimate

Sea State	Wind Speed (Knots)	Significant Wave Height (m)	Fatigue Life (Years)
3	11–16	0.5–1.25	27
6	28–33	4–6	9.47

The following relationship gives the value of damage:

$$\sum_{i=1}^{m} \frac{n}{N} \tag{5.38}$$

where m is the total number of stress bins. For the example case being discussed, the damage is estimated as $3.0149 \text{x} 10^{-6}$ for 900 seconds. Hence, in a year, the damage would be estimated as below:

$$\text{Damage in year} = \frac{3600\,x\,24\,x\,365\,x\,3.0149\,x\,10^{-6}}{900} = 0.1056 \tag{5.39}$$

This damage would be equivalent to 1 in 9.47 years, which would amount to the platform's service life, supporting the wind turbine under a rough sea state (sea state-6). Under the operating sea state (sea state–3), this value is estimated as 1 in 27 years. Table 5.5 summarizes these values.

EXERCISES

1. Write short notes on wind energy harvesting and the structural systems suitable to harvest wind energy.
2. What is a triceratops? Draw a neat sketch and explain its structural action.
3. What are the advantages and disadvantages of offshore wind energy?
4. Discuss various support systems for wind turbines.
5. How are offshore wind turbines classified?
6. List various components of floating wind turbine and explain their working briefly.
7. Write a brief note on SPAR-type wind turbine.
8. Write a brief note on TLP-type wind turbine.
9. Compare the structural response of a barge-type and semi-submersible-type wind turbine.
10. List various loads that act on offshore wind turbines.
11. Write a brief note on the blade momentum theory and its applicability in the analysis of offshore wind turbines.

12. Compare the blade element theory with blade momentum.
13. What is the wind shear effect? How is this handled on design?
14. Why are the responses of a wind turbine different in parked and operating conditions? Which is critical from a design perspective?
15. How does dynamic tension variation in tethers contribute to failure assessment of wind turbines?

References

Agarwal A. P. and Manuel L. 2011. Incorporating Irregular Nonlinear Waves in Coupled Simulation and Reliability Studies of Offshore Wind Turbines, *Applied Ocean Research*, 33(3): 215–227.

AGMA 1984. Gear Design Data Handbook, American Gear Manufacturers Association, Washington, DC.

Ahamed, R., McKee, K. and Howard, I. 2020. Advancements of Wave Energy Converters Based on Power Take Off (PTO) Systems: A Review, *Ocean Engineering*: 107248.

Alain Clément, Pat McCullen, António Falcão. 2002. Wave Energy in Europe: Current Status and Perspectives, *Renewable and Sustainable Energy Reviews*, 6: 405–431.

Amarkarthik, A., Chandrasekaran, S. Sivakumar, K., and Sinhmar, H. 2012. Laboratory Experiments Using Non-Floating Body to Generate Electrical Energy from Water Waves, *Front Energy*: 5–15.

Ambühl, S. et al. 2015. Operation and Maintenance Strategies for Wave Energy Converters, *Proceedings of Institution of Mechanical Engineers, Part O: Journal of Risk and Reliability*, 229(5): 417–441.

Anleitner, M. A. 2011. The Power of Deduction – Failure Modes and Effects Analysis for Design, *SDMIMD Journal of Management*, DOI: 10.15533/sdm/2011/v2i2/28794.

António, A. F. 2007. Modelling and Control of Oscillating-Body Wave Energy Converters with Hydraulic Power Take-Off and Gas Accumulator, *Ocean Engineering*, 34(14–15): 2021–2032.

Antonio, F. de O. Falcao. 2010. A Wave Energy Utilization: Review of Technologies, *Renewable and Sustainable Energy Reviews*, 14(3): 899–918.

API. 2000. API RP 2A-WSD, Recommended Practice for Planning, Designing and Constructing Fixed Offshore Platforms – Working Stress Design. American Petroleum Institute.

Atcheson, M., Garrad, A., Cradden, L., Henderson, A., Matha, D., Nichols, J., Roddier, D., and Sandberg, J. 2016. *Floating Offshore Wind Energy*, Springer, New York.

Athanasia, A. and Genachte, A. B. 2013. Deep Offshore and New Foundation Concepts, *Energy Procedia*, 35: 198–209.

Aubault A., Cermelli C., and Roddier D. 2009. WindFloat: A Floating Foundation for Offshore Wind Turbines Part III: Structural Analysis, Proceedings of International Conference on Ocean, Offshore and Arctic Engineering, Honolulu, HI, May 31–June 5.

Babarit, A. 2015. A Database of Capture Width Ratio of Wave Energy Converters, *Renewable Energy*, 80: 610–628.

Babarit, A., Duclos, G., and Clément, A. H. 2004. Comparison of Latching Control Strategies for a Heaving Wave Energy Device in Random Sea, *Applied Ocean Research*, 26(5): 227–238.

Babarit, A., Hals, J., Muliawan, M. et al. 2012. Numerical Benchmarking Study of a Selection of Wave Energy Converters, *Renewable Energy*, 41: 44–63.

Bae Y. H. and Kim M. H. 2011. Rotor-Floater-Mooring Coupled Dynamic Analysis of Mono-column-TLP-type FOWT (Floating Offshore Wind Turbine), *Ocean Systems Engineering*, 1(1): 95–111.

Birk, L. 2009. Application of Constrained Multi-Objective Optimization to the Design of Offshore Structure Hulls, *Journal of Offshore Mechanics and Arctic Engineering*, 131(1): 1–9.

Boaghe, O. M., Billings, S. A., and Stansby, P. K. 1998. Spectral Analysis for Nonlinear Wave Forces, *Applied Ocean Research*, 20: 199–212.

Bretschneider, C. L. 1960. A Theory of Waves of Finite Height, Proceedings of Seventh Conference of Coastal Engineering, The Hague: 146–183.

Bryne B., Houlsby G., Martin C., and Fish P. 2002. Suction Caisson Foundation for Offshore Turbine, *Journal of Wind Engineering*, 26(3): 145–155.

Buckingham, E. 1988. *Analytical Mechanics of Gears*, Dover Publications, New York.

Budar, K. and Falnes, J. 1975. A Resonant Point Absorber of Ocean-Wave Power, *Nature*, 256(5517): 478–479.

Butterfield, S., Musial, W., and Boone, A. 2004. Feasibility of Floating Platform Systems for Wind Turbines, Proceedings of ASME Wind Energy Symposium, Reno, NV, January 5–7: 476–486.

Butterfield, S., Musial, W., Jonkman, J., Sclavounos, P. 2007. *Engineering Challenges for Floating Offshore Wind Turbines*, National Renewable Energy Lab, Golden, CO.

Butterfield, S., Musial, W., Jonkman, J., Sclavounos, P., and Wayman, L. 2005. Engineering Challenges for Floating Offshore Wind Turbines, Proceedings of Copenhagen Offshore Wind 2005 Conference, Copenhagen, October 26–28.

Cargo, C. J., Plummer, A., Hillis, A. et al. 2012. Determination of Optimal Parameters for a Hydraulic Power Take-Off Unit of a Wave Energy Converter in Regular Waves, *Proceedings of Institution of Mechanical Engineers, Part A: Journal of Power and Energy*, 226(1): 98–111.

Cargo, C. J., Hillis, A. J., and Plummer, A. R. 2014. Optimisation and Control of a Hydraulic Power Take-Off Unit for a Wave Energy Converter in Irregular Waves, *Proceedings of Institution of Mechanical Engineers, Part A: Journal of Power and Energy*, 228(4): 462–479.

Cermelli C., Roddier D., and Aubault A. 2009. WindFloat: A Floating Foundation for Offshore Wind Turbines Part III: Structural Analysis, Proceedings of International Conference on Ocean, Offshore and Arctic Engineering, Honolulu, HI, May 31–June 5.

Chakrabarti, S. K. 1982. Transverse Forces on Vertical Tube Array in Waves, *Journal of Waterway, Port, Coastal and Ocean Division*, 108(1): 1–15.

Chakrabarti, S. K. and Tam, W. A. 1975. Wave Height Distribution around Vertical Cylinder, *Journal of Waterways Harbors and Coastal Engineering Division*, ASCE, 101: 225–230.

Chakrabarti, S. K., Wolbert, Allan L., and Tam, A. William. 1976. Wave Force on the Vertical Cylinder, Journal Waterways, Harbors and Coastal Engineering Division, *ASCE*, 10(2): 203–221.

Chandrasekaran, S. 2015. *Advanced Marine Structures*, CRC Press, Boca Raton, FL.

Chandrasekaran, S. 2016a. *Health, Safety and Environmental Management for Offshore and Petroleum Engineers*, John Wiley and Sons, Chichester.

Chandrasekaran, S. 2016b. *Offshore Structural Engineering: Reliability and Risk Assessment*, CRC Press, Boca Raton, FL.

Chandrasekaran, S. 2017. *Dynamic Analysis and Design of Ocean Structures*, 2nd Ed., Springer, Singapore.

Chandrasekaran, S. 2019a. *Advanced Design of Steel Structures*, CRC Press, Boca Raton, FL.

Chandrasekaran, S. 2019b. *Structural Health Monitoring with Application to Offshore Structures*, World Scientific Publishing, Singapore.

Chandrasekaran, S. 2020. *Offshore Semi-Submersible Platform Engineering*, CRC Press, Boca Raton, FL.

Chandrasekaran, S. and Abhishek, S. 2010. Potential Flow Based Numerical Study for the Response of Floating Offshore Structures with Perforated Columns, *Ships and Offshore Structures*, 5(4): 327–336.

Chandrasekaran, S. and Harender, N. A. 2012. Power Generation Using Mechanical Wave Energy Converter, *International Journal of Ocean and Climate Systems*, 3(1): 57–70.

Chandrasekaran, S. and Harender, N. A. 2014. Proposed Mechanism of Mechanical Wave Energy Converter, *Ship and Offshore*, 4: 101.

Chandrasekaran, S. and Harender, N. A. 2015. Failure Mode and Effects Analysis of Mechanical Wave Energy Converters, *International Journal of Intelligent Engineering Informatics*, 3(1): 57.

Chandrasekaran, S. and Harinder. 2011. Design and Efficiency Analysis of Mechanical Wave Energy Converter. Proceedings of 30th International Conference on Ocean, Offshore and Arctic Engineering, Rotterdam, June 19–24.

Chandrasekaran, S. and Jain, A. K. 2001. Dynamic Behaviour of Square and Triangular Offshore Tension Leg Platforms under Regular Wave Loads, *Ocean Engineering*, 29(3): 279–313.

Chandrasekaran, S. and Jain, A. K. 2002. Triangular Configuration Tension Leg Platform Behaviour under Random Sea Wave Loads, *Ocean Engineering*, 29(15): 1895–1928.

Chandrasekaran, S. and Jain, A. K. 2016. *Ocean Structures: Construction, Materials and Operations*, CRC Press, Boca Raton, FL.

Chandrasekaran, S. and Kiran, P. A. 2018. Mathieu Stability of Offshore Triceratops under Postulated Failure, *Ships and Offshore Structures*, 13(2): 143–148.

Chandrasekaran, S. and Madhavi, N. 2014a. Variations of Water Particle Kinematics of Offshore TLP's with Perforated Members: Numerical Investigations, Proceedings of Structural Engineering Convention, IIT, Delhi, December 22–24.

Chandrasekaran, S and Madhavi, N. 2014b. Retrofitting of Offshore Structural Member Using Perforated Cylinders, *SFA Newsletter*, 13: 10–11.

Chandrasekaran, S. and Madhavi, N. 2014c. Hydrodynamic Performance of Retrofitted Structural Member under Regular Waves, *International Journal of Forensic Engineering, Inder Science*, 2(2): 100–121.

Chandrasekaran, S. and Madhavi, N. 2014d. Variation of Water Particle Kinematics with Perforated Cylinder under Regular Waves, Proceedings of 24th International Society of Offshore and Polar Engineers, ISOPE, Busan, June 15–20.

Chandrasekaran, S. and Madhavi, N. 2015a. Flow Field around an Outer Perforated Circular Cylinder under Regular Waves: Numerical Study, *International Journal of Marine System and Ocean Technology*, 10: 91–100.

Chandrasekaran, S. and Madhavi, N. 2015b. Variation of Flow Field around Twin Cylinders with and without Outer Perforated Cylinder: Numerical Studies, *China Ocean Engineering, Springer*, 30(5): 763–771.

Chandrasekaran, S. and Madhavi, N. 2015c. Retrofitting of Offshore Cylindrical Structures with Different Geometrical Configuration of Perforated Outer Cover, *International Journal of Shipbuilding Progress*, 62(1–2): 43–56.

Chandrasekaran, S. and Madhavi, N. 2015d. Design Aids for Offshore Structures with Perforated Members, *Ship and Offshore Structures*, 10(2): 183–203.

Chandrasekaran, S. and Madhavi, N. 2016. Numerical Study on Geometrical Configurations of Perforated Cylindrical Structures, *Journal of Performance of Constructional Facilities, American Society of Civil Engineers*, 30(1): 04014185.

Chandrasekaran, S. and Madhuri, S. 2012a. Stability Studies of Offshore Triceratops, *International Journal of Research and Development*, 1(10): 398–404.

Chandrasekaran, S. and Madhuri, S. 2012b. Free Vibration Response of Offshore Triceratops: Experimental and Analytical Investigations, Proceedings of 3rd Asian Conf. on Mechanics of Functional Materials and Structures, IIT, Delhi, December 8–9: 965–968.

Chandrasekaran, S. and Madhuri, S. 2012c. Stability Studies on Offshore Triceratops, Proceedings of Conference on Technology of the Sea, Tech Samudhra, Indian Maritime University, Vishakapatnam, 1(10): 398–404.

Chandrasekaran, S. and Mayank, S. 2017. Dynamic Analyses of Stiffened Triceratops under Regular Waves: Experimental Investigations, *Ships and Offshore Structures*, 12(5): 697–705.

Chandrasekaran, S. and Nagavinotihini, R. 2017. Analysis and Design of Offshore Triceratops under Ultra-Deep Waters, *International Journal of Architectural, Civil and Construction Sciences*, 10(11): 1505–1513.

Chandrasekaran, S. and Nagavinothini, R. 2018a. Tether Analyses of Offshore Triceratops under Wind, Wave and Current, *Journal of Marine Systems and Ocean Technology*, 13: 34–42.

Chandrasekaran, S. and Nagavinothini, R. 2018b. Dynamic Analyses and Preliminary Design of Offshore Triceratops in Ultra-Deep Waters, *International Journal of Innovative Infrastructure Solutions*, 3(1): 16.

Chandrasekaran, S. and Nagavinothini, R. 2019a. Ice-induced Response of Offshore Triceratops, *Ocean Engineering*, 180: 71–96.

Chandrasekaran, S. and Nagavinothini, R. 2019b. Tether Analyses of Offshore Triceratops under Ice Loads due to Continuous Crushing, *International Journal of Innovative Infrastructure Solutions*, 4: 25.

Chandrasekaran, S. and Nagavinothini, R. 2020. *Offshore Compliant Platforms: Analysis, Design and Experimental Studies*, Wiley, Chichester.

Chandrasekaran, S. and Nassery, J. 2015. Springing and Ringing Response of Offshore Triceratops, Proceedings of 34th Conference on Ocean, Offshore and Arctic Engineering, St. John's, NL, May 31–June 5.

Chandrasekaran, S. and Nassery, J. 2017. Nonlinear Response of Stiffened Triceratops under Impact and Non-Impact Waves, *International Journal of Ocean Systems Engineering*, 7(3): 179–193.

Chandrasekaran, S. and Raghavi, B. 2015. Design, Development, and Experimentation of Deep-Ocean Wave Energy Converter System, *Energy Procedia*, 634–640.

Chandrasekaran, S. and Roy, A. 2004. Computational Wave Theories for Deep Water Compliant Offshore Structures. Proceedings of International Conference on Environmental Fluid Mechanics, IIT, Guwahati, March 2–3: 138–145.

Chandrasekaran, S. and Roy, A. 2005. Phase Plane Study of Offshore Structures Subjected to Nonlinear Hydrodynamic Loading, Proceedings of International Convention of Structural Engineering, IISc Bangalore: 397.

Chandrasekaran, S. and Sricharan, V. V. S. 2019. Improved Efficiency of a Floating Wave Energy Converter under Different Wave-Approach Angles: Numerical and Experimental Investigations, *Journal of Ocean Engg and Marine Energy*, 5(1): 41–50.

Chandrasekaran, S. and Sricharan, V. V. S. 2020a. Indian Institute of Technology Madras: A System for Wave Energy Conversion, India. Patent. 202041005812. Feb. 2020 [Filed].

Chandrasekaran, S. and Sricharan, V. V. S. 2020b. Indian Institute of Technology Madras. Bean-Shaped Multi-body Floating Wave Energy Converter, India. Design application/patente326437-001. Feb. 2020 [Granted].

Chandrasekaran, S. and Sricharan, V. V. S. 2020c. Numerical Analysis of a New Multi-Body Floating Wave Energy Converter with a Linear Power Take-Off System, *Renewable Energy*, 159: 250–271.

Chandrasekaran, S. and Sricharan, V. V. S. 2021. Numerical Study of Bean-Float Wave Energy Converter with Float Number Parametrization using WEC-Sim in Regular Waves with the Levelized Cost of Electricity Assessment for Indian Sea States, *Ocean Engineering*, 237.

Chandrasekaran, S. and Srinath, V. 2013. Experimental Investigations of Dynamic Response of Tension Leg Platforms with Perforated Members, Proceedings of 32nd International Conference on Ocean, Offshore and Arctic Engineering, Nantes, June 9–14.

Chandrasekaran, S. and Thomas, M. 2016. Suppression System for Offshore Cylinders under Vortex Induced Vibration, Proceedings of 21st Vibro-Engineering Conference, Berno, August 31–September 1, 7: 1–6.

Chandrasekaran, S., Jain, A. K., and Chandak, N. R. 2004. Influence of Hydrodynamic Coefficients in the Response Behavior of Triangular TLPs in Regular Waves, *Ocean Engineering*, 31(17–18): 2319–2342.

Chandrasekaran, S., Jain, A., Gupta, A., and Srivastava, A. 2007a. Response Behaviour of Triangular Tension Leg Platforms under Impact Loading, *Ocean Engineering*, 34(1): 45–53.

Chandrasekaran, S., Jain, A. K., and Chandak, N. R. 2007b. Response Behavior of Triangular Tension Leg Platforms under Regular Waves Using Stokes Nonlinear Wave Theory, *Journal of Waterway, Port, Coastal, and Ocean Engineering*, 133(3): 230–237.

Chandrasekaran, S., Jain, A. K., and Gupta, A. 2007c. Influence of Wave Approach Angle on TLP's Response, *Ocean Engineering*, 34(8–9): 1322–1327.

Chandrasekaran, S., Madhuri Seeram, A. K., and Jain, G. 2010. Dynamic Response of Offshore Triceratops under Environmental Loads. Proceedings of International Conference of Marine Technology, Dhaka, December 11–12: 61–66.

Chandrasekaran, S., Sundaravadivelu, R., Pannerselvam, R., and Madhuri, S. 2011. Experimental Investigations of Offshore Triceratops under Regular Waves, Proceedings of 30th International Conference on Ocean, Offshore and Arctic Engineering, Rotterdam, June 19–24.

Chandrasekaran, S., Jain, A. K., and Madhuri, S. 2013a. Aerodynamic Response of Offshore Triceratops, *Ships and Offshore Structures*, 8(2): 123–140.

Chandrasekaran, S., Amarkathik, A., Sivakumar, K., Selvamuthukumaran, D., and Sidney, S. 2013b. Experimental Investigation and ANN Modeling on Improved Performance of an Innovative Method of Using Heave Response of a Non-Floating Object for Ocean Wave Energy Conversion, *Front Energy*, DOI: 10.1007/s11708-013-0268-4.

Chandrasekaran, S., Natarajan, M., and Saravanakumar, S. 2013c. Hydrodynamic Response of Offshore Tension Leg Platforms with Perforated Members, *International Journal of Ocean and Climate Systems*, 4(3): 182–196.

Chandrasekaran, S., Amarkarthik, A., Sivakumar, K., and Selvamuthukumaran, D. 2013d. Experimental Investigation and ANN Modeling on Improved Performance of an Innovative Method of Using Heave Response of a Non-Floating Object for Ocean Wave Energy Conversion, *Frontiers in Energy*, 7(3): 279–287.

Chandrasekaran, S., Madhavi, N., and Natarajan, C. 2014a. Variations of Hydrodynamic Characteristics with the Perforated Cylinder, Proceedings of 33rd International Conference of Ocean, Offshore and Arctic Engineering, San Francisco, CA, June 8–13.

Chandrasekaran, S., Madhavi, N., and Sampath, S. 2014b. Force Reduction on Ocean Structures with Perforated Members, Proceedings of Structural Engineering Convention, IIT, Delhi, December 22–24.

Chandrasekaran, S., Raphel, D. C., and Saishri, P. 2014c. Deep Ocean Wave Energy Systems: Experimental Investigations, *Journal of Naval Archaeology and Marine Engineering*, 11(2): 139–146.

Chandrasekaran, S., Madhavi, N., and Sampath, S. 2015a. Advances in Structural Engineering, in V. Matsagar (Ed.), *Force Reduction on Ocean Structures with Perforated Members*, Springer, New York, pp. 647–661.

Chandrasekaran, S., Mayank, S., and Jain, A. 2015b. Dynamic Response Behavior of Stiffened Triceratops under Regular Waves: Experimental Investigations, Proceedings of 34th International Conference on Ocean, Offshore and Arctic Engineering, St. John's, NL, May 31–June 5.

Chandrasekaran, S., Jain, A. K., Shafiq, N., Mubarak, M., and Wahab, A. 2021. *Design Aids for Offshore Platforms under Special Loads*, CRC Press, Boca Raton, FL.
Christensen, L., Madsen, F., and Kofoed, J. P. 2005. The Wave Energy Challenge, Proceedings of Conference on POWER-GEN Europe, Milan, June 28–30.
Chung, J. S., Whitney, A. K., Lezius, D., and Conti, R. J. 1994. Flow-Induced Torsional Moment and Vortex Suppression for a Circular-Cylinder with Cables, Proceedings of 4th International Offshore and Polar Engineering Conference, Osaka, 3: 447–459.
Clauss, G. F. and Birk, L. 1996. Hydrodynamic Shape Optimization of Large Offshore Structures, *Applied Ocean Research*, 18(4): 157–171.
Clément, A., McCullen, P., Falcao, A. et al. 2002. Wave Energy in Europe: Current Status and Perspectives, *Renewable and Sustainable Energy Reviews*, 6(5): 405–431.
Cruz, J. 2008. *Ocean Wave Energy-Current Status and Future Perspective*, Springer New York.
Cummins, W. 1962. The Impulse Response Function and Ship Motions, Proceedings of Conference in Ship Theory, Institut flir Schiffbau der Universitat, Hamburg.
Davenport, A. G. 1961. The Spectrum of Horizontal Gushiness near the Ground in High Winds, *Journal of Royal Meteorological Society*, 87: 194–211.
De, S. C. 1955. Contributions to the Theory of Stokes Waves, *Proceedings of Cambridge Philosophical Society*, 51(4): 713–736.
Diaconu, S. and Rusu, E. 2013. The Environmental Impact of a Wave Dragon Array Operating in the Black Sea, *Scientific World Journal*, DOI: 10.1155/2013/498013.
Díaz, H. and Guedes Soares, C. 2020. Review of the Current Status, Technology and Future Trends of Offshore Wind Farms, *Ocean Engineering*, 209: 107381.
DNV, G. L. 2018. DNVGL-ST-0119: Floating Wind Turbine Structures. DNV GL.
Dunnett, J. S. W. 2009. Electricity Generation from Wave Power, *Renewable Energy*, 34: 179–195.
Eidsmoen, H. 1998. Tight-Moored Amplitude-Limited Heaving-Buoy Wave-Energy Converter with Phase Control, *Applied Ocean Research*, 20(3): 157–161.
Evans, D. V. 1976. A Theory for Wave-Power Absorption by Oscillating Bodies, *Journal of Fluid Mechanics*, 77(1): 1–25.
Falcão, A. F. D. O. 2010. Wave Energy Utilization: A Review of the Technologies, *Renewable and Sustainable Energy Reviews*, 14(3): 899–918.
Falnes, J. 2001. Optimum Control of Oscillation of Wave-Energy Converters, Proceedings of 11th International Offshore and Polar Engineering Conference, Stavanger, June 17–22.
Falnes, J. 2002. Optimum Control of Oscillation of Wave-energy Converters, *International Journal of Offshore and Polar Engineering*, 12(2): 147–155.
Falnes, J. 2004. *Ocean Waves and Oscillating Systems: Linear Interactions Including Wave-Energy Extraction*, Cambridge University Press, New York.
Fenton, J. D. 1985. A Fifth Order Stokes Theory for Steady Mass, *Journal of Waterways, Port, Coastal and Ocean Engineering, ASCE*, 111(2): 216–234.
Folley, M. (Ed). 2016. *Numerical Modelling of Wave Energy Converters*, Elsevier, New York.
French, M. J. and Bracewell, R. 1987. PS Frog: A Point-Absorber Wave Energy Converter Working in a Pitch/Surge Mode. 5th International Conference on Energy Options, Reading.
Fulton, G. R., Malcolm, D. J., and Moroz, E. 2006. Design of a Semi-Submersible Platform for a 5MW Wind Turbine, Proceedings of 44th Aerospace Sciences Meeting and Exhibit, Reno, NV, January 9–12.
Gaspar, J. F., Calvario, M., and Kamarlouei, M. 2016. Power Take-Off Concept for Wave Energy Converters Based on Oil-Hydraulic Transformer Units, *Renewable Energy*, 86: 1232–1246.
Gerhard PahlWolfgang BeitzJörg FeldhusenKarl-Heinrich Grote. 2006. *Engineering Design: A Systematic Approach*, Springer, New York.

Goggins, J. and Finnegan, W. 2014. Shape Optimisation of Floating Wave Energy Converters for a Specified Wave Energy Spectrum, *Renewable Energy*, 71: 208–220.

Hang S. Choi, Hyun S. Shin, Park I. K., and Jun B. Rho. 2003. An Experimental Study for Mooring Effects on the Stability of the SPAR Platforms, Proceedings of 13th International Offshore and Polar Engineering Conference, Honolulu, HI, 1: 285–288.

Hansen, R. H. and Kramer, M. M. 2011. Modelling and Control of the Wavestar Prototype, Proceedings of 9th European Wave and Tidal Energy Conference, Aalborg, September 5–9: 1–10.

Harris, R. I. 1971. The Nature of Wind and Modern Design of Wind-Sensitive Structures, Proceedings of Construction Industry Research and Information Association, London.

Haskind, M. D. 1957. The Exciting Forces and Wetting of Ships in Waves. *Izvest Akademii Nauk SSSR, Otdelenie Tekhnicheskikh Nauk*, 7: 65–79.

Henderson, A. R. and Morgan, C. S. 2003. Offshore Wind Energy in Europe: A Review of the State-of-the-Art, *Wind Energy*, 6(1): 35–52.

Henderson, A. R. and Patel, M. H. 2003. On the Modelling of a Floating Offshore Wind Turbine, *Wind Energy*, 6(1): 53–86.

Henderson, A. R. and Witcher, D. 2010. Floating Offshore Wind Energy: A Review of the Current Status and an Assessment of the Prospects, *Wind Engineering*, 34: 1–16.

Hogben, N. and Standing, R. G. 1974. Wave Loads on Large Bodies, Proceedings of Symposium on Dynamics of Marine Vehicles and Structures in Waves, University College, London: 258–277.

IEC 61400–1. 2005. *Wind Turbines, Part 1: Design Requirements*, International Electrotechnical Commission, Geneva.

IEC 61400–3. 2006. *Wind Turbines, Part 3: Design Requirements for Offshore Wind Turbines*, International Electrotechnical Commission, Geneva.

Ikoma, T., Nakamura, M., Moritsu, S., Aida, Y., Masuda, K., and Eto, H. 2019. Effects of Four Moon Pools on a Floating System Installed with Twin-VAWTs, Proceedings of 38th International Conference on Offshore Mechanics and Arctic Engineering, University of Strathclyde, London, February 18: V001T001A016.

Ikoma, T., Tan, L., Moritsu, S., Aida, Y., and Masuda, K. 2021. Motion Characteristics of a Barge-Type Floating Vertical-Axis Wind Turbine with Moonpools. *Ocean Engineering*, 230: 109006.

INCOSE. 2015. *INCOSE Systems Engineering Handbook: A Guide for System Life Cycle Processes and Activities*, 4th Ed., John Wiley & Sons, Chichester.

Isaacs, J. D., Castel, D., and Wick, G. L. 1976. Utilization of the Energy in Ocean Waves, *Ocean Engineering*, 3: 175–187.

Isaacson M., Baldwin, J., Allyn, N., and Cowdell, S. 2000. Wave Interactions with Perforated Breakwater, *Journal of Waterway, Port, Coastal and Ocean Engineering*, 126(5): 229–235.

Islam, A. B., Soeb, M. R., and Jumaat, M. Z. 2017. Floating SPAR Platform as an Ultra-Deepwater Structure in Oil and Gas Exploration, *Ships and Offshore Structures*, 12(7): 923–936.

Ittyavirah, D. and Philip, V. 2016. Dynamic Response of Offshore Triceratops- Supporting 5 MW Wind Turbine, *International Journal of Engineering Research*, 5(6): 708–714.

Ivanova, I. A., Bernhoff, H., Agren, O. et al. 2005. Simulated Generator for Wave Energy Extraction in Deep Water, *Ocean Engineering*, 32(14–15): 1664–1678.

James, R., Weng, W.-Y., Spradbery, C., Jones, J., Matha, D., Mitzlaff, A., Ahilan, R., Frampton, M., and Lopes, M. 2018. Floating Wind Joint Industry Project: Phase I Summary Report. Carbon Trust Technical Report, 19: 2–20.

Jarlan, G. E. 1961. A Perforated Vertical Breakwater, *Dock and Harbour Authority*, 41(486): 394–398.

Jin, S., Patton, R. J., and Guo, B. 2019. Enhancement of Wave Energy Absorption Efficiency via Geometry and Power Take-Off Damping Tuning, *Energy*, 169: 819–832.

Jonkman, J. M. and Buhl, Jr, M. L., 2005. *FAST User's Guide*, National Renewable Energy Laboratory, USA, Golden, CO.

Joseph, A., Mangal, L., and George, P. S. 2009. Coupled Dynamic Response of a Three-Column TLP, *Journal of Naval Archaeology and Marine Engineering*, 6(2): 52–61.

Jun Liu, Gao Lin, and Jianbo Li. 2012. Short-Crested Waves Interaction with a Concentric Cylindrical Structure with Double-Layered Perforated Walls, *Ocean Engineering*, 40: 76–90.

Jusoh, M. A., Ibrahim, M. Z., Daud, M. Z. et al. 2019. Hydraulic Power Take-Off Concepts for Wave Energy Conversion System: A Review, *Energies*, 12(23): 1–23.

Kaimal, J. C. 1972. Special Characteristics of Surface Layer Turbulence, *Journal of Royal Meteorological Society*, 98: 563–589.

Karimirad, M. 2010. Dynamic Response of Floating Wind Turbine, Centre for Ships and Ocean Structures (CeSOS), *Mechanical Engineering*, 17(2): 146–156.

Karimirad, M. 2014. *Floating Offshore Wind Turbines*, Springer International, Cham.

Kenny, C. J., Findlay, D., Lazakis, I. et al. 2017. Development of a Condition Monitoring System for an Articulated Wave Energy Converter by FMEA, Proceedings of 26th European Safety and Reliability Conference, Strathclyde, September 28: 184. DOI: 10.1201/9781315374987-173.

Kim, S. J., Koo, W., and Shin, M. J. 2019. Numerical and Experimental Study on a Hemispheric Point-Absorber-Type Wave Energy Converter with a Hydraulic Power Take-Off System, *Renewable Energy*, 135: 1260–1269.

Kolios, A., Mytilinou, V., Lozano-Minguez, E., and Salonitis, K. 2016. A Comparative Study of Multiple-Criteria Decision-Making Methods under Stochastic Inputs, *Energies*, 9: 566.

Larsen, T. J. and Hanson, T. D. 2007. A Method to Avoid Negative Damped Low Frequent Tower Vibrations for a Floating, Pitch Controlled Wind Turbine, Proceedings of 2nd Conference on the Science of Making Torque from Wind, Journal of Physics: Conference Series, Copenhagen, August 28–31.

Lee, C. H. and Newmann, J. N. 1997. *WAMIT User Manual, Ver 6.0*, Massachusetts Institute of Technology, Cambridge, MA.

Lefebvre, S. and Collu, M. 2011. Preliminary Design of a Floating Support Structure for a 5 MW Offshore Wind Turbine, *Ocean Engineering*, 40: 15–26.

Liu, Y., Li, S., Yi, Q., and Chen, D. 2016. Developments in Semi-Submersible Floating Foundations Supporting Wind Turbines: A Comprehensive Review, *Renewable and Sustainable Energy Reviews*, 60: 433–449.

López, M., Taveira-Pinto, F., and Rosa-Santos, P. 2017. Influence of the Power Take-Off Characteristics on the Performance of CECO Wave Energy Converter, *Energy*, 120: 686–697.

Lundgren, H. 1963. Wave Trust and Wave Energy Level, International Association of Hydraulic Research, London: 147–151.

Ma, K. T. 2019. *Mooring for Floating Wind Turbines*, Gulf Professional Publishing, Houston, TX, pp. 299–315.

Mast, E., Rawlinson, R., and Sixtensson, C. 2015. *Market Study Floating Wind in the Netherlands: Potential of Floating Offshore Wind*, DNV, Oslo.

McCormick, M. E. 1982. An Experimental Study of Wave Power Conversion by a Heaving, Vertical, Circular Cylinder in Restricted Waters, *Applied Ocean Research*, 4: 107–112.

Mei, C. C. 2012. Hydrodynamic Principles of Wave Power Extraction, *Philosophical Transactions of Royal Society A: Mathematical, Physical and Engineering Sciences*, 370(1959): 208–234.

Merigaud, A. and Ringwood, J. V. 2018. Free-Surface Time-Series Generation for Wave Energy Applications, *IEEE Journal of Oceanic Engineering*, 43(1): 19–35.

Moo-Hyun Kim. 2013. *SPAR Platforms: Technology and Analysis Methods*, American Society of Civil Engineers, Reston, VA.

Mott, R. L. 2004. *Machine Elements in Mechanical Design*, Pearson Prentice Hall, Hoboken, NJ.

Mueller, M. A. 2002. Electrical Generators for Direct Drive Wave Energy Converters, *IEE Proceedings: Generation, Transmission and Distribution*, 149(4): 446–456.

Musial, W. and Butterfield, S. 2004. Future for Offshore Wind Energy in the United States, Energy Ocean Proceedings, Palm Beach, FL, June 28–29.

Musial, W., Butterfield, S., and Ram, B. 2006. Energy from Offshore Wind, Proceedings of Offshore Technology Conference, Houston, TX, May 1–4.

Musial, W., Butterfield, S., and Boone, A. 2004. Feasibility of Floating Platform Systems for Wind Turbines, Proceedings of 42nd AIAA Aerospace Sciences Meeting and Exhibit, Reno, NV, January 5–8: 1007.

Nagavinothini, R. and Chandrasekaran, S. 2019. Dynamic Analyses of Offshore Triceratops in Ultra-Deep Waters under Wind, Wave and Current, *Structures*, 20: 279–289. https://doi.org/10.1016/j.istruc.2019.04.009.

Nagavinothini, R. and Chandrasekaran, S. 2020. Dynamic Response of Offshore Triceratops with Elliptical Buoyant Legs, *International Journal of Innovative Infrastructure Solutions*, 5(47): 1–14.

Newman, J. N. 1962. The Exciting Forces on Fixed Bodies in Waves, *Journal of Ship Research*, 6(4): 10–17.

Newman, J. N. 1979. The Theory of Ship Motions, *Advances in Applied Mechanics*: 221–283, DOI: 10.1016/S0065-2156(08)70268-0.

Nielsen, F. G., Hanson, T. D., and Skaare, B. 2006. Integrated Dynamic Analysis of Floating Offshore Wind Turbines, Proceedings of 25th International Conference on Offshore Mechanical and Arctic Engineering, Hamburg, June 4–9: 671–679.

Niemann, G. 1978. *Machine Elements: Design and Calculation in Mechanical Engineering*, Springer, New York.

Nygaard, T., Vaal, J., Pierella, F., Oggiano, L., and Stenbro, R. 2016. Development, Verification and Validation of 3DFloat: Aero-Servo-Hydro-Elastic Computations of Offshore Structures, *Energy Procedia*, 94: 425–433.

Ogilvie, T. F. 1964. Recent Progress Toward the Understanding and Prediction of Ship Motions, Proceedings of 5th Symposium on Naval Hydrodynamics, Bergen, September 10–12: 3–80.

Panicker, N. N. 1976. Power Resource Estimate of Ocean Surface Waves, *Ocean Engineering*, 3: 429–439.

Raphael, D. C., Kumaraswamy, S., Rao, M. M., and Ravindran, M. 2004. Knowledge Based Approach Towards Design of Mechanical Systems and Components for Deep Sea Mining, Proceedings of 23rd International Conference on Offshore Mechanics and Arctic Engineering, Vancouver, June 20–25: 875–881.

Raphael, D. C., Kumaraswamy, S., Rao, M. M., and Ravindran, M. 2007. Ocean Engineering: Design Synthesis Studies for the Development of Deep Ocean Manganese Nodule Collecting Systems, Proceedings of 26th International Conference on Offshore Mechanics and Arctic Engineering, San Diego, CA, June 10–15.

Reliability Analysis Center. 1968. Failure Mode, Effects, and Criticality Analysis, *Microelectronics Reliability*, 7(3): 268.

Rhinefrank, K., Schacher, A., Prudell, J. et al. 2012. Comparison of Direct-Drive Power Takeoff Systems for Ocean Wave Energy Applications, *IEEE Journal of Oceanic Engineering*, 37(1): 35–44.

Ringwood, J. V. and Bacelli, G. 2014. Energy-Maximizing Control of Wave-Energy Converters: The Development of Control System Technology to Optimize Their Operation, *IEEE Control Systems*, 34(5): 30–55.

Roddier, D., Cermellli, C., and Weinstein, A. 2009. WindFloat: A Floating Foundation for Offshore Wind Turbines, Part I: Design Basis and Qualification Process, Proceedings of 28th International Conference on Ocean, Offshore and Arctic Engineering, Honolulu, HI, May 31–June 5 .

Roddier, D., Cermelli, C., Aubault, A., and Weinstein, A. 2010. WindFloat: A Floating Foundation for Offshore Wind Turbines, *Journal of Renewable and Sustainable Energy*, 2: 033104.

Rodríguez, C. A., Rosa-Santos, P., and Taveira-Pinto, F. 2019. Assessment of Damping Coefficients of Power Take-Off Systems of Wave Energy Converters: A Hybrid Approach, *Energy*, 169: 1022–1038.

Saaty, T. L. 1990. *Multicriteria Decision Making: The Analytic Hierarchy Process*. RWS Publications, Pittsburgh, PA.

Salter, S. H. 1974. Wave Power, *Nature*, 249(5459): 720–724.

Santo, H., Taylor, P. H., and Stansby, P. K. 2020. The Performance of the Three-Float M4 Wave Energy Converter off Albany, on the South Coast of Western Australia, Compared to Orkney (EMEC) in the UK, *Renewable Energy*, 146: 444–459.

Sclavounos, P. 2008. Floating Offshore Wind Turbines, *Journal of Marine Technology Society*, 42: 39–43.

Sergiienko, N. Y., Neshat, M., da Silva, L., Alexander, B., and Wanger, M. 2020. Design Optimisation of a Multi-Mode Wave Energy Converter, *Ocean Renewable Energy*, 9.

Sharmila, N., Jalihal, P., Swamy, A. et al. 2004. Wave Powered Desalination System, *Energy*, 29(11): 1659–1672.

Sheng, W. and Lewis, A. 2012. Assessment of Wave Energy Extraction from Seas: Numerical Validation, *Journal of Energy Resources Technology*, 134(4), DOI: 10.1115/1.4007193.

Simiu, E. 1971. Wind Spectra and Dynamic Along-Wind Response, *Journal of Structural Engineering Division, ASCE*, 100(9): 1897–1910.

Skaare, B., Hanson, T. D., and Nielsen, F. G. 2007. Importance of Control Strategies on Fatigue Life of Floating Wind Turbines, Proceedings of 26th International Conference on Offshore Mechanical and Arctic Engineering, San Diego, CA, June 10–15.

Skaare, B., Nielsen, F. G., Hanson, T. D., Yttervik, R., Havmøller, O., and Rekdal, A. 2015. Analysis of Measurements and Simulations from the Hywind Demo Floating Wind Turbine. *Wind Energy*, 18: 1105–1122.

Skjelbreia, L. and Hendrickson, J. 1960. Fifth Order Gravity Wave Theory, Proceedings of 7th Conference Coastal Engineering, Aug 1–2, The Hague: 184–196.

Sricharan, V. V. S. and Chandrasekaran, S. 2021. Time-Domain Analysis of a Bean-Shaped Multi-body Floating Wave Energy Converter with a Hydraulic Power Take-Off Using WEC-Sim, *Energy*, 223: 119985.

Stansby, P., Carpintero Moreno, E., and Stallard, T. 2017. Large Capacity Multi-Float Configurations for the Wave Energy Converter M4 Using a Time-Domain Linear Diffraction Model, *Applied Ocean Research*, 68: 53–64.

Stokes, G. G. 1847. On the Theory of Oscillatory Waves, *Transactions of the Cambridge Philosophical Society*, 8: 441–445.

Stokes, G. G. 1880. Appendics to the Paper on the theory of Solitary Waves, *Math Phys Papers*, 1: 314–326.

Suh, N. P. 2001. *Axiomatic Design: Advances and Applications*, MIT-Pappalardo, Cambridge, MA.

Taboada, J. V. 2015. Comparative Analysis Review on Floating Offshore Wind Foundations (FOWF), Proceedings of 54th Naval Engineering and Maritime Industry Congress, Ferrol, October 14–16.

Terret, F. L., Osorio, J. D. C., and Lean, G. H. 1968. Model Studies of a Perforated Breakwater, Proceedings of 11th International Conference on Coastal Engineering, September 1–2, London: 1104–1120.

Tong, K. 1998. Technical and Economic Aspects of a Floating Offshore Wind Farm, *Journal of Wind Engineering and Industrial Aerodynamics*, 74: 399–410.

Utsunomiya, T., Sato, T., Matsukuma, H., and Yago, K. 2009. Experimental Validation for Motion of a SPAR-Type Floating Offshore Wind Turbine Using 1/22.5 Scale Model, Proceedings of 28th International Conference on Offshore Mechanics and Arctic Engineering, Honolulu, HI, May 31–June 5: 951–959.

Vantorre, M., Banasiak, R., and Verhoeven, R. 2004. Modelling of Hydraulic Performance and Wave Energy Extraction by a Point Absorber in Heave, *Applied Ocean Research*, 26(1–2): 61–72.

Vivek, P. and Ittyavah, D. 2016. Dynamic Analysis of 5 MW Offshore Turbine Mounted on Triceratops, *International Journal of Engineering Research and Technology*, 5(6).

Wang, K.-H. and Ren, X. 1994. Wave Interaction with a Concentric Porous Cylinder System, *Ocean Engineering*, 21(4): 343–360.

Wayman, E. N., Sclavounos, P. D., Butterfield, S., Jonkman, J., and Musial, W. 2006. Coupled Dynamic Modeling of Floating Wind Turbine Systems, Proceedings of Offshore Technology Conference, Houston, TX, May 1–4.

WEC-Sim. 2020. WEC-Sim Tool. Available at: https://wec-sim.github.io/WEC-Sim/.

Wehausen, J. V. 1971. The Motion of Floating Bodies, *Annual Review of Fluid Mechanics*, 3(1): 237–268.

William, A. N. and Li, W. 1998. Wave Interaction with a Semi-Porous Cylindrical Breakwater Mounted on a Storage Tank, *Ocean Engineering*, 25(2–3): 195–219.

Williams, A. N. and Li, W. 2000. Water Wave Interaction with an Array of Bottom-Mounted Surface-Piercing Porous Cylinders, *Ocean Engineering*, 27(8): 841–866.

Williams, A. N., Li, W. and Wang, K.-H. 2000. Water Wave Interaction with a Floating Porous Cylinder, *Ocean Engineering*, 27(1): 1–28.

Yemm, R. et al. 2012. Pelamis: Experience from Concept to Connection, *Philosophical Transactions of Royal Society, A: Mathematical, Physical and Engineering Sciences*, 370(1959): 365–380.

Zaheer, MM. and Islam, N. 2008. Fluctuating Wind-Induced Response of Double-Hinged Articulated Loading Platform. Proceedings of 27th International Conference on Offshore Mechanical and Arctic Engineering, Estori, June 15–20: 723–731.

Zaheer, MM. and Islam, N. 2012. Stochastic Response of a Double-hinged Articulated Leg Platform under Wind and Waves, *Journal of Wind Engineering and Industrial Aerodynamics*, 111: 53–60.

Zaheer, MM. and Islam, N. 2017. Dynamic Response of Articulated Towers under Correlated Wind and Waves, *Ocean Engineering*, 132: 114–125.

Zambrano, T., MacCready, T., Kiceniuk, T., Jr., Roddier, D. G., and Cermelli, C. A. 2006. Dynamic Modeling of Deepwater Offshore Wind Turbine Structures in Gulf of Mexico Storm Conditions, Proceedings of 25th International Conference on Offshore Mechanical and Arctic Engineering, Hamburg, June 4–9.

Zhao Hong-Jun, Song Zhi-Yao, LI Ling, Kohg Jun, Wang Le-Qiang, Yang Jie. 2016. On the Fifth Order Stokes Solution for Steady Water Waves, *China Ocean Engineering*, 30(5): 794–810.

Index

Note: Figures are indicated by page numbers in *italic*, and tables are indicated by page numbers in **bold**.

3DFloat 220–221

A

ADAM (Automatic Dynamic Analysis of Mechanical Systems) 220
aero-elastic model 227–229
aerofoil theory 225–227
Airy's linear wave theory 2–4
  Chakrabarti's modifications 4, *5*
  Wheeler's modifications 3–4, *5*
Airy's two-dimensional small-amplitude linear wave theory 4, *5*, *6*
American Petroleum Institute (API) spectrum 22–28, 29, *30*, 31
Ansys AQWA 91, 207–208, 221
  frequency domain-specific workflow 69–70
Aqua Boy 143
attenuator wave-energy devices 62–63

B

Baltic Sea, offshore wind energy extraction 203
barge-type floating wind turbines 209, 213–214
beach tourism 187, *194–196*, 197, **198**
Bean Floating Wave Energy Converter (BFWEC) 77–91
Beaufort scale 22
blade element theory 225–227
blade momentum theory 222–225
Blue H prototype TLP platform 213
boundary value problem 70
Brazil offshore, wave spectrum model for 21
breakwater 64
Bredsneidger spectrum 21
buoyancy of floating offshore structures 1

C

capture width ratio (CWR) 64
CETO-FWEC 63
climate change mitigation, role of wave energy 144
Cnoidal theory 4
coastal erosion and accretion 197, **198**
coastal protection 64, 187, *193*, 197, **198**, 199
corrosion on offshore structures 217
Cummins equation 75, 88
current
  definition of 31
  tidal current 31–32
  wind-generated current 31–32
current load, effect on offshore wave energy devices 1, 22

D

Davenport spectrum 22, 28, 29, *30*
deep-ocean wave-energy converters
  conceptual design 163
  cylindrical float 172–176
  cylindrical float integrated with fin 176, **177–184**
  experimental investigations 168–176, **177–184**
  geometric design of the device 163
  working principle 166–168
Denmark, offshore wind farms 207
desalination plants 64, 196, 199, **198**
diffraction theory 1
double-rack mechanical wave-energy converter (DRMWEC) 154–158
  advantages of 156
  conceptual design 154–156
  experimental studies 156–158
Douglas sea state 22

E

earthquakes, impact on offshore structures 1, *220*, 222
environmental loads on offshore structures 1

F

failure assessment 158–162
failure mode and effects analysis (FMEA) 158–162
  floating wave energy converters 134–141
  mechanical wave-energy converters 162–163, **164–165**
FAST (Fatigue, Aerodynamics, Structures and Turbulence) code 208, 220, 227–228
FAST2 227
FAST3 227
FLOATAGEN 214
floating offshore structures
  form-dominant design 1
  positive buoyancy 1
  wave-structure interaction 1

floating wave-energy converters (FWECs) 61–141
advantages of floating point absorbers 64
analysis with hydraulic PTO system 104–109
applications of point-absorbers 64
arrays of 64
attenuators 62–63
Bean Floating Wave Energy Converter (BFWEC) 77–91
capture width ratio (CWR) 64
classification of 62–63
computational tools 77, *78*
definition of 61
design features 61
effective length of floats **99**
failure assessment 134–141
frequency-domain modeling 69–74
hydraulic PTO systems 65
hydrodynamic coefficients 91–93
influence of the geometric shape of the floats 64
influence of wave theories on design 2–15
integration with other systems 64
linear time-domain model (LTD) 75–77
mathematical system for 62, *63*
modeling and analyses 65–66
multibody FWECs 77–91
novel hydraulic PTO system 124–132, **133–134**
numerical modeling 66–69
numerical studies on hydraulic PTO systems 110–115, **116**
oscillating water column (OWC) 63
overtopping devices 63
performance measurement 64
point absorbers (PA) 62–63
power extraction parameters **99**
power take-off (PTO) systems 65
practical guide to design of hydraulic PTO system 109–110
preliminary design stages 65–66
research on harnessing ocean energy 61
response without PTO system 116–123
Salter's duck 62
site-specific design 61–62
spring-mass-damper system 66–69
submerged pressure differentials 63
surge or oscillating pitch converters 63
terminators 62–63
three stages in wave energy conversion 62
time-domain modeling 74–76
virtual PTO system performance curves 100–104
wave power 94–99
wave-to-wire transfer 65–66
WEC-Sim (Wave Energy Converter Simulator) 77, *78*
flooding prevention 187, *189*
frequency-domain modeling 69–74
freshwater production units, wave-powered 64
Froude-Krylov force 71–72, 73, 77
Fukushima Daiichi nuclear power plant, flooding caused by a tsunami 187, 199
Fukushima FORWARD semi-submersible type floating wind turbine 215

G

GICON-SOF Pilot 213
Gulf of Mexico, wave spectrum model 21

H

Harris spectrum 28, 29–30
health tourism 187, *194–196*, 197, **198**, 199
hydraulic PTO systems 65
analysis of FWEC with hydraulic PTO 104–109
novel hydraulic PTO system 124–132, **133–134**
numerical studies on 110–115, **116**
practical guide to design of 109–110
hydrodynamic coefficients of FWECs 91–93
Hywind Scotland floating wind power project 212

I

ice loads on offshore floating wind turbines *220*, 221–222
IECKAI (IEC Kaimal model) 230
IECVKM (IEC Von Karman Isotropic Model) 230
impulse response function (IRF) 76, 77
India
combating coastal erosion near Chennai 199
Kalpakkam nuclear power plant 199
Kudankulam nuclear power plant 187, 199
potential contribution of wave energy 61
sources of renewable energy 61
Indian Institute of Technology, Madras 78
Indian renewable energy, Bean Floating Wave Energy Converter (BFWEC) 77–91
International Ship Structures Congress (ISSC) spectrum (two parameters) 16–17, 18, 20, 21
irregular waves 4
wave spectra 20
wave parameters **99**
wave power 95–99

J

Joint North Sea Wave Project (JONSWAP) spectrum (five parameters) 16, 17–19, 20, 21

K

Kaimal spectrum 25, 28, 30–31, 230
Kane method 228

L

life-cycle engineering 185
linear diffraction force 71, 73
linear time-domain model (LTD) 75–77

M

MATLAB® program 77, 101, 104, 116, 135
plotting wind spectra 29–31
wave spectra plot 18–20
mechanical wave-energy converters (MWECs) 143
advantages of 143–144
double-rack mechanical wave-energy converter (DRMWEC) 154–158
equation of motion 145–147
experimental studies 152–155
failure mode and effect analysis (FMEA) 162–163, **164–165**
high energy density of wave energy 143–144
ideal locations for 144
power take-off system design 147–152
preliminary geometric design 144–155
meteorological measuring devices, integrated with FWECs 64
MetService 22
modified PM spectrum (two parameters) 16, 18, 19
Morison's equation 1, 32, 73, 207, 208
multibody floating wave-energy converters 77–91
multi-utility development devices 185–201
advantages of 187
applications for 187, 196–199
multi-objective design synthesis approach *185*, 185–187
TSUSUCA-DOLPHIN device 187–201

N

National Renewable Energy Laboratory (NREL) 208, 220
wind spectral models 230
NEMOH software tool 70
Netherlands, offshore wind farms 207
Newfoundland offshore, wave spectrum model 21
North Sea
Joint North Sea Wave Project (JONSWAP) spectrum (five parameters) 16, 17–19, 20, 21
offshore wind energy extraction 203
nuclear power plants, flooding prevention 187–201

O

ocean energy, research on harnessing 61
ocean environment 1
ocean living simulation 187
Ochi-Hubble spectrum 21
offshore floating wind turbines
aero-elastic model 227–229
aerofoil theory 225–227
barge (pontoon) type 209, 213–214
blade element theory 225–227
blade momentum theory 222–225
components 209, *210*
dynamic tension variation of tethers 238–241
effects of earthquakes *220*, 222
environmental loads on 220–222
experimental and numerical analyses 220–222
fatigue life estimation 239–241
horizontal axis wind turbines (HAWT) 209
ice loads *220*, 221–222
Kane method 228
mathematical background 222–227
normal turbulence model 229–230
numerical analysis of triceratops-supported wind turbine 230–235
pontoon (barge) type 209, 213–214
responses under operable loads and parked conditions 236–238
semi-submersible type 209, 210, 214–215
service life estimation 239–241
SPAR type 209–212
tension-leg platform (TLP) type 209, 210, 212–213
triceratops type 209, 215–220
types of 209–220
vertical axis wind turbines (VAWT) 209
wind load determination 228–229
wind shear effect 229
wind spectrum 230
see also offshore wind turbines
offshore oil and gas platforms
base support for offshore wind turbines 206
deep-ocean wave-energy converters 166–168
integrated FWECs 64
wave-energy devices as power source 143
offshore wind turbines 203–241
advantages 203
classification 208–209
developments around the world 203, 204
disadvantages 203
evolution of wind turbine technology 206–207
fixed-bottom installations 204
floating platforms 204
floating wind turbines 208–209
locations of offshore wind farms in Europe 206–207
numerical tools for aerodynamic analysis 207–208
potential of offshore wind energy 203

shallow-water wind turbines 208
SPAR (single point anchor reservoir) platforms 204, *205*, 205, 208
support systems for 204–206
tension-leg platforms (TLPs) 204, 205, 208
transitional wind turbines 208
triceratops platforms 204–205, *206*, 208
turbine output capacities 207
water depths 204
wind and current effects 1
see also offshore floating wind turbines
oil and gas industry *see* offshore oil and gas platforms
oscillating pitch converter wave-energy devices 63
oscillating water column (OWC) wave-energy devices 63
overtopping wave-energy devices 63
Oyster wave-energy device 63

P

Palmgren-Miners rule 240
Pelamis 65, 143
perforated cylinders 33–59
effect of annular spacing on force reduction 37–40
effect of perforation parameters on force reduction 41–47
effect of perforation ratio on force reduction 37–41
effect of twin perforated cylinders 47–59
force reduction in the inner cylinder 33–37
purpose of 33
retrofitting coastal and offshore structure 33
wave force reduction effects 33
Pierson-Moskowitz (PM) spectrum 16, 18, 19, 21, 97
point-absorber wave-energy devices 62–63
pontoon-type floating wind turbines 209, 213–214
Portugal, floating offshore wind turbine 207
power take-off (PTO) systems 65
for mechanical wave-energy converters 147–152
virtual PTO system performance curves 100–104
see also hydraulic PTO systems
Prime Minister's Research Fellowship (PMRF) 78

R

radiation/diffraction theory 207
random waves *see* irregular waves
reef generating devices, wave-powered 64
regular waves
models 2–15
wave parameters **98**
wave power 94–95, **98**
renewable energy
research on harnessing ocean energy 61
waves as a source of 143–144
requirements engineering approaches 185
Response Amplitude Operator (RAO) 4, 73–74
retrofitting perforated cylinders 33
Reynolds-averaged Navier–Stoke (RANS) turbulence 36
risk priority number (RPN) 134–141
Riso smooth-terrain model 230

S

Salter's duck 62
Scotland
Hywind Scotland floating wind power project 212
Pelamis wave energy converter 65, 143
sea state
characteristics of random sea states 20–22
definition of 16
description of 21–22
Douglas sea state 22
typical wave energy PM spectra 21
seaquakes 1
semi-submersible type floating wind turbines 209, 210, 214–215
ship motion simulator *196*
Simscape 104
Simulink 104
solitary wave theory 4
SPAR (single point anchor reservoir) platforms 78, 204, *205*, 205, 208, 209–212
spring-mass-damper system 66–69
Stokes' fifth-order theory 4–15
stream function theory 4
submerged pressure differentials wave-energy devices 63
surge converter wave-energy devices 63
sustainable energy, waves as a source of 143–144
sustainable engineering 185

T

Tapchan 63
tension-leg platforms (TLPs) 78, 204, 205, 209, 209, 210, 212–213
terminator wave-energy devices 62–63
tidal current 31–32
time-domain modeling 74–76
triceratops-type floating wind turbines 204–205, *206*, 208, 209, 215–219
numerical analysis 230–235
studies 219–220
tsunami flooding prevention, TSUSUCA-DOLPHIN device 187–201

TSUSUCA-DOLPHIN device 187–201
cost benefits 199
non-patent citations **199**
patent citations **200**
range of applications for 187–199
workable alternative designs 197

U

United Kingdom, offshore wind farms 206–207

V

value engineering 185

W

WAMIT software tool 70
water particle kinematics, describing 2–4
wave analysis
design wave approach 2
irregular (random) waves 4
regular waves 2–15
statistical approach 2
wave classification according to relative depth 3
Wave Dragon 63, 143
wave energy
benefits of 144
concentrated form of wind energy 143
high energy density 143–144
source of clean and sustainable energy 143–144
wave-energy conversion, three stages of 62
Wave Energy Converter Simulator (WEC-Sim) 77, *78*, 104
wave-energy devices
advantages of mechanical wave-energy converters 143–144
comparison of different technologies 143
multi-utility development devices 185–201
wave force reduction, use of perforated cylinders 33–59
wave loads on floating offshore structures 1
wave power
irregular waves 95–99
range of applications for 64
regular waves 94–95, **98**
wave spectra 2, 16–22
Bredsneidger spectrum 21
comparison of *18*
definition of the sea state 16
definition of the wave spectrum 16
International Ship Structures Congress (ISSC) spectrum (two parameters) 16–17, 18, 20, 21
irregular (random) waves 20
JONSWAP (Joint North Sea Wave Project) spectrum (five parameters) 16, 17–19, 20, 21
Modified PM spectrum (two parameters) 16, 18, 19
Ochi-Hubble spectrum 21
Pierson-Moskowitz (PM) spectrum (one parameter) 16, 18, 19, 21
selection of the wave spectrum 16
suitability for different regions 21
wave-structure interaction 1
wave theories 2–15
Airy's linear wave theory 2–4
Airy's two-dimensional small-amplitude linear wave theory 4, *5*, *6*
Cnoidal theory 4
Solitary wave theory 4
Stokes' fifth-order theory 4–15
wave-wind-PV integrated systems 64
Wavestar 65
WEC-Sim (Wave Energy Converter Simulator) 77, *78*, 104
Western Africa offshore, wave spectrum model 21
Western Australia offshore, wave spectrum model 21
wind, effects on floating offshore structures 1
wind components 26
gust component 26–27
mean wind component 26–27
wind energy
potential of 203
see also offshore wind turbines
wind farms, locations of offshore wind farms in Europe 206–207
wind-generated current 31–32
wind load
determination 228–229
effect on offshore wave energy devices 22–31
wind power 206–207
wind shear effect 229
wind spectra 22, 28–31, 230
American Petroleum Institute (API) spectrum 22–28, 29, *30*, 31
Davenport spectrum 22, 28, 29, *30*
Harris spectrum 28, 29–30
IECVKM (IEC Von Karman Isotropic Model) 230
Kaimal spectrum 25, 28, 30–31, 230
NREL site-specific models 230
Riso smooth-terrain model 230
WindFloat semi-submersible type floating wind turbine 207, 215
World Meteorological Organization 22